Gauge Theory and the Early Universe

NATO ASI Series

Advanced Science Institutes Series

A Series presenting the results of activities sponsored by the NATO Science Committee, which aims at the dissemination of advanced scientific and technological knowledge, with a view to strengthening links between scientific communities.

The Series is published by an international board of publishers in conjunction with the NATO Scientific Affairs Division

A	**Life Sciences**	Plenum Publishing Corporation
B	**Physics**	London and New York
C	**Mathematical and Physical Sciences**	Kluwer Academic Publishers Dordrecht, Boston and London
D	**Behavioural and Social Sciences**	
E	**Applied Sciences**	
F	**Computer and Systems Sciences**	Springer-Verlag
G	**Ecological Sciences**	Berlin, Heidelberg, New York, London,
H	**Cell Biology**	Paris and Tokyo

Series C: Mathematical and Physical Sciences - Vol. 248

Gauge Theory and the Early Universe

edited by

P. Galeotti
Istituto di Cosmogeofisica del CNR,
Torino, Italy

and

David N. Schramm
Astronomy and Astrophysics Center,
University of Chicago, Chicago, IL, U.S.A.

Kluwer Academic Publishers

Dordrecht / Boston / London

Published in cooperation with NATO Scientific Affairs Division

Proceedings of the NATO Advanced Study Institute on
Gauge Theory and the Early Universe
Erice, Italy
20–30 May 1986

Library of Congress Cataloging in Publication Data

NATO Advanced Study Institute (1986 : Erice, Sicily)
Gauge theory and the early universe.

(NATO ASI series. Series C, Mathematical and Physical Sciences ; no. 248
"Proceedings of the NATO-Advanced Study Institute held in Erice, Italy, 20-30 May 1986.
1. Cosmology--Congresses. 2. Gauge fields (Physics)--Congresses. 3. Astrophysics--Congresses. I. Galeotti, Piero, 1942- . II. Schramm, David N. III. Title. IV. Series.
QB980.N37 1986 523.1 88-27388

ISBN-13: 978-94-010-7876-4 e-ISBN-13: 978-94-009-3059-9
DOI: 10.1007/ 978-94-009-3059-9

Published by Kluwer Academic Publishers,
P.O. Box 17, 3300 AA Dordrecht, The Netherlands.

Kluwer Academic Publishers incorporates the publishing programmes of
D. Reidel, Martinus Nijhoff, Dr W. Junk, and MTP Press.

Sold and distributed in the U.S.A. and Canada
by Kluwer Academic Publishers,
101 Philip Drive, Norwell, MA 02061, U.S.A.

In all other countries, sold and distributed
by Kluwer Academic Publishers Group,
P.O. Box 322, 3300 AH Dordrecht, The Netherlands.

Table of Contents

The Erice School is supported by NATO, NSS, Ettore Majorana Centre, the Italian Ministry of Education, Scientific and Technological Research, and the first Erice School which will meet every other year.

GAUGE THEORY AND THE EARLY UNIVERSE

P.Galeotti[(1)], D.N.Schramm[(2)]
(1) Istituto di Cosmo-geofisica del CNR and Università di Torino, Italy.
(2) University of Chicago and Fermilab, USA

The link between cosmology and particle physics has became closer and closer in recent years, and new ideas and developments in one field are strictly connected with the new ideas and developments in the other field. Roughly speaking, many different subdisciplines of physics seem to converge in cosmology, the study of the early evolution of the universe. The range of parameters (dimensions, temperatures, densities, etc...) to be taken into account in cosmology is so vast that one is forced to use theories or experimental data of wide ranging origin. This is demonstrated in the following few examples.

Astronomy. The astronomical results clearly indicate that the universe is expanding, and that the scale factor varies according to the Hubble law $\dot{R} = HR$, where $H = 100\ h_o\ \mathrm{Km\ s^{-1}\ Mpc^{-1}}$ (with $0.4 \lesssim h_o \lesssim 1$ from the observational uncertainties).

From the Hubble law and General Relativity, with zero cosmological constant for simplicity, one immediately obtains the relationship between the average density of the universe and the curvature K, where $K > 0$ is a closed 3-sphere, $K < 0$ is an open 3-hyperboloid and $K = 0$ is flat space:

$$\rho = \rho_{cr}\left(1 + \frac{K}{R^2 H^2}\right) \qquad (1)$$

The critical density, ρ_{cr}, at present, has the value:

$$\rho_{cr} = \frac{3H^2}{8\pi G} \simeq 2\cdot 10^{-29} h_o^{\ 2}\ \mathrm{gr\ cm^{-3}} \simeq 10\ \mathrm{KeV}\ h_o^{\ 2}\ \mathrm{cm^{-3}} \qquad (2)$$

P. Galeotti and D. N. Schramm (eds.), Gauge Theory and the Early Universe, 1–4.

Thus, from simple astronomical considerations, one immediately asks the question is the universe flat (K = 0) and is there missing mass (or better missing light), since the observational data on the matter associated with the light-emitting galaxies indicate $\Omega_{light} = (\rho/\rho_{cr}) \lesssim 0.01$, and astronomical abundances when compared to Big Bang nucleosynthesis argue that the density in baryons $\Omega_b = \rho_b/\rho_{cr} \lesssim 0.1$. Thus to get $\Omega = 1$ would require non-baryonic matter.
Recent astronomical results on the large scale structure of the universe, namely matter distribution and motion of clusters of galaxies, also indicate the existence of streaming motion (of order of 600 km s^{-1}) and voids (defined as regions where galaxies are not visible to observational limits) on scale lenghts of order 50 Mpc or larger. Obviously, the nature of these structures should reflect, in some way, the initial conditions in the past universe when matter formation and galaxy condensation have occured.
Atomic physics. The problem of the horizon in the universe occurs when one analyses the experimental data of the background radiation. Indeed assuming a blackbody distribution one gets the present temperature of the universe $T = 2.73 \pm 0.04$ K, and observes a high degree of isotropy and uniformity to within $\Delta T/T \lesssim 10^{-5}$, once the dipole motion of the observer is removed.
Thus one knows that the universe was hotter in the past, at least as hot as $\sim 3\ 10^3$ K, and that the background radiation was produced at the time $t \sim 10^5$ years, when the dimensions of what became todays knowable universe, according to standard models, were of order 10^7 light years. The problem of horizon simply consists on the fact that it is hard to reconcile the high degree of isotropy and uniformity of the background radiation with its emission from many regions physically disconnected from each other (some by almost 100 horizon lengths at the time when the radiation last interacted with matter).
Nuclear physics. The methods of nuclear physics are used to produce the abundances of chemical elements in the universe. In particular, the Big Bang nucleosynthesis of light elements is a powerfull tool to study the physical conditions of the early universe, at times of order 1 to 10^2 s. From the excellent agreement of the observed abundances with the theory we know that the universe should have been as hot as 10^{10} K in the past.
In addition, the abundance ratios among light elements firmly constrain the number, N_ν, of neutrino species to $N_\nu \lesssim 4$, with a best fit to $N_\nu = 3$; so we know that there could at most be only one additional neutrino flavour besides the three known. This bound is much stronger than that obtained with the accelerators, $N_\nu \lesssim 5.5$ (up until recently accelerator limits were at $N \lesssim 1000$), by measuring the width of the

Z^o particle, which is related to the number of neutrino species by $\Gamma(Z^o) \simeq 2.8 + 0.2\ (N_\nu - 3)$ GeV. The fact that accelerator limits are converging on the cosmological prediction is a dramatic proof that cosmological arguments affect fundamental physics.

This result, connected with the number density of neutrinos in the universe

$$n_\nu = \frac{7}{11} n_\gamma \simeq 300\ cm^{-3}$$

contraints the neutrino rest mass to a few 10's of eV, if one requires that the energy density of neutrinos in the universe does not exceed the upper limits on the cosmological density.

Particle physics. From the experimental point of view, the past evolution of the universe can be studied up to a time $t \sim 10^{-10}$ s from the Big Bang, where electroweak processes decoupled. The recent results of UA1 and UA2 at CERN give a clear measure of the physics at that time, through the values of $W^\pm$ and Z^o masses: 83.5 and 93.4 GeV respectively.

For earlier times we still have no direct experimental evidence but only a few indirect arguments from cosmic ray studies and proton decay limits. Beyond the hypothetical desert, in which no "oasis" has been found so far, particle physics joins cosmology at the GUT's era ($t \sim 10^{-35}$ s) through studying proton decay processes and magnetic monopoles. Searches for monopoles are however still inconclusive because no experiment has reached so far the Parker flux bound from the survival of galactic magnetic fields with monopoles present. Experimental searchs for proton decay seem to give some indications: minimal SU(5) has been ruled out because there is no evidence of the predicted dominant channel $p \rightarrow e^+ \pi^o$. This result agrees with the earlier astrophysical arguments from the baryons to photons density ratio $n_b/n_\gamma \simeq 10^{-10}$, which excludes minimal SU(5) and requires a larger group at the GUT era. In any case, we'll get better information from new huge underground experiments, able to test the SUSY dominant channel $p \rightarrow K^+ \tilde{\nu}$ up to a life-time of order 10^{32} years.

Condensed matter physics. The early universe presumably went through several phase transitions. One of these at the GUT epoch may have produced a rapid expansion known as "inflation". Another may have produced remnant cosmic strings which led to the formation of galaxies and the large scale structure.

Quantum gravity. An important phase transition in the early universe is believed to have occured at the Plank time ($t \sim 10^{-43}$ s), where gravitation decoupled from the other fundamental interactions. Since a quantum field theory introduced into the Einstein gravitation theory encounters a disaster at the Plank time, unless primordial point structures are abandoned; one possibility is to introduce multiple dimensional structures (superstrings) with 10 dimensions. Such models lead to many new and exciting possibility, including a possible shadow world. Such a superstring theory may indeed be the ultimate theory of every thing (T.o.E.).

Aim of course. The aim of this course was to bring together physicists from a variety of fields and backgrounds to enable students to have a more complete perspective on the origin of the universe.

TOWARD THE INFLATIONARY PARADIGM: LECTURES ON INFLATIONARY COSMOLOGY

Michael S. Turner
Theoretical Astrophysics
Fermi National Accelerator Laboratory
Batavia, Illinois 60510
and
Departments of Physics and Astronomy and Astrophysics
The University of Chicago
Chicago, Illinois 60637

1. OVERVIEW

Guth's inflationary Universe scenario has revolutionized our thinking about the very early Universe. The inflationary scenario offers the possibility of explaining a handful of very fundamental cosmological facts—the homogeneity, isotropy, and flatness of the Universe, the origin of density inhomogeneities and the origin of the baryon asymmetry, while at the same time avoiding the monopole problem. It is based upon microphysical events which occurred early ($t \leq 10^{-34}$ sec) in the history of the Universe, but well after the planck epoch ($t \geq 10^{-43}$ sec). While Guth's original model was fundamentally flawed, the variant based on the slow-rollover transition proposed by Linde, and Albrecht and Steinhardt (dubbed 'new inflation') appears viable. Although old inflation and the earliest models of new inflation were based upon first order phase transitions associated with spontaneous-symmetry breaking (SSB), it now appears that the inflationary transition is a much more generic phenomenon, being associated with the evolution of a weakly-coupled scalar field which for some reason or other was initially displaced from the minimum of its potential. Models now exist which are based on a wide variety of microphysics: SSB, SUSY/SUGR, compactification of extra dimensions, R^2 gravity, induced gravity, and some random, weakly-coupled scalar field. While there are several models which successfully implement the inflation, none is particularly compelling and all seem somewhat *ad hoc*. The common distasteful feature of all the successful models is the necessity of a small dimensionless number in the model—usually in the form of a dimensionless coupling of order 10^{-15}. And of course, all inflationary scenarios rely upon the assumption that vacuum energy (or equivalently a cosmological term) was once dynamically very significant, whereas today there exists every evidence that it is not (although we have no understanding why it is not). For these reasons I have entitled these lectures *Toward the Inflationary Paradigm*. I have divided my lectures into the following sections: Successes of the standard cosmology; Shortcomings of the standard cosmology; New inflation—the slow–rollover transition; Scalar field dynamics; Origin of density inhomogeneities; Specific models, I. Interesting failures; Lessons learned—prescription for successful inflation; Two models that work; The Inflationary paradigm; Loose ends; and Inflation confronts observation.

P. Galeotti and D. N. Schramm (eds.), Gauge Theory and the Early Universe, 5–69.

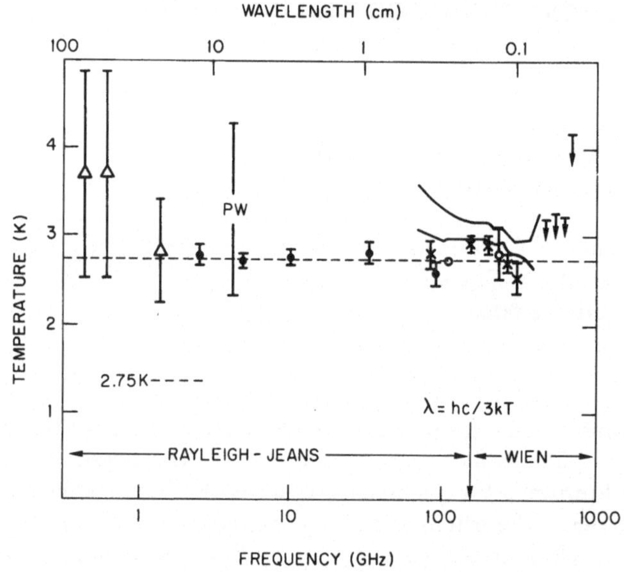

Figure 1: Summary of microwave background temperature measurements from $\lambda \simeq 0.05$ to 80 cm (see refs. 4). Measurements indicate that the background radiation is well described as a 2.75 ± 0.05K blackbody. PW denotes the discovery measurement of Penzias and Wilson.

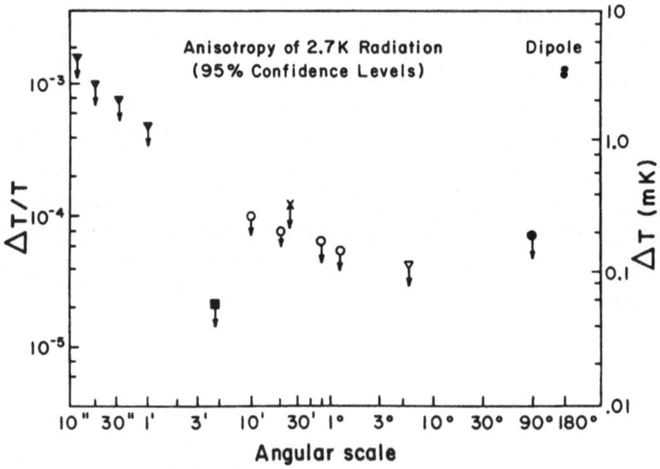

Figure 2: Summary of microwave background anisotropy measurements on angular scales from 10″ to 180° (see ref. 5). With the exception of the dipole measurements, the rest are 95% confidence upper limits to the anisotropy.

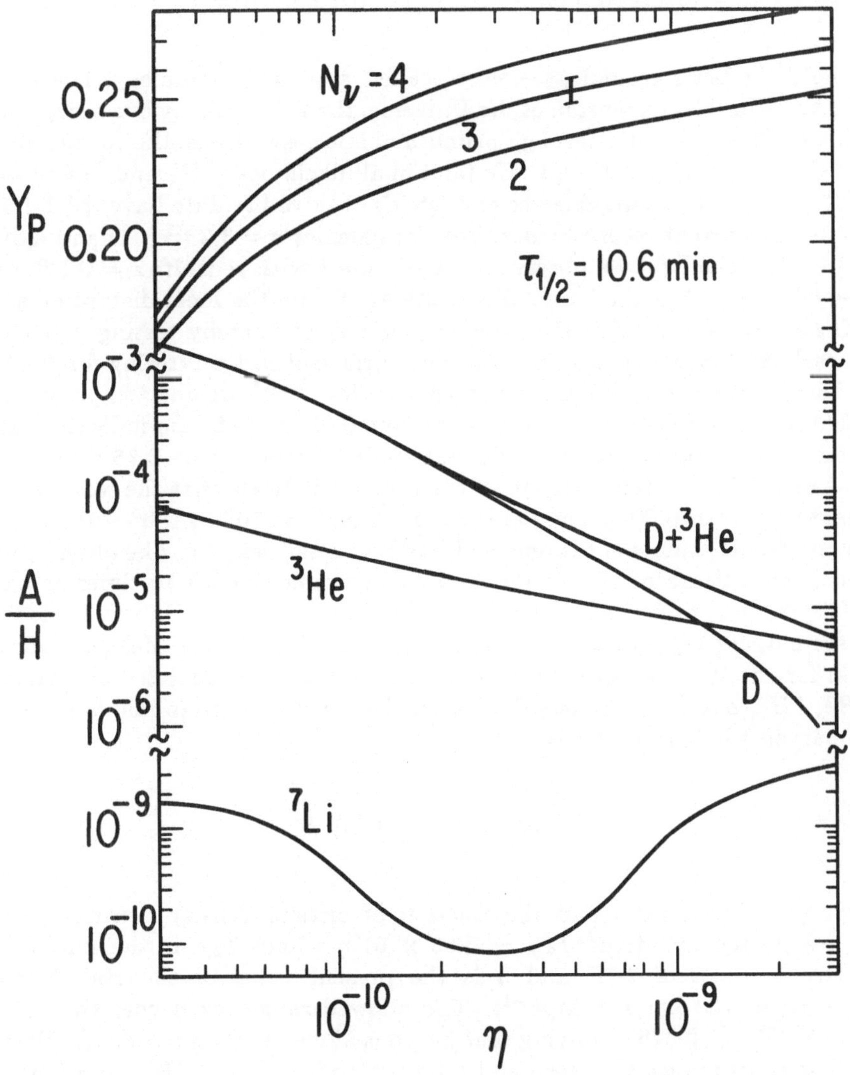

Figure 3: Big bang nucleosynthesis predictions for the primordial abundances of D, ^{3}He, ^{4}He, and ^{7}Li. Y_p = mass fraction of ^{4}He, shown for N_ν = 2, 3, 4 light neutrino species. Present observational data suggest: $0.23 \leq Y_p \leq 0.25$, $(D/H)_p \geq 1\times10^{-5}$, $[(D+^3He)/H]_p \leq 10^{-4}$, and $(^7Li/H)_p \simeq (1.1\pm0.4)\times10^{-10}$. Concordance requires $\eta \simeq (4-7)\times10^{-10}$. For further discussion see ref. 6.

2. THE STANDARD COSMOLOGY AND ITS SUCCESSES

The hot, big bang cosmology—the so-called standard cosmology, neatly accounts for the (Hubble) expansion of the Universe, the 2.7 K microwave background radiation (see Figs. 1,2), and through primordial nucleosynthesis, the cosmic abundances of the light elements D and ^4He (and in all likelihood, ^3He and ^7Li as well; see Fig. 3). The most distant galaxies and QSO's observed to date have redshifts in excess of 3—the current record holders are: for galaxies $z = 3.2$ (ref. 1) and QSO's $z = 4.0$ (ref. 2). The light we observe from an object with redshift $z = 3$ left that object only 1–2 Byr after the bang. Observations of even the most distant galaxies and QSO's are consistent with the standard cosmology, thereby testing it back to times as early as 1 Byr (see, e.g., ref. 3). The surface of last scattering for the microwave background is the Universe at an age of a few $\times 10^5$ yrs and temperature of about 3000 K. Measurements at wavelengths from 0.05 cm to 80 cm indicate that it is consistent with being radiation from a blackbody of temperature 2.75 K $\pm$ 0.05 K (see Fig. 1 and ref. 4). Measurements of the isotropy indicate that the temperature is uniform to a part in 1000 on angular scales ranging from 10″ to 180°—to a part in 10^4 after the dipole component is removed (see Fig. 2 and ref. 5). The observations of the microwave background test the standard cosmology back to times as early as 100,000 yrs. According to the standard cosmology, when the Universe was 0.01 sec–300 sec old, corresponding to temperatures of $\simeq$ 10 MeV–0.1 MeV, conditions were right for the synthesis of a number of light nuclei. The predicted abundances of D, ^3He, ^4He, and ^7Li are consistent with their observed abundances provided that the baryon-to-photon ratio is

$$\eta \equiv n_b/n_\gamma \simeq (4-7) \times 10^{-10} \tag{1}$$

The baryon-to-photon ratio and the fraction of critical density contributed by baryons are related by: $\Omega_b h^2/T_{2.7}^3 \simeq 3.53 \times 10^7 \eta$ where $T_{2.7}$ is the microwave temperature in units of 2.7K and h is the present value of the Hubble constant in units of 100 km s^{-1} Mpc^{-1}. The allowed range for η corresponds to: $0.014 \lesssim \Omega_b h^2/T_{2.7}^3 \lesssim 0.025$, implying that baryons alone cannot provide the closure density. The concordance of theory and observation for D and ^{4}He is particularly compelling evidence in support of the standard cosmology as there are no known contemporary astrophysical sites which can simultaneously account for the primordial abundances of both these isotopes (see Fig. 3; see ref. 6 for further discussion of primordial nucleosynthesis). In sum, all the available evidence indicates that the standard cosmology provides an accurate accounting of the evolution of the Universe from 0.01 sec after the bang until today, some 15 or so Byr late—quite a remarkable achievement!

I will now briefly review the standard cosmology (more complete discussions of the standard cosmology are given in ref. 3). Throughout I will use high energy physics units, where $\hbar = k = c = 1$. The following conversion factors may be useful.

$$\begin{aligned}
1GeV^{-1} &= 0.197 \times 10^{-13} cm \\
1GeV^{-1} &= 0.658 \times 10^{-24} sec \\
1GeV &= 1.160 \times 10^{13} K \\
1GeV^4 &= 2.32 \times 10^{17} gcm^{-3} \\
1M_\odot &= 1.99 \times 10^{33} g \simeq 1.2 \times 10^{57} baryons \\
1pc &= 3.26 light-year \simeq 3.09 \times 10^{18} cm \\
1Mpc &= 3.09 \times 10^{24} cm \\
G_N &= 6.673 \times 10^{-8} cm^3 g^{-1} sec^{-2} \equiv m_{pl}^{-2} \\
& (m_{pl} = 1.22 \times 10^{19} GeV)
\end{aligned}$$

On large scales ($\gg 100Mpc$) the Universe is isotropic and homogeneous, as evidenced by the uniformity of the 2.7 K background radiation, the x-ray background, and counts of galaxies and radio sources, and so the standard cosmology is based on the maximally-symmetric Robertson–Walker line element

$$ds^2 = -dt^2 + R^2(t)[dr^2/(1-kr^2) + r^2 d\theta^2 + r^2 sin^2\theta d\phi^2] \tag{2}$$

where ds^2 is the square of the proper separation between two space-time events, k is the curvature signature (and can, by a suitable rescaling of R, be set equal to -1, 0, or +1), and $R(t)$ is the cosmic scale factor. The expansion of the Universe is embodied in $R(t)$—as $R(t)$ increases all proper (i.e., physical—as measured by meter sticks) distances scale with $R(t)$. The coordinates r, θ,and φ are comoving coordinates: test particles initially at rest will have constant comoving coordinates, and the velocity of NR test particles moving with respect to the comoving coordinates decrease ($\propto R(t)^{-1}$). The distance between two objects comoving with the expansion, e.g., two galaxies, simply scales up with $R(t)$. The momentum of any freely-propagating particle decreases as $1/R(t)$. In particular, the wavelength of a photon $\lambda \propto R(t)$, i.e., is redshifted by the expansion of the Universe

The coordinate distance at which curvature effects become noticeable is $|k|^{-1/2}$, which corresponds to the physical (or proper) distance

$$R_{curv} \simeq R(t)|k|^{-1/2} \tag{3}$$

—which one might call the curvature radius of the Universe. Note that R_{curv} also just scales with the cosmic scale factor $R(t)$.

The evolution of the cosmic scale factor and of the stress energy in the Universe are governed by the Friedmann equations:

$$H^2 \equiv (\dot{R}/R)^2 = 8\pi G\rho/3 - k/R^2 \tag{4}$$

$$d(\rho R^3) = -pd(R^3) \tag{5}$$

where ρ is the total energy density and p is the isotropic pressure. [The assumption of isotropy and homogeneity require that the stress-energy tensor take on the perfect fluid form: $T^{\mu}_{\nu} = diagonal(-\rho, p, p, p)$.] Because $\rho \propto R^{-n}$ ($n = 3$ for matter, $n = 4$ for radiation) it follows from Eqn.(4) that model Universes with $k < 0$ expand forever, while those with $k > 0$ must necessarily recollapse.

The expansion rate H (also known as the Hubble parameter) sets the characteristic timescale for the growth of $R(t)$: H^{-1} is the e-folding time for R. The present value of H is

$$H = 100h\ km\ sec^{-1}Mpc;$$

where the observational data strongly suggest that $0.4 \leq h \leq 1$ (ref. 7).

The sign of the spatial curvature k—and the ultimate fate of the Universe can be determined from measurements of ρ and H:

$$\begin{aligned} k/H^2R^2 &= \rho/(3H^2/8\pi G) - 1 \\ &\equiv \Omega - 1 \end{aligned} \tag{6}$$

where $\Omega = \rho/\rho_{crit}$ and $\rho_{crit} = 1.88h^2 \times 10^{-29} gcm^{-3} = 1.05 \times 10^4 h^2 eVcm^{-3}$. The curvature radius, R_{curv}, is related to Ω by

$$(R_{curv}/H^{-1})^2 = 1/(\Omega - 1) \tag{7}$$

A reliable and definitive determination of Ω has thus far eluded cosmologists. Based upon the luminous matter in the Universe (which is relatively easy to keep track of) we can set a lower bound to Ω

$$\Omega \geq \Omega_{LUM} \simeq 0.01$$

Based on dynamical techniques—which all basically involve Kepler's third law in one guise or another, the observational data seem to indicate that the material that clusters with visible galaxies on scales $\lesssim$ 10-30 Mpc accounts for

$$\Omega_{GAL} \simeq 0.1 - 0.3$$

Although Ω can, in principle, be determined by measurements of the deceleration parameter q_0

$$\begin{aligned} q_0 &\equiv -(\ddot{R}/R)/H^2, \\ &= \Omega(1 + 3p/\rho)/2, \end{aligned} \tag{8}$$

because of the difficulty of reliably determining q_0, the observations probably only restrict Ω to be less than a few[7]. [For a more thorough discussion of the amount of matter in the Universe see ref. 8.]

The best upper limit to Ω comes from the age of the Universe. The age of the Universe is related to the Hubble time H^{-1} by

$$t_u = f(\Omega)H_0^{-1} \tag{9}$$

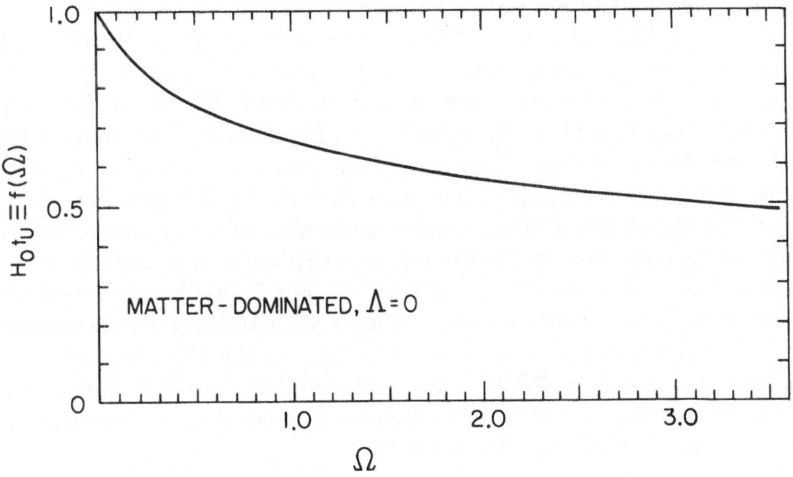

Figure 4a: The age of a matter-dominated, $\Lambda = 0$ model universe in Hubble units, $f(\Omega) = H_0 t_u$, as a function of Ω.

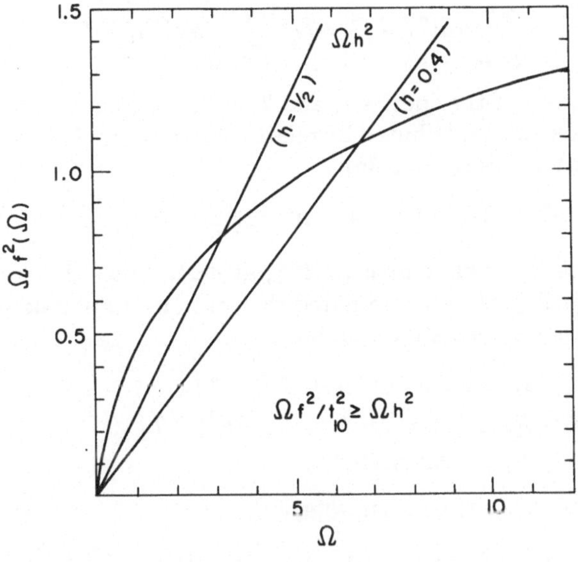

Figure 4b: The functions Ωf^2 and Ωh^2 ($h = 0.4$ and 0.5). The function $\Omega f^2/t_{10}^2$ bounds Ωh^2 from above. For $t_{10} \geq 1$ and $h \geq 0.4(0.5)$, this implies $\Omega h^2 \leq 1.1(0.8)$. The age of the Universe $t_u = t_{10} 10$ Gyr.

where $f(\Omega)$ is a monotonically decreasing function of Ω; $f(0) = 1$ and $f(1) = 2/3$ for a matter-dominated Universe and 1/2 for a radiation-dominated Universe. The dating of the oldest stars and the elements strongly suggest that the Universe is at least 10 Byr old—the best estimate being around 15 Byr old[9]. From Eqn(9) and $t_u \geq t_{10}10$ Byr it follows that $\Omega f^2/t_{10}^2 \geq \Omega h^2$. The function Ωf^2 is monotonically increasing and bounded above by $\pi^2/4$, implying that *independent* of h, $\Omega h^2 \leq 2.5/t_{10}^2$. Requiring $h \geq 0.4$ and $t_{10} \geq 1$, it follows that $\Omega h^2 \leq 1.1$ (see Fig. 4).

The energy density of the Universe quite naturally splits up into that contributed by relativistic particles—today the microwave photons and cosmic neutrino backgrounds, and that contributed by non-relativistic particles—baryons and whatever else! The energy density contributed by non-relativistic particles decreases as $R(t)^{-3}$—just due to the increase in the proper volume of the Universe, while that of relativistic particles varies as $R(t)^{-4}$—the additional factor of R being due to the fact that the momenta of relativistic particles are redshifted by the expansion. [Both of these results follow directly from Eqn(5).]

The energy density contributed by relativistic particles at temperature T is

$$\rho_R = g_*(T)\frac{\pi^2}{30}T^4 \tag{10}$$

where $g_*(T)$ counts the effective number of degrees of freedom (weighted by their temperature) of all the relativistic particle species (those with $m \ll T$):

$$g_*(T) = \sum_{Bose} g_B(T_i/T)^4 + 7/8 \sum_{Fermi} g_F(T_i/T)^4, \tag{11}$$

here T_i is the temperature of the species i, and T is the photon temperature.

Today the energy density contributed by relativistic particles (photons and three neutrino species) is very small ($g_* = 3.36$)

$$\Omega_{\gamma\ 3\nu}h^2 \simeq 4 \times 10^{-5}T_{2.7}^4$$

However, because $\rho_R \propto R^{-4}$, while $\rho_{NR} \propto R^{-3}$, at early times the energy density contributed by relativistic particles dominated that of non-relativistic particles. To be specific, the Universe was radiation-dominated for

$$\begin{aligned} t &\leq t_{EQ} \simeq 4 \times 10^{10} sec(\Omega h^2)^{-2}T_{2.7}^6, \\ R &\leq R_{EQ} \simeq 4 \times 10^{-5} R_{today}(\Omega h^2)^{-1}T_{2.7}^4, \\ T &\geq T_{EQ} \simeq 5.8 eV\, \Omega h^2 T_{2.7}^3. \end{aligned}$$

Therefore, at very early times Eqn(4) simplifies to

$$\begin{aligned} H &= (\dot{R}/R) = (4\pi^3 g_*/45)^{1/2}T^2/m_{pl}, \\ &= 1.66 g_*^{1/2}T^2/m_{pl} \end{aligned} \tag{12}$$

[Note since the curvature term varies as $R(t)^{-2}$ it too is negligible compared to the energy density in relativistic particles.] For reference, $g_*(\text{few}MeV) =$

10.75 (γ, $e^{\pm}$, $3\nu\bar{\nu}$); $g_*(100GeV) \simeq 110$ (γ, $8G$, $W^{\pm}Z$, 3 families of quarks and leptons, and 1 Higgs doublet).

So long as thermal equilibrium is maintained, the second Friedmann equation, Eqn(5), implies that the entropy per comoving volume, $S \propto sR^3$, remains constant. Here s is the entropy density which is dominated by the contribution from relativistic particles, and is

$$s = (\rho + p)/T \simeq (2\pi^2/45)g_* T^3. \tag{13}$$

The entropy density is just proportional to the number density of relativistic particles. Today the entropy density is just 7.04 times the number density of photons. The constancy of S means that $s \propto R^{-3}$, or that the ratio of any number density to s is just proportional to the number of that species per comoving volume. The baryon number-to-entropy ratio is

$$n_B/s \simeq (1/7)\eta \simeq (6-10) \times 10^{-11}$$

and since today the number density of baryons is much greater than that of antibaryons, this ratio is also the net baryon number per comoving volume—which is conserved so long as the rate of baryon-number non-conserving reactions is small.

The constancy of S implies that

$$T \propto g_*(T)^{-1/3} R(t)^{-1}. \tag{14}$$

Whenever g_* is constant, this means that $T \propto R(t)^{-1}$. Together with Eqn(12) this gives

$$\begin{aligned} R(t) &= R(t_0)(t/t_0)^{1/2}, \\ t &\simeq 1/2H^{-1} \simeq 0.3 g_*^{-1/2} m_{pl}/T^2, \\ &\simeq 2.4 \times 10^{-6} sec\; g_*^{-1/2}(T/GeV)^{-2}. \end{aligned} \tag{15}$$

Finally, let me mention one more important feature of the standard cosmology, the existence of particle horizons. In the standard cosmology the distance a photon could have traveled since the bang is finite, meaning that at a given epoch the Universe is comprised of many causally-distinct domains. Photons travel on paths characterized by $ds^2 = 0$; for simplicity and without loss of generality consider a trajectory with $d\varphi = d\theta = 0$. The coordinate distance traversed by a photon since 'the bang' is

$$\int_0^t dt'/R(t')$$

which corresponds to the physical distance (measured at time t)

$$d_H(t) = R(t) \int_0^t dt'/R(t'). \tag{16}$$

If $R(t) \propto t^n$ and $n < 1$, then the horizon distance $d_H(t)$ is finite and $d_H(t) = t/(1-n) = nH^{-1}/(1-n) \simeq t$.

Note that even if $d_H(t)$ diverges (e.g., if $R(t) \propto t^n$ with $n > 1$), the Hubble radius H^{-1} still sets the scale of the 'Physics Horizon'. All physical distances scale with $R(t)$. Thus microphysical processes operating on a timescale $\gtrsim H^{-1}$ will have their effects distorted by the expansion, strongly suggesting that a coherent microphysical process can only operate over a time interval of order H^{-1}. Then, causally-coherent microphysical processes can only operate on distances $\leq$ the Hubble radius, H^{-1}. The intuitive notion that the Hubble radius acts as the 'Physics Horizon' is borne out quantitatively time and time again, and so it is useful to think of H^{-1} as the maximum scale for microphysical processes.

During the radiation-dominated era $n = 1/2$ and $d_H(t) = 2t$; the entropy and baryon number within the horizon at a given time are easily computed:

$$\begin{aligned} S_{HOR} &= (4\pi/3)t^3 s, \\ &\simeq 0.05 g_*^{-1/2} (m_{pl}/T)^3, \\ N_{B-HOR} &= (n_B/s) S_{HOR}, \\ &\simeq 10^{-12} (m_{pl}/T)^3, \\ &\simeq 10^{-2} M_\odot (T/MeV)^{-3}. \end{aligned}$$

We can compare these numbers to the entropy and baryon number contained within the present horizon volume:

$$\begin{aligned} S_U &\simeq 10^{88}, \\ N_{BU} &\simeq 10^{78}. \end{aligned}$$

Evidently, in the standard cosmology the comoving volume which corresponds to the part of the Universe which is presently observable contained many, many horizon volumes at early times. This is an important point to which we shall return shortly.

3. SHORTCOMINGS OF THE STANDARD COSMOLOGY

The standard cosmology is very successful—it provides us with a reliable framework for describing the history of the Universe as early as 10^{-2} sec after the bang (when the temperature was about 10 MeV) and perhaps as early as 10^{-43} sec after the bang (see Fig. 5). In sum, the standard cosmology is a great achievement. [There is nothing in our present understanding of physics that would indicate that it is incorrect to extrapolate the standard cosmology back to times as early as 10^{-43} sec—the fundamental constituents of matter, quarks and leptons, are point-like particles and their known interactions should remain 'weak' up to energies as high as 10^{19} GeV—justifying the dilute gas approximation made in writing $\rho_r \propto T^4$. (This fact was first pointed out by Collins and Perry[9a]). However, at times earlier than 10^{-43} sec, corresponding to temperatures greater than 10^{19} GeV, quantum corrections to general relativity—a classical theory, should become very significant.]

However, it is not without its shortcomings. There are a handful of very important and fundamental cosmological facts which, while it can accommodate, it in no way elucidates. I will briefly review these puzzling facts.

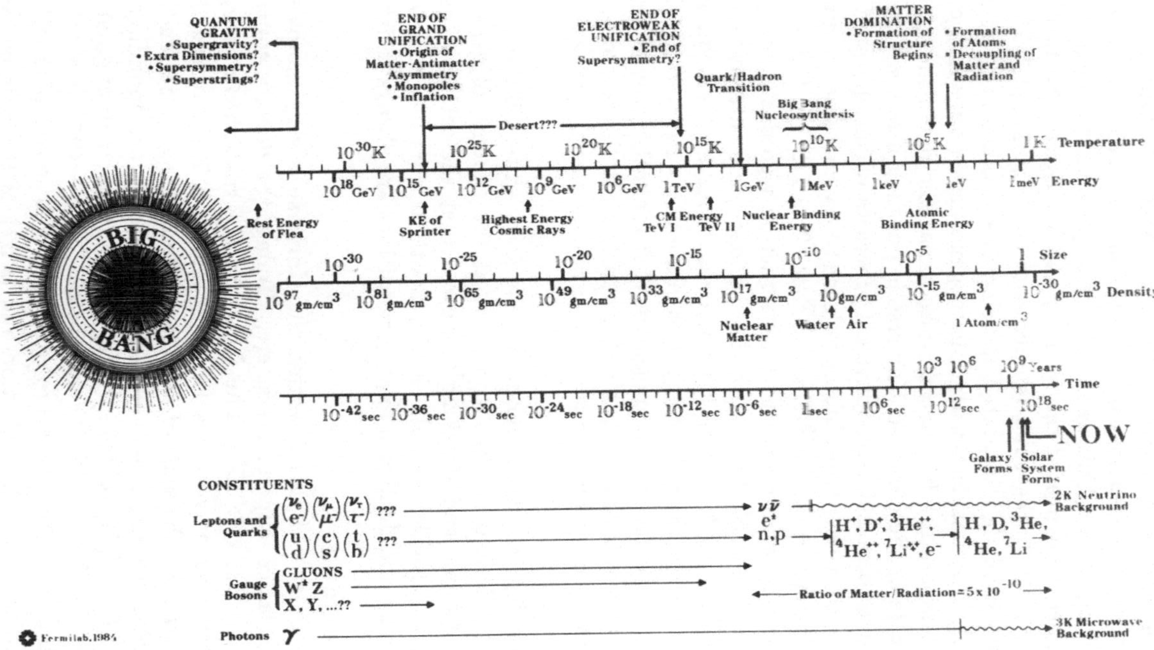

Figure 5: History of the Universe according to the standard cosmology and currently fashionable ideas in particle theory.

(1–2) Large-scale Isotropy and Homogeneity

The observable Universe (size $\simeq H^{-1} \simeq 10^{28} cm \simeq 3000$ h^{-1} Mpc) is to a high degree of precision isotropic and homogeneous on the largest scales, say $> 100 Mpc$. [Of course, our knowledge of the Universe outside our past light cone is very limited; see ref. 10.] The best evidence for the isotropy and homogeneity is provided by the uniformity of the cosmic background temperature (see Fig. 2): $(\delta T/T) < 10^{-3}$ (10^{-4} if the dipole anisotropy is interpreted as being due to our motion relative to the cosmic rest frame). Large-scale density inhomogeneities or anisotropic expansion would result in temperature fluctuations of comparable magnitude (see refs. 11, 12). The smoothness of the observed Universe is puzzling if one wishes to understand it as being due to causal, microphysical processes which operated during the early history of the Universe. Our Hubble volume today contains an entropy of about 10^{88}. At decoupling ($t \simeq 6 \times 10^{12} (\Omega h^2)^{-1/2} sec$, $T \simeq 1/3 eV$), the last epoch when matter and radiation were known to be interacting vigorously and particle interactions might have been able to smooth the radiation, the entropy within the horizon was only about 8×10^{82}; that is, the comoving volume which contains the presently-observable Universe, then was comprised of about 2×10^5 causally-distinct regions. How is it that they came to be homogeneous? Put another way, the particle horizon at decoupling only subtends an angle of about $1/2°$ on the sky today—how is it that the cosmic background temperature is so uniform on angular scales much greater than this?

The standard cosmology can accommodate these facts—after all the FRW cosmology is exactly isotropic and homogeneous, but at the expense of very special initial data. In 1973 Collins and Hawking[13] showed that the set of initial data which evolve to a Universe which globally is as smooth as ours has measure zero (provided that the strong and dominant energy conditions are always satisfied).

(3) Small-scale Inhomogeneity

As any real astronomer will gladly testify, the Universe is very lumpy—stars, galaxies, clusters of galaxies, superclusters, etc. Today, the density contrast on the scale of galaxies is: $\delta\rho/\rho \simeq 10^5$. The fact that the microwave background radiation is very uniform even on very small angular scales ($\ll 1°$) indicates that the Universe was smooth even on the scale of galaxies at decoupling. [The relationship between the angle on the sky and mass contained within the corresponding length scale at decoupling is: $\theta \simeq 1'(M/10^{12} M_\odot)^{1/3} \Omega^{-1/3} h^{1/3}$.] On small angular scales: $\delta T/T \simeq c(\delta\rho/\rho)_{dec}$, where the numerical constant $c \simeq 10^{-1} - 10^{-2}$ (see ref. 12 for further details). Whence came the structure which today is so conspicuous?

Once matter decouples from the radiation and is free of the pressure support provided by the radiation, any density inhomogeneities present will grow via the Jeans (or gravitational instability)—in the linear regime, $\delta\rho/\rho \propto R(t)$. [If the mass density of the Universe is dominated by a collisionless particle species, e.g., a light, relic neutrino species or relic axions, density perturbations in these particles can begin to grow as soon as the Universe becomes matter-dominated.] In order to account for the present structure, density perturbations of amplitude $\sim 10^{-3}$ or so at decoupling are necessary on the scale of galaxies. The standard cosmology sheds no light as to the origin or nature (spectrum and type—adiabatic or isothermal) of the primordial density perturbations so crucial for understanding the structure observed in the Universe today. [For a review of the formation of structure in the Universe according to the gravitational instability picture, see ref. 14.]

(4) Flatness (or Oldness) of the Universe

The observational data suggest that

$$0.01 \leq \Omega \leq \text{few}.$$

Ω is related to both the expansion rate of the Universe and the curvature radius of the Universe:

$$\Omega = 8\pi G\rho/3H^2 \equiv H^2_{crit}/H^2, \tag{17}$$

$$|\Omega - 1| = (H^{-1}/R_{curv})^2, \tag{18}$$

The fact that Ω is not too different from unity today implies that the present expansion rate is close to the critical expansion rate and that the curvature radius of the Universe is comparable to or larger than the Hubble radius. As the Universe expands Ω does not remain constant, but evolves away from 1

$$\Omega = 1/(1 - x(t)), \tag{19}$$

$$x(t) = (k/R^2)/(8\pi G\rho/3), \tag{20}$$

$$x(t) \propto \begin{cases} R(t)^2 & \text{radiation} - \text{dominated} \\ R(t) & \text{matter} - \text{dominated} \end{cases}$$

That Ω is still of order unity means that at early times it was equal to 1 to a very high degree of precision:

$$|\Omega(10^{-43}sec) - 1| \simeq O(10^{-60}),$$
$$|\Omega(1sec) - 1| \simeq O(10^{-16}).$$

This in turn implies that at early times the expansion rate was equal to the critical rate to a high degree of precision and that the curvature of the Universe was much, much greater than the Hubble radius. If it were not, i.e., suppose that $|(k/R^2)/(8\pi G\rho/3)| \simeq O(1)$ at $t \simeq 10^{-43}sec$, then the Universe would have collapsed after a few Planck times ($k > 0$) or would have quickly become curvature-dominated, ($k < 0$), in which case $R(t) \propto t$ and $t(T = 3K) \leq 10^{-11}$ sec! Why was this so?

The so-called flatness problem has sometimes been obscured by the fact that it is conventional to rescale $R(t)$ so that $k = -1$, 0, or $+1$, making it seem as though there are but three FRW models. However, that clearly is not the case; there are an infinity of models, specified by the curvature radius $R_{curv} = R(t)|k|^{-1/2}$ at some given epoch, say the planck epoch. Our model corresponds to one with a curvature radius that exceeds its initial Hubble radius by 30 orders-of-magnitude. Again, this fact can be accommodated by FRW models, but the extreme flatness of our Universe is in no way explained by the standard cosmology. [The flatness problem and the naturalness of the $k = 0$ model have been emphasized by Dicke and Peebles.[14a]]

(5) Baryon Number of the Universe

There is ample evidence (see ref. 15) for the dearth of antimatter in the observable Universe. That fact together with the baryon-to-photon ratio ($\eta \simeq 4 - 7 \times 10^{-10}$) means that our Universe is endowed with a net baryon number, quantified by the baryon number-to-entropy ratio

$$n_B/s \simeq (6 - 10) \times 10^{-11},$$

which in the absence of baryon number non-conserving interactions or significant entropy production is proportional to the constant net baryon number per comoving volume which the Universe has always possessed. Until five or so years ago this very fundamental number was without explanation. Of course it is now known that in the presence of interactions that violate B, C, and CP a net baryon asymmetry will evolve dynamically. Such interactions are predicted by Grand Unified Theories (or GUTs) and 'baryogenesis' is one of the great triumphs of the marriage of grand unification and cosmology. [See ref. 16 for a review of grand unification.] If the baryogenesis idea is correct, then the baryon asymmetry of the Universe is subject to calculation just as the primordial Helium abundance is. Although the idea is very attractive and certainly appears to be qualitatively correct, a precise calculation of the baryon number-to-entropy ratio cannot be performed until *The* Grand Unified Theory is known. [Baryogenesis is reviewed in ref. 17.]

(6) The Monopole Problem

If the great success of the marriage of GUTs and cosmology is baryogenesis, then the great disappointment is 'the monopole problem'. 't Hooft-Polyakov monopoles[18] are a generic prediction of GUTs. In the standard cosmology (and for the simplest GUTs) monopoles are grossly overproduced during the GUT symmetry-breaking transition, so much so that the Universe would reach its present temperature of 3K at the very tender age of 30,000 yrs! [For a detailed discussion of the monopole problem, see refs. 19, 20.] Although the monopole problem initially seemed to be a severe blow to the Inner Space/Outer Space connection, as it has turned out it provided us with a valuable piece of information about physics at energies of order 10^{14} GeV and the Universe at times as early as 10^{-34} sec—the standard cosmology and the simplest GUTs are definitely incompatible! As it turned out, it was the search for a solution to the monopole problem which in the end led Guth to come upon the inflationary Universe scenario[21,22].

(7) The Smallness of the Cosmological Constant

With the possible exception of supersymmetry/supergravity and superstring theories, the absolute scale of the scalar potential $V(\phi)$ is not specified (here ϕ represents the scalar fields in the theory, be they fundamental or composite). A constant term in the scalar potential is equivalent to a cosmological term (the

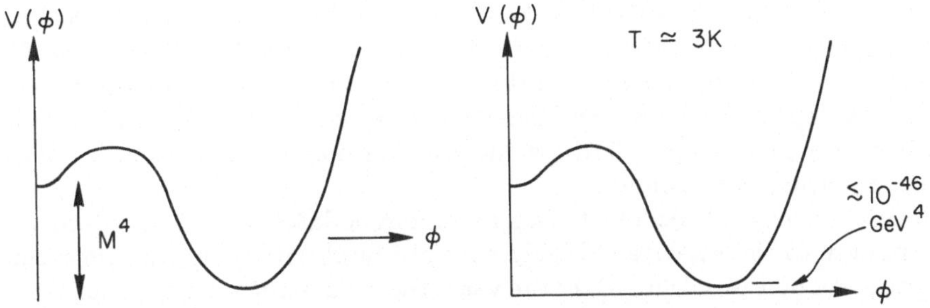

Figure 6: In gauge theories the vacuum energy is a function of one or more scalar fields (here denoted collectively as ϕ); however, the absolute energy scale is not set. Vacuum energy behaves like a cosmological term; the present expansion rate of the Universe constrains the value of the vacuum energy today to be $\leq 10^{-46}$ GeV^4.

scalar potential contributes a term $Vg_{\mu\nu}$ to the stress energy of the Universe[23]). At low temperatures (say temperatures below any scale of spontaneous symmetry-breaking) the constant term in the potential receives contributions from all the stages of SSB—chiral symmetry breaking, electroweak SSB, GUT SSB, etc. The observed expansion rate of the Universe ($H = 100h\ km\ sec^{-1}Mpc^{-1}$) limits the total energy density of the Universe to be

$$\rho_{TOT} \leq O(10^{-46} GeV^4).$$

Making the seemingly very reasonable assumption that all stress energy self-gravitates (which is dictated by the equivalence principle) it follows that the vacuum energy of our $SU(3) \times U(1)$ vacuum must be less than $10^{-46} GeV^4$. Compare this to the scale of the various contributions to the scalar potential: $0(M^4)$ for physics associated with a symmetry breaking scale of M

$$V_{today}/M^4 \leq \rho_{TOT}/M^4 \leq \begin{cases} 10^{-122} & M = m_{pl} \\ 10^{-102} & M = 10^{14} GeV \\ 10^{-56} & M = 300 GeV \\ 10^{-46} & M = 1 GeV \end{cases}$$

At present there is no explanation for the vanishingly small value of the energy density of our very unsymmetrical vacuum. It is easy to speculate that a fundamental understanding of the smallness of the cosmological constant will likely involve an intimate link between gravity and quantum field theory.

Today we can be certain the vacuum energy is small and plays a minor role in the dynamics of the expansion of the Universe (compared to the potential role that it could play). If we accept this as an empirical determination of the absolute scale of the scalar potential $V(\phi)$, it then follows that the energy density associated with an expectation value of ϕ near zero is enormous—of order M^4 (see Fig. 6) and therefore could have played an important role in the dynamics of the very early Universe. Accepting this ***empirical determination*** of the zero of vacuum energy—which is a very great leap of faith, is the starting point for inflation. In fact, the rest of the journey is downhill.

All of these cosmological facts can be accommodated by the standard model, but seemingly at the expense of highly special initial data (the possible exception being the monopole problem). Over the years there have been a number of attempts to try to understand and/or explain this apparent dilemma of initial data. Inflation is the most recent attempt and I believe shows great promise. Let me begin by briefly mentioning the earlier attempts:

- Mixmaster Paradigm–Starting with a solution with a singularity which exhibits the features of the most general singular solutions known (the so-called mixmaster model) Misner and his coworkers hoped that they could show that particle viscosity would smooth out the geometry. In part because horizons still effectively exist in the mixmaster solution 'the chaotic cosmology program' has proven unsuccessful (for further discussion see refs. 24).
- Nature of the Initial Singularity–Penrose[25]explored the possibility of explaining the observed smoothness of the Universe by restricting the kinds of initial singularities which are permitted in Nature (those with vanishing Weyl curvature). In a sense his approach is to postulate a law of physics governing allowed initial data.
- Quantum Gravity Effects–The first two solutions involve appealing to classical gravitational effects. A number of authors have suggested that quantum gravity effects might be responsible for smoothing out the space-time geometry (deWitt[26];Parker[27]; Zel'dovich[28]; Starobinskii[29]; Anderson[30]; Hartle and Hu[31]; Fischetti et al.[32]). The basic idea being that anisotropy and/or inhomogeneity would drive gravitational particle creation, which due to back reaction effects would eliminate particle horizons and smooth out the geometry. Recently, Hawking and Hartle[33] have advocated the Quantum Cosmology approach to actually compute the initial state. All of these approaches necessarily involve events at times $\lesssim 10^{-43} sec$ and energy densities $\gtrsim m_{pl}^4$.
- Anthropic Principle–Some (see, e.g., refs. 34) have suggested (or in some cases even advocated) 'explaining' many of the puzzling features of the Universe around us (and in some cases, even the laws of physics!) by arguing that unless they were as they are intelligent life would not have been able to develop and observe them! Hopefully we will not have to resort to such an explanation.

The approach of inflation is somewhat different from previous approaches. Inflation (at least from my point-of-view) is based upon well-defined and reasonably well-understood microphysics (albeit, some of it very speculative). That microphysics is:

- Classical Gravity (general relativity), at least as an effective, low-energy theory of gravitation.

⋆ 'Modern Particle Physics'–grand unification, supersymmetry /supergravity, field theory limit of superstring theories, etc. at energy scales $\lesssim m_{pl}$

As I will emphasize, in all viable models of inflation the inflationary period (at least the portion of interest to us) takes place well after the planck epoch, with the energy densities involved being far less than m_{pl}^4 (although semi-classical quantum gravity effects might have to be included as non-renormalizable terms in the effective Lagrangian). Of course, it could be that a resolution to the cosmological puzzles discussed above involves both 'modern particle physics' and quantum gravitational effects in their full glory (as in a fully ten dimensional quantum theory of strings).

I will not take the time or the space here to review the historical development of our present view of inflation; I refer the interested reader to the interesting paper on this subject by Lindley[35]. It suffices to say that Guth's very influential paper of 1981 was the 'shot heard 'round the world' which initiated the inflation revolution,[22] and that Guth's doomed original model (see Guth and Weinberg[36]; Hawking et al[36]) was revived by Linde's[37] and Albrecht and Steinhardt's[38] variant, 'new inflation'. I will focus all of my attention on the present status of the 'slow-rollover' model of Linde[37] and Albrecht and Steinhardt[38].

4. NEW INFLATION—THE SLOW-ROLLOVER TRANSITION

The basic idea of the inflationary Universe scenario is that there was an epoch when the vacuum energy density dominated the energy density of the Universe. During this epoch $\rho \simeq V \simeq$ constant, and thus $R(t)$ grew exponentially ($\propto exp(Ht)$), allowing a small, causally-coherent region (initial size $\leq H^{-1}$) to grow to a size which encompasses the region which eventually becomes our presently-observable Universe. In Guth's original scenario[22], this epoch occurred while the Universe was trapped in the false ($\phi = 0$) vacuum during a strongly, first-order phase transition. In new inflation, the vacuum-dominated, inflationary epoch occurs while the region of the Universe in question is slowly, but inevitably, evolving toward the true, SSB vacuum. Rather than considering specific models in this section, I will try to discuss new inflation in the most general context. For the moment I will however assume that the epoch of inflation is associated with a first-order, SSB phase transition, and that the Universe is in thermal equilibrium before the transition. As we shall see later new inflation is more general than these assumptions. But for definiteness (and for historical reasons), let me begin by making these assumptions.

Consider a SSB phase transition characterized by an energy scale M. For $T \geq T_c \simeq 0(M)$ the symmetric ($\phi = 0$) vacuum is favored, i.e., $\phi = 0$ is the global minimum of the finite temperature effective potential $V_T(\phi)$ (=free energy density). As T approaches T_c a second minimum develops at $\phi - 0$, and at $T - T_c$, the two minima are degenerate. At temperatures below T_c the SSB ($\phi = \sigma$) minimum is the global minimum of $V_T(\phi)$ (see Fig. 7). However, the Universe does not instantly make the transition from $\phi = 0$ to $\phi = \sigma$; the details and time required are a question of dynamics. [The scalar field ϕ is the order parameter for the SSB transition under discussion; in the spirit of generality ϕ might be a gauge singlet field or might have nontrivial transformation properties under the gauge group,

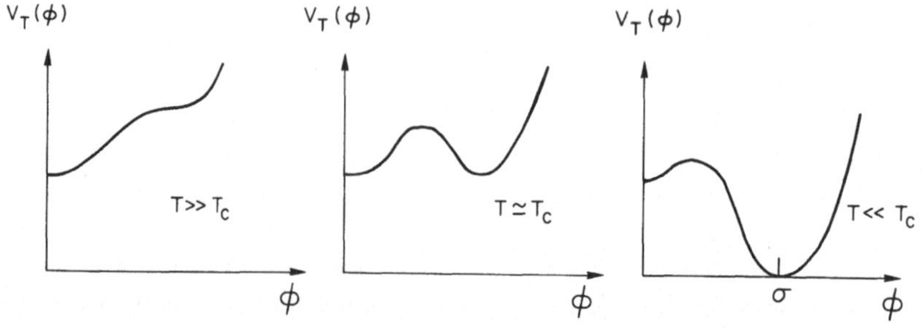

Figure 7: The finite temperature effective potential as a function of T (schematic). The Universe is usually assumed to start out in the high temperature, symmetric minimum ($\phi = 0$) of the potential and must eventually evolve to the low temperature, asymmetric minimum ($\phi = \sigma$). The evolution of ϕ from $\phi = 0$ to $\phi = \sigma$ can prove to be very interesting—as in the case of an inflationary transition.

possibly even responsible for the SSB of the GUT.] Once the temperature of the Universe drops below $T_c \simeq O(M)$, the potential energy associated with ϕ being far from the minimum of its potential, $V \simeq V(0) \simeq M^4$, dominates the energy density in radiation ($\rho_r < T_c^4$), and causes the Universe to expand exponentially. During this exponential expansion (known as a deSitter phase) the temperature of the Universe decreases exponentially causing the Universe to supercool. The exponential expansion continues so long as ϕ is far from its SSB value. Now let's focus on the evolution of ϕ.

Assuming a barrier exists between the false and true vacua, thermal fluctuations and/or quantum tunneling must take ϕ across the barrier. The dynamics of this process determine when and how the process occurs (bubble formation, spinodal decomposition, etc.) and the value of ϕ after the barrier is penetrated. If the action for bubble nucleation remains large, $S_b \gg 1$, then the barrier will be overcome by the nucleation of Coleman-deLuccia bubbles;[39] on the other hand if the action for bubble nucleation becomes of order unity, then the Universe will undergo spinodal decomposition, and irregularly-shaped fluctuation regions will form (see Fig. 8; for a more detailed discussion of the barrier penetration process see refs. 38–40). For definiteness suppose that the barrier is overcome when the temperature is T_{MS} and that after the barrier is penetrated the value of ϕ is ϕ_0. From this point the journey to the true vacuum is downhill (literally). For the moment let us assume that the evolution of ϕ is adequately described by semi-classical equations of motion:

$$\ddot{\phi} + 3H\dot{\phi} + \Gamma\dot{\phi} + V' = 0, \tag{21}$$

where ϕ has been normalized so that its kinetic term in the Lagrangian is $1/2\partial_\mu\phi\partial^\mu\phi$, and prime indicates a derivative with respect to ϕ. The subscript T on V has been dropped; for $T \ll T_c$ the temperature dependence of V_T can be neglected and the zero temperature potential ($\equiv V$) can be used. The $3H\dot{\phi}$ term

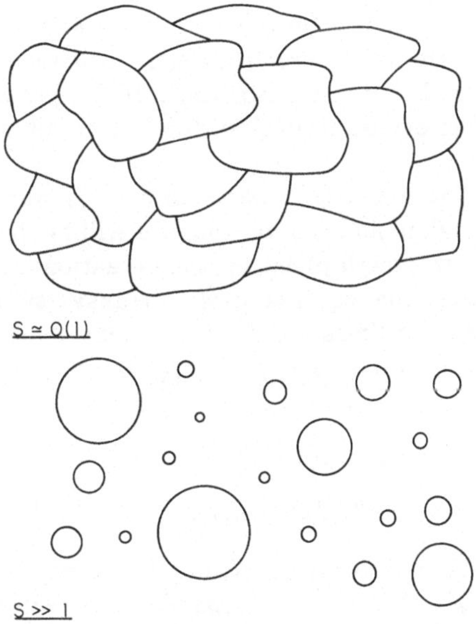

Figure 8: If the tunneling action is large ($S \gg 1$), barrier penetration will proceed via bubble nucleation, while in the case that it becomes small ($S \simeq O(1)$), the Universe will fragment into irregularly-shaped fluctuation regions. The very large scale (scale $\gg$ bubble or fluctuation region) structure of the Universe is determined by whether $S \simeq O(1)$—in which case the Universe is comprised of irregularly-shaped domains, or $S \gg O(1)$—in which case the Universe is comprised of isolated bubbles.

acts like a frictional force, and arises because the expansion of the Universe 'red-shifts away' the kinetic energy of $\phi(\propto R^{-3})$. The $\Gamma\dot{\phi}$ term accounts for particle creation due to the time-variation of ϕ[refs. 41, 42]. The quantity Γ is determined by the particles which couple to ϕ and the strength with which they couple ($\Gamma^{-1} \simeq$ lifetime of a ϕ particle). As usual, the expansion rate H is determined by the energy density of the Universe:

$$H^2 = 8\pi G\rho/3, \tag{22}$$

$$\rho \simeq 1/2\dot{\phi}^2 + V(\phi) + \rho_r, \tag{23}$$

where ρ_r represents the energy density in radiation produced by the time variation of ϕ. [For $T_{MS} \ll T_c$ the original thermal component makes a negligible contribution to ρ.] The evolution of ρ_r is given by

$$\dot{\rho}_r + 4H\rho_r = \Gamma\dot{\phi}^2, \tag{24}$$

where the $\Gamma\dot{\phi}^2$ term accounts for particle creation by ϕ.

In writing Eqns.(21–24) I have implicitly assumed that ϕ is spatially homogeneous. In some small region (inside a bubble or a fluctuation region) this will be a good approximation. The size of this smooth region will turn out to be unimportant; take it to be of order the 'Physics Horizon', H^{-1}—certainly, it is not likely to be larger. Now follow the evolution of ϕ within the small, smooth patch of size H^{-1}.

If $V(\phi)$ is sufficiently flat somewhere between $\phi = \phi_0$ and $\phi = \sigma$, then ϕ will evolve very slowly in that region, and the motion of ϕ will be 'friction-dominated' so that $3H\dot{\phi} \simeq -V'$ (in the slow growth phase particle creation is not important[43]). If V is sufficiently flat, then the time required for ϕ to transverse the flat region can be long compared to the expansion timescale H^{-1}; for definiteness say, $\tau_\phi = 100H^{-1}$. During this slow growth phase $\rho \simeq V(\phi) \simeq V(\phi = 0)$; both ρ_r and $1/2\dot{\phi}^2$ are $\ll V(\phi)$. The expansion rate H is then just

$$\begin{aligned} H &\simeq (8\pi V(0)/3m_{pl}^2)^{1/2} \\ &\simeq O(M^2/m_{pl}), \end{aligned} \tag{25}$$

where $V(0)$ is assumed to be of order M^4. While $H \simeq$ constant, R grows exponentially: $R \propto exp(Ht)$; for $\tau_\phi = 100H^{-1}$, R expands by a factor of e^{100} during the slow rolling period, and the physical size of the smooth region increases to $e^{100}H^{-1}$.

As the potential steepens, the evolution of ϕ quickens. Near $\phi = \sigma$, ϕ oscillates around the SSB minimum with frequency m_ϕ : $m_\phi^2 \simeq V''(\sigma) \simeq O(M^2) \gg H^2 \simeq M^4/m_{pl}^2$. As ϕ oscillates about $\phi = \sigma$ its motion is damped both by particle creation and the expansion of the Universe. If $\Gamma^{-1} \ll H^{-1}$, then coherent field energy density $(V + 1/2\dot{\phi}^2)$ is converted into radiation in less than an expansion time ($\Delta t_{RH} \simeq \Gamma^{-1}$), and the patch is reheated to a temperature $T \simeq 0(M)$—the vacuum energy is efficiently converted into radiation ('good reheating'). On the other hand, if $\Gamma^{-1} \gg H^{-1}$, then ϕ continues to oscillate and the coherent field energy redshifts away with the expansion: $(V + 1/2\dot{\phi}^2) \propto R^{-3}$—the coherent energy behaves like non-relativistic matter. Eventually, when $t \simeq \Gamma^{-1}$ the energy in radiation begins to dominate that in coherent field oscillations, and the patch is reheated to a temperature $T \simeq (\Gamma/H)^{1/2}M \simeq (\Gamma m_{pl})^{1/2} \ll M$ ('poor reheating'). The evolution of ϕ is summarized schematically in Fig. 9. In the next section I will discuss the all-important scalar field dynamics in great detail.

For the following discussion let us assume 'good reheating' ($\Gamma \gg H$). After reheating the patch has a physical size $e^{100}H^{-1}$ ($\simeq 10^{17}$cm for $M \simeq 10^{14}$GeV), is at a temperature of order M, and in the approximation that ϕ was initially constant throughout the patch, the patch is exactly smooth. From this point forward the region evolves like a radiation-dominated FRW model. How have the cosmological conundrums been 'explained'?

First, *the homogeneity and isotropy*; our observable Universe today ($\simeq 10^{28}$cm) had a physical size of about 10cm ($= 10^{28}$cm $\times$ $3K/10^{14}$GeV) when T was 10^{14}GeV—thus it lies well within one of the smooth regions produced by the inflationary epoch. Put another way, inflation has resulted in a smooth patch which

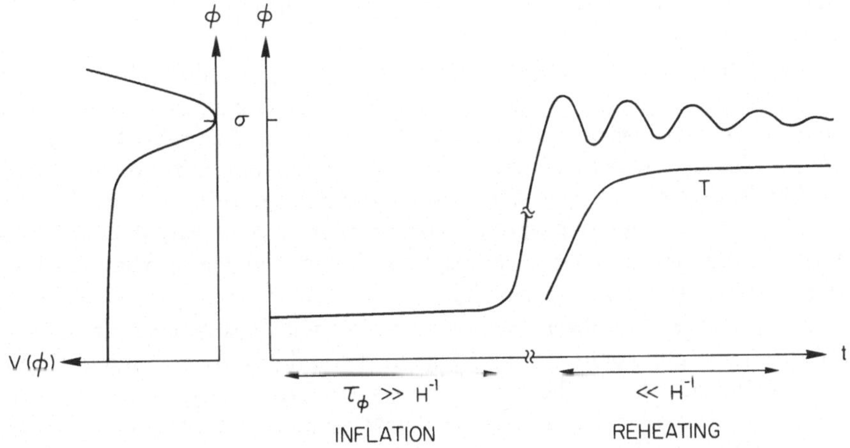

Figure 9: Evolution of ϕ and the temperature inside the bubble or fluctuation region (schematic). Early on ϕ evolves slowly (relative to the expansion timescale), then as the potential steepens ϕ evolves rapidly (on the expansion timescale). The oscillations of ϕ are damped by particle creation, which leads to the reheating of the bubble or fluctuation region.

contains an entropy of order $(10^{17}\mathrm{cm})^3 \times (10^{14}\mathrm{GeV})^3 \simeq 10^{134}$, which is much, much greater than that within the presently-observed Universe ($\simeq 10^{88}$). Before inflation that same volume contained only a very small amount of entropy, about $(10^{-23}cm)^3(10^{14}GeV)^3 \simeq 10^{14}$. The key to inflation then is the highly non-adiabatic event of reheating (see Fig. 10). The very large-scale cosmography depends upon the state of the Universe before inflation and how inflation was initiated (bubble nucleation or spinodal decomposition); see ref. 45 for further discussion.

Since we have assumed that ϕ is spatially constant within the bubble or fluctuation region, after reheating the patch in question is precisely uniform, and at this stage *the inhomogeneity puzzle* has not been solved. Inflation *has* produced a smooth manifold on which small fluctuations can be impressed. Due to deSitter space produced quantum fluctuations in ϕ, ϕ is not exactly uniform even in a small patch. Later, I will discuss the density inhomogeneities that result from the quantum fluctuations in ϕ.

The flatness puzzle involves the smallness of the ratio of the curvature term to the energy density term. This ratio is exponentially smaller after inflation: $x_{\text{after}} \simeq e^{-200}x_{\text{before}}$ since the energy density before and after inflation is $0(M^4)$, while k/R^2 has exponentially decreased (by a factor of e^{200}). Since the ratio x is reset to an exponentially small value, the inflationary scenario predicts that today Ω should be $1 \pm 0(10^{-BIG\#})$.

If the Universe is reheated to a temperature of order M, a *baryon asymmetry* can evolve in the usual way, although the quantitative details may be different[17,43]. If the Universe is not efficiently reheated ($T_{RH} \ll M$), it may be possible for n_B/s to be produced directly in the decay of the coherent field oscillations[41–44] (which behave just like NR ϕ particles); this possibility will be discussed later. In any case, it is absolutely necessary to have baryogenesis occur after reheating since any baryon number (or any other quantum number) present before inflation is diluted by a factor of $(M/T_{MS})^3\ exp(3H\tau_\phi)$—the factor by which the total entropy increases. Note that if C, CP are violated spontaneously, then ϵ (and n_B/s) could have a different sign in different patches—leading to a Universe which on the very largest scales ($\gg e^{100}H^{-1}$) is baryon symmetric.

Since the patch that our observable Universe lies within was once (at the beginning of inflation) causally-coherent, the Higgs field could have been aligned throughout the patch (indeed, this is the lowest energy configuration), and thus there is likely to be ≤ 1 monopole within the entire patch which was produced as a topological defect. *The glut of monopoles* which occurs in the standard cosmology does not occur. [The production of other topological defects (such as domain walls, etc.) is avoided for similar reasons.] Some monopoles will be produced after reheating in rare, very energetic particle collisions[46a]. The number produced is both exponentially small and exponentially uncertain. [In discussing the resolution of the monopole problem I am tacitly assuming that the SSB of the GUT is occurring during the SSB transition in question, or that it has already occurred in an earlier SSB transition; if not then one has to worry about the monopoles produced in the subsequent GUT transition.] Although monopole production is intrinsically small in inflationary models, the uncertainties in the number of monopoles produced are exponential and of course, it is also possible that monopoles might be produced as topological defects in a subsequent phase transition[46b] (although it may be difficult to arrange that they not be overproduced).

Finally, the inflationary scenario sheds no light upon *the cosmological constant puzzle.* Although it can potentially successfully resolve all of the other puzzles in my list, inflation is, in some sense, a house of cards built upon the cosmological constant puzzle.

5. SCALAR FIELD DYNAMICS

The evolution of the scalar field ϕ is key to understanding new inflation. In this section I will focus on the semi-classical dynamics of ϕ. Later, I will return to the question of the validity of the semi-classical approach. Much of what I will discuss here is covered in more detail in ref. 47.

Stated in the most general terms, the current view of inflation is that it involves the dynamical evolution of a very weakly-coupled scalar field (hereafter referred to as ϕ) which is, for one reason or another, initially displaced from the minimum of its potential (see Fig. 11). While it is displaced from its minimum, and is slowly-evolving toward that minimum, its potential energy density drives the rapid (exponential) expansion of the Universe, now known as inflation.

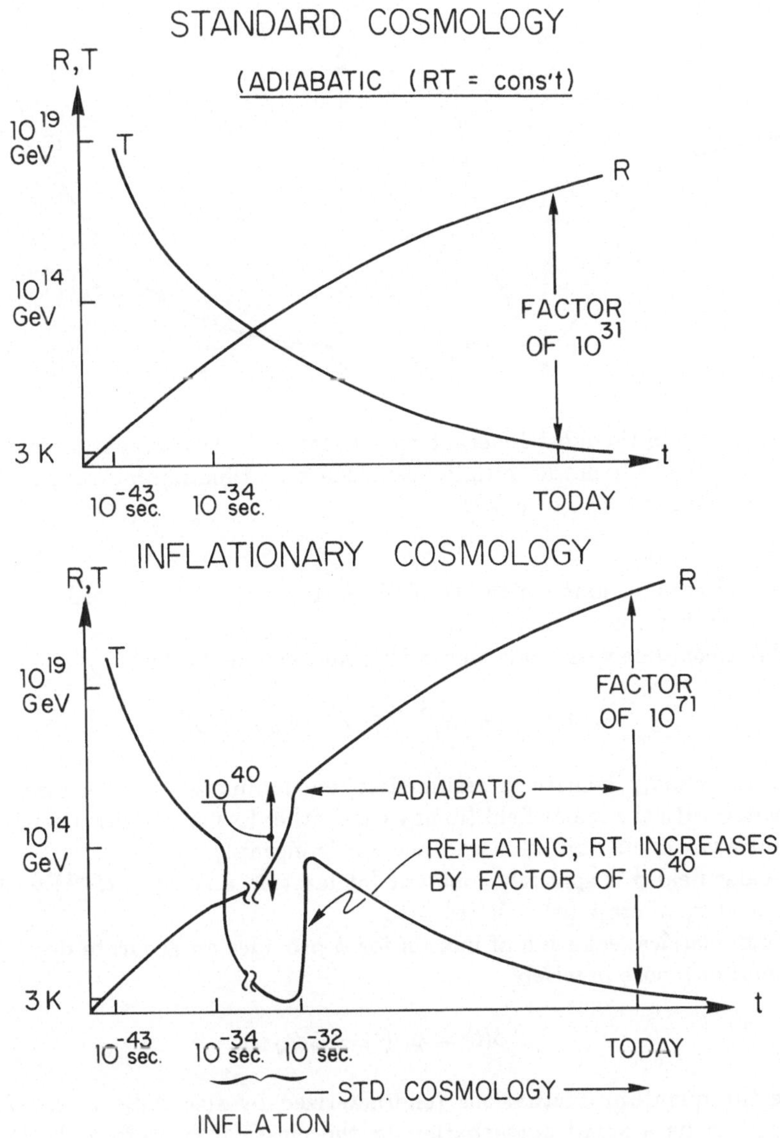

Figure 10: Evolution of the scale factor R and temperature T of the Universe in the standard cosmology and in the inflationary cosmology. The standard cosmology is always adiabatic ($RT \simeq$ const), while the inflationary cosmology undergoes a highly non-adiabatic event (reheating) after which it is adiabatic.

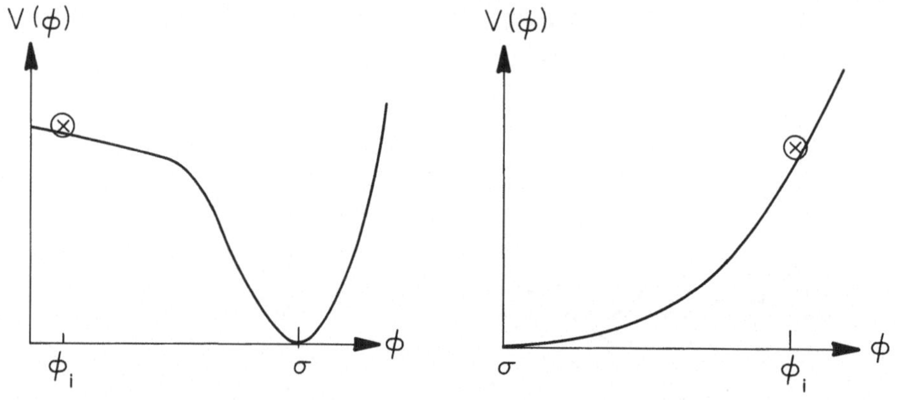

Figure 11: Stated in the most general terms, inflation involves the dynamical evolution of a scalar field which was initially displaced from the minimum of its potential, be that minimum at $\sigma = 0$ or $\sigma \neq 0$.

The usual assumptions which are made (often implicitly) in order to analyze the scalar field dynamics inflation are:

⋆ A FRW spacetime with scale factor $R(t)$ and expansion rate

$$H^2 \equiv (\dot{R}/R)^2 = 8\pi\rho/3m_{pl}^2 - k/R^2 \tag{26}$$

where the energy density is assumed to be dominated by the stress energy associated with the scalar field (in any case, other forms of stress energy rapidly redshift away during inflation and become irrelevant).

⋆ The scalar field ϕ is spatially constant (at least on a scale $\gtrsim H^{-1}$) with initial value $\phi_i \neq \sigma$, where $V(\sigma) = V'(\sigma) = 0$.

⋆ The semi-classical equation of motion for ϕ provides an accurate description of its evolution; more precisely,

$$\phi(t) = \phi_{cl}(t) + \Delta\phi_{QM}$$

where the quantum fluctuations (characterized by size $\Delta\phi_{QM} \simeq H/2\pi$) are assumed to be a small perturbation to the classical trajectory $\phi_{cl}(t)$. From this point forward I will drop the subscript 'cl'. I will return later to these assumptions to discuss how they have been or can be relaxed and/or justified.

Consider a classical scalar field (minimally coupled) with lagrangian density given by

$$\mathcal{L} = -\frac{1}{2}\partial_\mu\phi\partial^\mu\phi - V(\phi). \tag{27}$$

For now I will ignore the interactions that ϕ must necessarily have with other fields in the theory. As it will turn out they must be weak for inflation to work, so that this assumption is a reasonable one. The stress-energy tensor for this field is then

$$T_{\mu\nu} = -\partial_\mu\phi\partial_\nu\phi - \mathcal{L}g_{\mu\nu} \tag{28}$$

Assuming that in the region of interest ϕ is spatially-constant, $T_{\mu\nu}$ takes on the perfect fluid form with energy density and pressure given by

$$\rho = \frac{1}{2}\dot{\phi}^2 + V(\phi)(+(\nabla\phi)^2/2R^2), \tag{29a}$$

$$p = \frac{1}{2}\dot{\phi}^2 - V(\phi)(-(\nabla\phi)^2/6R^2), \tag{29b}$$

where I have included the spatial gradient terms for future reference. [Note, once inflation begins the spatial gradient terms decrease rapidly, $(\nabla\phi)^2/R^2 \propto R^{-2}$, for wavelengths $\gtrsim H^{-1}$, and quickly become negligible.] That the spatial gradient term in ϕ be unimportant is crucial to inflation; if it were to dominate the pressure and energy density, then $R(t)$ would grow as t (since $p = -\frac{1}{3}\rho$) and not exponentially.

The equations of motion for ϕ can be obtained either by varying the action or by using $T^{\mu\nu}_{;\nu} = 0$. In either case the resulting equation is:

$$\ddot{\phi} + 3H\dot{\phi}(+\Gamma\dot{\phi}) + V'(\phi) = 0. \tag{30}$$

I have explicitly included the $\Gamma\dot{\phi}$ term which arises due to particle creation. The $3H\dot{\phi}$ friction term arises due to the expansion of the Universe; as the scalar field gains momentum, that momentum is redshifted away by the expansion.

This equation, which is analogous to that for a ball rolling with friction down a hill with a valley at the bottom, has two qualitatively different regimes, each of which has a simple, approximate, analytic solution. Fig. 12 shows schematically the potential $V(\phi)$.

(1) The slow-rolling regime

In this regime the field rolls at terminal velocity and the $\ddot{\phi}$ term is negligible. This occurs in the interval where the potential is very flat, the conditions for sufficient flatness being[11]:

$$|V''| \leq 9H^2, \tag{31a}$$

$$|V'm_{pl}/V| \lesssim (48\pi)^{1/2}, \tag{31b}$$

Condition (31a) usually subsumes condition (31b), so that condition (31a) generally suffices. During the slow-rolling regime the equation of motion for ϕ reduces to

$$\dot{\phi} \simeq -V'/3H. \tag{32}$$

During the slow-rolling regime particle creation is exponentially suppressed[43] because the timescale for the evolution of ϕ (which sets the energy/momentum scale of

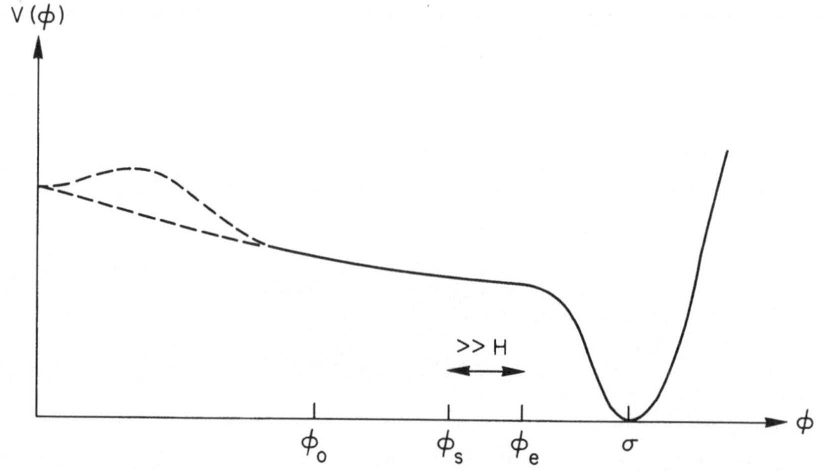

Figure 12: Schematic plot of the potential required for inflation. The shape of the potential for $\phi \ll \sigma$ determines how the barrier between $\phi = 0$ and $\phi = \sigma$ (if one exists) is penetrated. The value of ϕ after barrier penetration is taken to be ϕ_0; the flat region of the potential is the interval $[\phi_s,\ \phi_e]$.

the particles created) is much greater than the Hubble time (which sets the physics horizon), i.e., any particles radiated would have to have wavelengths much larger than the physics horizon, which results in the exponential suppression of particle creation during this epoch. Thus, the $\Gamma\dot{\phi}$ term can be neglected during the 'slow roll'.

Suppose the interval where conditions (31a,b) are satisfied is $[\phi_s,\ \phi_e]$, then the number of e-folds of expansion which during the time ϕ is evolving from $\phi = \phi_s$ to $\phi = \phi_e$ $(\equiv N)$ is

$$N \equiv \int H dt \simeq -3 \int_{\phi_s}^{\phi_e} H^2 d\phi / V'(\phi) \simeq -(8\pi/m_{pl}^2) \int V(\phi) d\phi / V'(\phi). \tag{33}$$

[Note that $R_e/R_s \equiv exp(N)$ since $\dot{R}/R \equiv H$.] Taking H^2/V' to be roughly constant over this interval and approximating V' as $\simeq \phi V''$ (which is approximately true for polynominal potentials) it follows that

$$N \approx 3H^2/V'' \geq 3.$$

If there is a region of the potential where the evolution is friction-dominated, then N will necessarily be greater than 1 (by condition (31a)).

(2) Coherent field oscillations

In this regime

$$|V''| \gg 9H^2,$$

and ϕ evolves rapidly, on a timescale $\ll$ the Hubble time H^{-1}. Once ϕ reaches the bottom of its potential, it will oscillate with an angular frequency equal to $m_\phi \equiv V''(\sigma)^{1/2}$. In this regime it proves useful to rewrite Eqn.(30) for the evolution of ϕ as

$$\dot{\rho}_\phi = -3H\dot{\phi}^2 - \Gamma\dot{\phi}^2. \tag{34}$$

where

$$\rho_\phi \equiv 1/2\dot{\phi}^2 + V(\phi).$$

Once ϕ is oscillating about $\phi = \sigma$, $\dot{\phi}^2$ can be replaced by its average over a cycle

$$< \dot{\phi}^2 >_{cycle} = \rho_\phi,$$

and Eqn.(34) becomes

$$\dot{\rho}_\phi = -3H\rho_\phi - \Gamma\rho_\phi \tag{35}$$

which is nothing else but the equation for the evolution of the energy density of zero momentum, massive particles with a decay width Γ.

Referring back to Eqn(29) we can see that the cycle average of the pressure (i.e., space-space components of $T_{\mu\nu}$) is zero—as one would expect for NR particles. The coherent ϕ oscillations are in every way equivalent to a very cold condensate of ϕ particles. The decay of these oscillations due to quantum particle creation is equivalent to the decay of zero-momentum ϕ particles.

The complete set of semi-classical equations for the reheating of the Universe is

$$\dot{\rho}_\phi = -3H\rho_\phi - \Gamma\rho_\phi, \tag{36a}$$
$$\dot{\rho}_r = -4H\rho_r + \Gamma\rho_\phi, \tag{36b}$$
$$H^2 = 8\pi G(\rho_r + \rho_\phi)/3, \tag{36c}$$

where $\rho_r = (\pi^2/30)g_* T^4$ is the energy density in the relativistic particles produced by the decay of the coherent field oscillations. [I have tacitly assumed that the decay products of ϕ rapidly thermalize; Eqn(36b) is correct whether or not the decay products thermalize, so long as they are relativistic.] The evolution for the energy density in the scalar is easy to obtain

$$\rho_\phi = M^4(R/R_e)^{-3}\exp[-\Gamma(t - t_e)], \tag{37}$$

where I have set the initial energy equal to M^4, the initial epoch being when the scalar field begins to evolve rapidly (at $R = R_e$, $\phi = \phi_e$, and $t = t_e$).

From $t = t_e$ until $t \simeq \Gamma^{-1}$, the energy density of the Universe is dominated by the coherent sloshings of the scalar field ϕ, set into motion by the initial vacuum energy associated with $\phi \ll \sigma$. During this phase

$$R(t) \propto t^{2/3}$$

that is, the Universe behaves as if it were dominated by NR particles—which it is!

Interestingly enough it follows from Eqn(36) that during this time the energy density in radiation is actually decreasing ($\rho_r \propto R^{-3/2}$—see Fig. 13). [During the first Hubble time after the end of inflation ρ_r does increase.] However, the all important entropy per comoving volume is increasing

$$S \propto R^{15/8}.$$

When $t \simeq \Gamma^{-1}$, the coherent oscillations begin to decay exponentially, and the entropy per comoving volume levels off—indicating the end of the reheating epoch. The temperature of the Universe at this time is,

$$T_{RH} \simeq g_*^{-1/4}(\Gamma m_{pl})^{1/2} \tag{38}$$

If Γ^{-1} is less than H^{-1}, so that the Universe reheats in less than an expansion time, then all of the vacuum is converted into radiation and the' Universe is reheated to a temperature

$$T_{RH} \simeq g_*^{-1/4}M \quad (\text{if } \Gamma \geq H) \tag{38'}$$

the so-called case of good reheating.

To summarize the evolution of the scalar field ϕ: early on ϕ evolves very slowly, on a timescale $\gg$ the Hubble time H^{-1}; then as the potential steepens (and $|V''|$ becomes $> 9H^2$) ϕ begins to evolve rapidly, on a timescale $\ll$ the Hubble time H^{-1}. As ϕ oscillates about the minimum of its potential the energy density in these oscillations dominates the energy density of the Universe and behaves like NR matter ($\rho_\phi \propto R^{-3}$); eventually when $t \simeq \Gamma^{-1}$, these oscillations decay exponentially, 'reheating' the Universe to a temperature of $T_{RH} \simeq g_*^{-1/4}(\Gamma m_{pl})^{1/2}$ (if $\Gamma > H$, so that the Universe does not e-fold in the time it takes the oscillations to decay, then $T_{RH} \simeq g_*^{-1/4}M$). Saying that the Universe reheats when $t \simeq \Gamma^{-1}$ is a bit paradoxical as the temperature has actually been *decreasing* since shortly after the ϕ oscillations began. However, the fact that the temperature of the Universe was actually once greater than T_{RH} for $t < \Gamma^{-1}$ is probably of no practical use since the entropy per comoving volume increases until $t \simeq \Gamma^{-1}$—by a factor of $(M^2/\Gamma m_{pl})^{5/4}$, and any interesting objects that might be produced (e.g., net baryon number, monopoles, etc.) will be diluted away by the subsequent entropy production. By any reasonable measure, T_{RH} is the reheat temperature of the Universe. The evolution of ρ_ϕ, ρ_r, and S are summarized in Fig. 13.

Armed with our detailed knowledge of the evolution of ϕ we are ready to calculate the precise number of e-folds of inflation necessary to solve the horizon and flatness problems and to discuss direct baryon number production. First consider the requisite number of e-folds required for sufficient inflation. To solve the homogeneity problem we need to insure that a smooth patch containing an entropy of at least 10^{88} results from inflation. Suppose the initial bubble or fluctuation region has a size $H^{-1} \simeq m_{pl}/M^2$—certainly it is not likely to be significantly larger than this. During inflation it grows by a factor of $\exp(N)$. Next, while the Universe is dominated by coherent field oscillations it grows by a factor of

$$(R_{RH}/R_e) \simeq (M^4/T_{RH}^4)^{1/3},$$

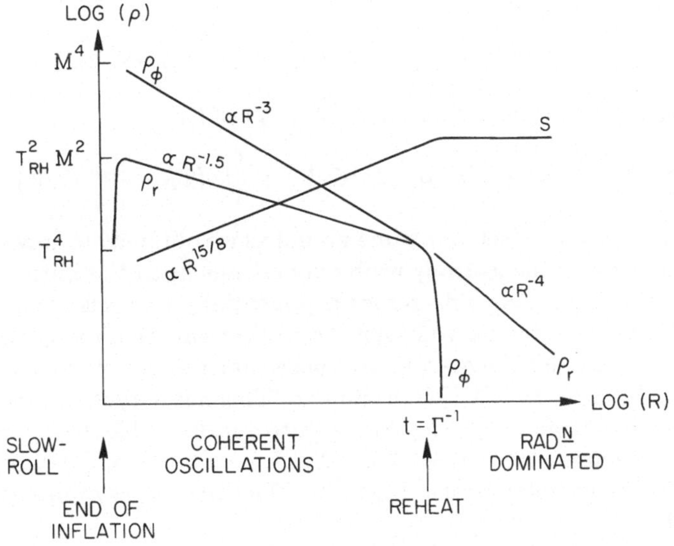

Figure 13: The evolution of ρ_ϕ, ρ_r, and S during the epoch when the Universe is dominated by coherent ϕ-oscillations. The reheat temperature $T_{RH} \simeq g_*^{-1/4}(\Gamma m_{pl})^{1/2}$. The maximum temperature achieved after inflation is actually greater, $T_{max} \simeq (T_{RH}M)^{1/2}$.

where T_{RH} is the reheat temperature. Cubing the size of the patch at reheating (to obtain its volume) and multiplying its volume by the entropy density ($s \approx T_{RH}^3$), we obtain

$$S_{patch} \simeq e^{3N} m_{pl}^3/(M^2 T_{RH}).$$

Insisting that S_{patch} be greater than 10^{88}, it follows that

$$N \geq 53 + \frac{2}{3} ln(M/10^{14}GeV) + \frac{1}{3} ln(T_{RH}/10^{10}GeV). \tag{39}$$

Varying M from $10^{19}GeV$ to $10^8 GcV$ and T_{RH} from $1GeV$ to $10^{19}GeV$ this lower bound on N only varies from 36 to 68.

The flatness problem involves the smallness of the ratio

$$x = (k/R^2)/(8\pi G\rho/3)$$

required at early times. Taking the pre-inflationary value of x to be x_i and remembering that

$$x(t) \propto \begin{cases} R^{-2} & \rho = \text{cons't} \\ R & \rho \propto R^{-3} \\ R^2 & \rho \propto R^{-4} \end{cases}$$

it follows that the value of x today is

$$x_{today} = x_i e^{-2N}(M/T_{RH})^{4/3}(T_{RH}/10eV)^2(10eV/3K).$$

Insisting that x_{today} be at most of order unity implies that

$$N \geq 53 + ln(x_i) + \frac{2}{3}ln(M/10^{14}GeV) + \frac{1}{3}ln(T_{RH}/10^{10}GeV)$$

—up to the term $ln(x_i)$, precisely the same bound as we obtained to solve the homogeneity problem. Solving the isotropy problem depends upon the initial anisotropy present; during inflation isotropy decreases exponentially (see refs. 48).

Finally, let's calculate the baryon asymmetry that can be directly produced by the decay of the ϕ particles themselves. Suppose that the decay of each ϕ particle results in the production of net baryon number ϵ. This net baryon number might be produced directly by the decay of a ϕ particle (into quarks and leptons) or indirectly through an intermediate state ($\phi \to X\bar{X}$; X, $\bar{X} \to$ quarks and leptons; e.g., X might be a superheavy, color triplet Higgs[49]). The baryon asymmetry produced per volume is then

$$n_B \simeq \epsilon n_\phi.$$

On the other hand we have

$$(g_* \pi^2/30)T_{RH}^4 \simeq n_\phi m_\phi.$$

Taken together it follows that[42,50]

$$n_B/s \simeq (3/4)\epsilon T_{RH}/m_\phi. \tag{40}$$

This then is the baryon number per entropy produced by the decay of the ϕ particles directly. If the reheat temperature is not very high, baryon number non-conserving interactions will not subsequently reduce the asymmetry significantly. Note that the baryon asymmetry produced only depends upon the *ratio* of the reheat temperature to the ϕ particle mass. This is important, as it means that a very low reheat temperature can be tolerated, so long as the ratio of it to the ϕ particle mass is not too small.

6. ORIGIN OF DENSITY INHOMOGENEITIES

To this point I have assumed that ϕ is precisely uniform within a given bubble or fluctuation region. As a result, each bubble or fluctuation region resembles a perfectly isotropic and homogeneous Universe after reheating. However, because of deSitter space produced quantum fluctuations, ϕ cannot be exactly uniform, even within a small region of space. It is a well-known result that a massless and non-interacting scalar field in deSitter space has a spectrum of fluctuations given by (see, e.g., ref. 51)

$$(\Delta\phi)^2 \equiv (2\pi)^{-3}k^3|\delta\phi_k|^2 \;=\; H^2/16\pi^3, \tag{41}$$

where

$$\delta\phi = (2\pi)^{-3}\int d^3k\delta\phi_k^{-ikx}, \tag{42}$$

and $\vec{x}$ and $\vec{k}$ are comoving quantities. This result is applicable to inflationary scenarios as the scalar field responsible for inflation must be very weakly-coupled and nearly massless. [That Universe is not precisely deSitter during inflation, i.e., $\rho + p = \dot{\phi}^2 \neq 0$, does not affect this result significantly; this point is addressed in ref. 52.] These deSitter space produced quantum fluctuations result in a calculable spectrum of adiabatic density perturbations. These density perturbations were first calculated by the authors of refs. 53-56; they have also been calculated by the authors of refs. 57 who addressed some of the technical issues in more detail. All the calculations done to date arrive at the same result. I will briefly describe the calculation in ref. 56; my emphasis here will be to motivate the result. I refer the reader interested in more details to the aforementioned references.

It is conventional to expand density inhomogeneities in a Fourier expansion

$$\delta\rho/\rho = (2\pi)^{-3}\int \delta_k e^{-ikx} d^3k. \tag{43}$$

The physical wavelength and wavenumber are related to comoving wavelength and wave number, λ and k, by

$$\begin{aligned}\lambda_{ph} &= (2\pi/k)R(t) \equiv \lambda R(t),\\ k_{ph} &= k/R(t).\end{aligned}$$

The quantity most people refer to as $\delta\rho/\rho$ on a given scale is more precisely the RMS mass fluctuation on that scale

$$(\delta\rho/\rho)_k \equiv <(\delta M/M)^2>_k \simeq \Delta_k^2 \equiv (2\pi)^{-3}k^3|\delta_k|^2, \tag{44}$$

which is just related to the Fourier component δ_k on that scale.

The cosmic scale factor is often normalized so that $R_{today} = 1$; this means that given Fourier components are characterized by the physical size that they have today (neglecting the fact that once a given scale goes non-linear ($\delta\rho/\rho \gtrsim 1$) objects of that size form bound 'lumps' that no longer participate in the universal expansion and remain roughly constant in size). The mass (in NR matter) contained within a sphere of radius $\lambda/2$ is

$$M(\lambda) \simeq 1.5\times 10^{11} M_\odot (\lambda/Mpc)^3\Omega h^2.$$

Although physics depends on physical quantities (k_{ph}, λ_{ph}, etc.), the comoving labels k, M, and λ are the most useful way to label a given component as they have the affect of the expansions already scaled out.

I want to emphasize at the onset that the quantity $\delta\rho/\rho$ is not gauge invariant (under general coordinate transformations). This fact makes life very difficult when

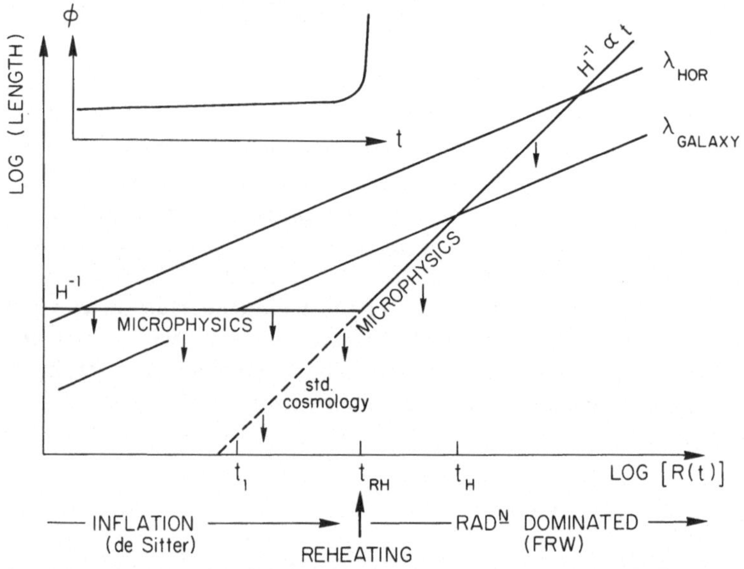

Figure 14: The evolution of the physical size of galactic- and (present) horizon-sized perturbations ($\lambda_{ph} \propto R$) and the size of the physics horizon H^{-1}. Causally-coherent microphysics operates only on scales $\leq H^{-1}$. In the standard cosmology a perturbation crosses the horizon but once as $H^{-1} \propto R^n (n > 1)$, making it impossible for microphysics to create density perturbations at early times. In the inflationary cosmology a perturbation crosses the horizon twice (since $H^{-1} \simeq$ const during inflation), and so microphysics can produce density perturbations at early times.

discussing Fourier components with wavelengths larger than the physics horizon (i.e., $\lambda_{ph} \geq H^{-1}$). The gauge non-invariance of $\delta\rho/\rho$ is not a problem when $\lambda_{ph} \leq H^{-1}$, as the analysis is essentially Newtonian. The usual approach is to pick a convenient gauge (e.g., the synchronous gauge where $g_{oo} = -1$, $g_{oi} = 0$) and work very carefully (see refs. 58, 59). The more elegant approach is to focus on gauge-invariant quantities; see ref. 60. I will gloss over the subtleties of gauge invariance in my discussion, which is aimed at motivating the correct answer.

The evolution of a given Fourier component (in the linear regime—$\delta\rho/\rho \ll 1$) separates into two qualitatively different regimes, depending upon whether or not the perturbation is inside or outside the physics horizon ($\simeq H^{-1}$). When $\lambda_{ph} \leq H^{-1}$, microphysical processes can affect its evolution—such processes include: quantum mechanical effects, pressure support, free-streaming of particles, 'Newtonian gravity', etc. In this regime the evolution of the perturbation is very dynamical. When a perturbation is outside the physics horizon, $\lambda_{ph} \geq H^{-1}$, microphysical processes do not affect its evolution; in a very real sense its evolution is kinematic—it evolves as a wrinkle in the fabric of space–time.

In the standard cosmology, a given Fourier component crosses the horizon only once, starting outside the horizon and crossing inside at a time (see Fig. 14)

$$t \simeq (M/M_{\odot})^{2/3} sec$$

(valid during the radiation-dominated epoch). For this reason it is not possible to create adiabatic (more precisely, curvature) perturbations by causal microphysical processes which operate at early times[59,60]. In the standard cosmology, if adiabatic perturbations are present, they must be present *ab initio*. The smallness of the particle horizon at early times relative to the comoving volume occupied by the observable Universe today strikes again!

[It is possible for microphysical processes to create isothermal, more precisely isocurvature, perturbations. Once such perturbations cross inside the horizon they are characterized by a spectrum

$$(\delta\rho/\rho) \propto (M/M_H)^{-1/2}$$

or steeper. Here M_H is the horizon mass when the perturbations were created. Thus the earlier the processes operate, the smaller the perturbations are on interesting scales. By an appropriate choice of gauge it is possible to view these isothermal perturbations as adiabatic perturbations with a very steep spectrum, $\delta\rho/\rho \propto M^{-7/6}$; however, as must be the case, they cross the horizon with the amplitude mentioned above. For more details, see refs. 59, 60.]

Because the distance to the physics horizon remains approximately constant during inflation, the situation is very different in the inflationary Universe. All interesting scales start inside the horizon, cross outside the horizon and once again come inside the horizon (at the usual epoch); see Fig. 14. This means that causal microphysical processes can set up curvature perturbations on astrophysically-interesting scales. [This point seems to have been first appreciated by Press.[61]]

Consider the evolution of a given Fourier component k. Early during the inflationary epoch $\lambda_{ph} \leq H^{-1}$, and quantum fluctuations in ϕ give rise to density perturbations on this scale. As the scale passes outside the horizon, say at $t = t_1$, microphysical processes become impotent, and $\delta\rho/\rho$ freezes out at a value,

$$\begin{aligned}(\delta\rho/\rho)_k &\simeq O(\dot{\phi} H \Delta\phi/M^4), \\ &\simeq O(\dot{\phi} H^2/M^4),\end{aligned} \tag{45}$$

as the scale leaves the horizon. From that point forward, the QM fluctuation is assumed to 'freeze in' and thereafter evolve classically. Note in the approximation that H and $\dot{\phi}$ are constant during the inflationary epoch the value of $\delta\rho/\rho$ as the perturbation leaves the horizon is independent of k. This scale independence of $\delta\rho/\rho$ when perturbations cross outside the horizon is of course traceable to the time translation invariance of deSitter space.

While outside the horizon the evolution of a perturbation is kinematical, independent of scale, and gauge dependent. There is a gauge independent quantity

($\equiv \varsigma$) which remains constant while the perturbation is outside the horizon, and which at horizon crossing ($t = t_1$ and t_H) is given by

$$\varsigma|_{\text{horizon crossing}} \simeq \delta\rho/(\rho+p),$$
$$\varsigma(t_1) = \varsigma(t_H)$$
$$\Rightarrow [\delta\rho/(\rho+p)]_{t=t_H} \simeq [\delta\rho/(\rho+p)]_{t=t_1}. \tag{46}$$

(see refs. 56 and 62 for more details). When the perturbation crosses back inside the horizon: $(\rho+p) = n\rho (n = 4/3-$ radiation-dominated; $n = 1$, matter-dominated) so that up to a numerical factor $(\delta\rho/\rho)|_{t_H} \simeq [\delta\rho/(\rho+p)]|_{t_H}$. During inflation, however, $\rho+p = \dot{\phi}^2 \ll \rho \simeq M^4$ so that $(\delta\rho/\rho)|_{t_1} \simeq (\dot{\phi}^2/M^4)[\delta\rho/(\rho+p)]|_{t_1}$. Note, $M^4/\dot{\phi}^2$ is typically a very large number. Eqns(45, 46) then imply

$$(\delta\rho/\rho)_H \equiv (\delta\rho/\rho)_{t=t_H} \simeq (M^4/\dot{\phi}^2)(\delta\rho/\rho)_{t_1} \simeq H^2/\dot{\phi}, \tag{47}$$

Note that in the approximation that $\dot{\phi}$ and H are are constant during inflation the amplitude of $\delta\rho/\rho$ at horizon crossing ($= (\delta\rho/\rho)_H$) is independent of scale. This fact is traceable to the time-translation invariance of the nearly-deSitter inflationary epoch and the scale-independent evolution of $(\delta\rho/\rho)$ while the perturbation is outside the horizon. The so-called scale-invariant or Zel'dovich spectrum of density perturbations was first discussed, albeit in another context, by Harrison[63] and Zel'dovich[64]. Scale-invariant adiabatic density perturbations are a generic prediction of inflation. [Because H and $\dot{\phi}$ are not precisely constant during inflation, the spectrum is not quite scale-invariant. However the scales of astrophysical interest, say $\lambda \simeq 0.1Mpc - 100Mpc$, cross outside the horizon during a very short interval, $\Delta N \simeq 6.9$, during which H, $\dot{\phi}$, and ϕ are very nearly constant. For most models of inflation the deviations are not expected to be significant; for further discussion see refs. 65, 66.] Although the details of structure formation are not presently sufficiently well understood to say what the initial spectrum of perturbations must have been, the Zel'dovich with an amplitude of about $10^{-4} - 10^{-5}$ is certainly a viable possibility.

Before moving on, let me be very precise about the amplitude of the inflation-produced adiabatic density perturbations. Perturbations which re-enter the horizon while the Universe is still radiation-dominated ($\lambda \leq \lambda_{eq} \simeq 13h^{-2}Mpc$), do so as a sound wave in the photons and baryons with amplitude

$$(\delta\rho/\rho)_H \equiv k^{3/2}|\delta_k|/(2\pi)^{3/2} \simeq H^2/(\pi^{3/2}\dot{\phi}) \tag{48a}$$

Perturbations in non-interacting, relic particles (such as massive neutrinos, axions, etc.), which by the equivalence principle must have the same amplitude at horizon crossing, do not oscillate, but instead grow slowly ($\propto \ln R$). By the epoch of matter-radiation equivalence they have an amplitude of 2–3 times that of the initial baryon–photon sound wave, or

$$(\delta\rho/\rho)_{MD} \simeq (2-3)(\delta\rho/\rho)_H \simeq (2-3)H^2/(\pi^{3/2}\dot{\phi}) \tag{48b}$$

It is this amplitude which must be of order $10^{-5} - 10^{-4}$ for successful galaxy formation.

Perturbations which re-enter the horizon when the Universe is already matter-dominated (scales $\lambda \geq \lambda_{eq} \simeq 13h^{-2}Mpc$) do so with amplitude

$$(\delta\rho/\rho)_H \simeq k^{3/2}|\delta_k|/(2\pi)^{3/2} \simeq (H^2/10)/(\pi^{3/2}\dot{\phi}) \tag{49}$$

Once inside the horizon they continue to grow (as $t^{2/3}$ since the Universe is matter-dominated).

When the structure formation problem is viewed as an initial data problem, it is the spectrum of density perturbations at the epoch of matter domination which is the relevant input spectrum. The shape of this spectrum has been carefully computed by the authors of ref. 67. Roughly speaking, on scales less than λ_{eq} the spectrum is almost flat, varying as $\lambda^{-3/4} \propto M^{-1/4}$ for scales around the galaxy scale ($\simeq 1Mpc$). On scales much greater than λ_{eq}, $(\delta\rho/\rho) \propto \lambda^{-2} \propto M^{-2/3}$ (in the synchronous gauge where adiabatic perturbations grow as $t^n; n = 2/3$ matter dominated, $n = 1$ radiation dominated. Since these scales have yet to re-enter the horizon they have not yet achieved their horizon-crossing amplitude).

In order to compute the amplitude of the inflation-produced adiabatic density perturbations we need to evaluate $H^2/\dot{\phi}$ when the astrophysically-relevant scales crossed outside the horizon. Recall, in the previous section we computed when the comoving scale corresponding to the present Hubble radius crossed outside the horizon during inflation—up to 'ln terms' $N \simeq 53$ or so e-folds before the end of inflation, cf., Eqn.(39). The present Hubble radius corresponds to a scale of about $3000Mpc$; therefore the scale λ must have crossed the horizon $ln(3000Mpc/\lambda)$ e-folds later:

$$N_\lambda \simeq N_{HOR} - 8 + ln(\lambda/Mpc) \simeq 45 + ln(\lambda/Mpc) + \frac{2}{3}ln(M/10^{14}GeV) + \frac{1}{3}ln(T_{RH}/10^{10}GeV).$$

Typically $H^2/\dot{\phi}$ depends upon N_λ to some power[65]; since N_λ only varies logarithmically ($\Delta N/N \simeq 0.14$ in going from $0.1Mpc$ to $3000Mpc$), the scale dependence of the spectrum is almost always very minimal.

As mentioned earlier, a generic prediction of the inflationary Universe is that today Ω should be equal to one to a high degree of precision. Equivalently, that means

$$|(k/R^2)/(8\pi G\rho/3)| \ll 1$$

since

$$\Omega = 1/[1 - (k/R^2)/(8\pi G\rho/3)].$$

Therefore one might conclude that an accurate measurement of Ω would have to yield 1 to extremely high precision. However, because of the adiabatic density perturbations produced during inflation that is not the case. Adiabatic density fluctuations correspond to fluctuations in the local curvature

$$\delta\rho/\rho \simeq \delta(k/R^2)/(8\pi G\rho/3)$$

This means that should we be able to very accurately probe the value of Ω (equivalently the curvature of space) on the scale of our Hubble volume, say by using the Hubble diagram, we would necessarily obtain a value for Ω which is dominated by the curvature fluctuations on the scale of the present horizon,

$$\Omega_{obs} \simeq 1 + \delta(k/R^2)/(8\pi G\rho/3) \simeq 1 \pm O(10^{-4} - 10^{-5}),$$

and so we would obtain a value different from 1 by about a part in 10^4 or 10^5.

Finally, let me briefly mention that isothermal density perturbations can also arise during inflation. [Isothermal density perturbations are characterized by $\delta\rho = 0$, but $\delta(n_i/n_\gamma) \neq 0$ in some components. They correspond to spatial fluctuations in the local pressure due to spatial fluctuations in the local equation of state.] Such perturbations can arise from the deSitter produced fluctuations in other quantum fields in the theory.

The simplest example occurs in the axion-dominated Universe[68,69,70]. Suppose that Peccei-Quinn symmetry breaking occurs before or during inflation. Until instanton effects become important ($T \simeq$ few $100MeV$) the axion field $a = f_a\theta$ is massless and θ is in general not aligned with the minimum of its potential: $\theta = \theta_1 \neq 0$ (I have taken the minimum of the axion potential to be $\theta = 0$). Once the axion develops a mass (equivalently, its potential develops a minimum) θ begins to oscillate; these coherent oscillations correspond to a condensate of very cold axions, with number density $\propto \theta^2$. [For further discussion of the coherent axion oscillations see refs. 71–73.] During inflation deSitter space produced quantum fluctuations in the axion field gave rise to spatial fluctuations in θ_1:

$$\delta\theta \simeq \delta a/f_a \simeq H/f_a$$

Once the axion field begins to oscillate, these spatial fluctuations in the axion field correspond to fluctuations in the local axion to photon ratio

$$\delta(n_a/n_\gamma)/(n_a/n_\gamma) \simeq 2\delta\theta/\theta_1 \simeq 2H/(f_a\theta_1)$$

More precisely

$$(\delta n_a/n_a) = k^{3/2}|\delta a(k)|/(2\pi)^{3/2} = H/(2\pi^{3/2} f_\lambda \theta_1), \tag{50}$$

where f_λ is the expectation value of f_a when the scale λ leaves the horizon (in some models the expectation value of the field which breaks PQ symmetry evolves as the Universe is inflating so that f_λ can be $< f_a$). It is possible that these isothermal axion fluctuations can be important for galaxy formation in an axion-dominated, inflationary Universe.[69]

7. SPECIFIC MODELS–PART I. INTERESTING FAILURES

(1) 'Old Inflation'

By old inflation I mean Guth's original model of inflation. In his original model the Universe inflated while trapped in the $\phi = 0$ false vacuum state. In order to inflate enough the vacuum had to be very metastable; however, that being the case, the bubble nucleation probability was necessarily small—so small that the bubbles that did nucleate never percolated, resulting in a Universe which resembled swiss cheese more than anything else[36]. The interior of an individual bubble was not suitable for our present Universe either. Because he was not considering flat potentials, essentially all of the original false vacuum energy resided in bubble walls rather than in vacuum energy inside the bubbles themselves. Although individual bubbles would grow to a very large size given enough time, their interiors would contain very little entropy (compared to the 10^{88} in our observed Universe). In sum, the Universe inflated all right, but did not 'gracefully exit' from inflation back to a radiation-dominated Universe—close, Alan, but no cigar!

(2) Coleman–Weinberg SU(5)

The first model of new inflation studied was the Coleman–Weinberg SU(5) GUT[37,38]. In this model the field which inflates is the 24-dimensional Higgs which also breaks SU(5) down to SU(3) × SU(2) × U(1). Let ϕ denote its magnitude in the SU(3) × SU(2) × U(1) direction. The one-loop, zero-temperature Coleman–Weinberg[74] potential is

$$\begin{aligned} V(\phi) &= 1/2B\sigma^4 + B\phi^4\{ln(\phi^2/\sigma^2) - 1/2\}, \\ B &= 25\alpha_{GUT}^2/16 \simeq 10^{-3} \\ \sigma &\simeq 2\times 10^{15} GeV \end{aligned} \tag{51}$$

Due to the absence of a mass term $(m^2\phi^2)$, the potential is very flat near the origin (SSB arises due to one-loop radiative corrections[74]); for $\phi \ll \sigma$:

$$\begin{aligned} V(\phi) &\simeq B\sigma^4/2 - \lambda\phi^4/4 \\ \lambda &\simeq |4Bln(\phi^2/\sigma^2)| \simeq 0.1 \end{aligned} \tag{52}$$

The finite temperature potential has a small temperature dependent barrier [height $O(T^4)$] near the origin [$\phi \simeq O(T)$]. The critical temperature for this transition is $O(10^{14} - 10^{15} GeV)$, however the $\phi = 0$ vacuum remains metastable. When the temperature of the Universe drops to $O(10^9 GeV)$ or so, the barrier becomes low enough that the finite temperature action for bubble nucleation drops to order unity and the $\phi = 0$ false vacuum becomes unstable[38]. In analogy with solid state phenomenon it is expected that at this the temperature of the Universe will undergo 'spinodal decomposition', i.e., will break up into irregularly shaped regions within which ϕ is approximately constant (so-called fluctuation regions). Approximately the potential by Eqn(52). It is easy to solve for the evolution of ϕ in the slow-rolling regime $[|V''| \leq 9H^2$ for $\phi^2 \leq \phi_e^2 \simeq \sigma^2(\pi\sigma^2/m_{pl}^2|ln(\phi^2/\sigma^2)|)]$

$$(H/\phi)^2 \simeq \frac{2\lambda}{3}N(\phi), \tag{53}$$

$$H^2 \simeq \frac{4\pi}{3}\frac{B\sigma^4}{m_{pl}^2}, \tag{54}$$

where $N(\phi) \equiv \int_{\phi}^{\phi_e} H dt$ is the number of e-folds of inflation the Universe undergoes while ϕ evolves from ϕ to ϕ_e. Clearly, the number of e-folds of inflation depends upon the initial value of $\phi(\equiv \phi_0)$; in order to get sufficient inflation ϕ_0 must be $\simeq O(H)$. Although one might expect ϕ_0 to be of this order in the fluctuation regions since $H \simeq 5 \times 10^9 GeV \simeq$ (temperature at which the $\phi = 0$ false vacuum loses its metastability), there is a fundamental difficulty. In using the semi-classical equations of motion to describe the evolution of ϕ one is implicitly assuming

$$\phi \simeq \phi_{cl} + \Delta\phi_{QM},$$
$$\Delta\phi_{QM} \ll \phi_{cl}$$

The deSitter space produced quantum fluctuations in ϕ are of order H. More specifically, it has been shown that[75,76]

$$\Delta\phi_{QM} \simeq (H/2\pi)(Ht)^{1/2}$$

Therein lies the difficulty—in order to achieve enough inflation the initial value of ϕ must be of the order of the quantum fluctuations in ϕ. At the very least this calls into question the semiclassical approximation.

The situation gets worse when we look at the amplitude of the adiabatic density perturbations:

$$(\delta\rho/\rho)_H \simeq (H^2/\pi^{3/2}\dot{\phi}) \tag{55}$$
$$\simeq (3/\pi^{3/2})(H^3/\lambda\phi^3) \tag{56}$$
$$(\delta\rho/\rho)_H \simeq (2/\pi)^{3/2}(\lambda/3)^{1/2}N^{3/2} \tag{57}$$

For galactic-scale perturbations $N \simeq 50$, implying that $(\delta\rho/\rho)_H \simeq 30$! Again, its clear that the basic problem is traceable to the fact that during inflation $\phi \leq H$.

The decay width of the ϕ particle is of order $\alpha_{GUT}\sigma \simeq 10^{13} GeV$ which is much greater than H (implying good reheating), and so the Universe reheats to a temperature of order $10^{14} GeV$ or so.

From Eqns(53,57) it is clear that by reducing λ both problems could be remedied—however $\lambda \lesssim 10^{-13}$ is necessary[56]. Of course, as long as the inflating field is a gauge non-singlet λ is set by the gauge coupling strength. From this interesting failure it is clear that one should focus on weakly-coupled, gauge singlet fields for inflation.

(3) Geometric Hierarchy Model

The first model proposed to address the difficulty mentioned above, was a supersymmetric GUT[77,78]. In this model ϕ is a scalar field whose potential at tree level is absolutely flat, but due to radiative corrections develops curvature. In the model ϕ is also responsible for the SSB of the GUT. The potential for ϕ is of the form

$$V(\phi) \simeq \mu^4[c_1 - c_2 ln(\phi/m_{pl})] \tag{58}$$

where $\mu \simeq 10^{12} GeV$ is the scale of supersymmetry breaking, and c_1 and c_2 are constants which depend upon details of the theory. This form for the potential is only valid away from its SSB minimum ($\sigma \simeq m_{pl}$) and for $\phi \gg \mu$. The authors presume that higher order effects will force the potential to develop a minimum for $\phi \simeq m_{pl}$. Since $V' \propto \phi^{-1}$ the potential gets flatter for large ϕ—which already sounds good.

The inflationary scenario for this potential proceeds as follows. The shape of the potential is not determined near $\phi = 0$; depending on the shape ϕ gets to some initial value, say $\phi = \phi_0$, either by bubble nucleation or spinodal decomposition. Then it begins to roll. During slow roll which begins when $|V''| \simeq 9H^2$ and $\phi_s \simeq (c_2/24\pi c_1)^{1/2} m_{pl}$,

$$H^2 \simeq \frac{8\pi}{3m_{pl}^2} c_1 \mu^4 \tag{59a}$$

$$(1 - \phi^2/m_{pl}^2) \simeq (c_2/4\pi c_1) N(\phi) \tag{59b}$$

$$(\delta\rho/\rho)_H \simeq (H^2/\pi^{3/2}\dot{\phi}), \tag{60a}$$

$$\simeq 8(8/3)^{1/2}(c_1^{3/2}/c_2)\mu^2\phi/m_{pl}^3. \tag{60b}$$

Note that during the slow roll ($\phi \geq \phi_s$)

$$\frac{\phi}{H} \geq \frac{\phi_s}{H} \simeq \frac{c_2^{1/2}}{c_1}\frac{1}{8\pi}\frac{m_{pl}^2}{\mu^2},$$

$$\simeq 10^{13} c_2^{1/2}/c_1 \gg 1,$$

thereby avoiding the difficulty encountered in the Coleman–Weinberg where $\phi \leq H$ was required to inflate. For $c_1 \simeq O(1)$, $c_2 \simeq 10^{-8}$—acceptable values in the model, $(\delta\rho/\rho)_H \simeq 10^{-5}$ and $N(\phi_s) \simeq 4\pi c_1/c_2 \simeq 10^9$. The number of e-folds of inflation is very large—10^9. This is quite typical of the very flat potentials required to achieve $(\delta\rho/\rho) \simeq 10^{-4} - 10^{-5}$.

Now for the bad news. In this model ϕ is very weakly coupled—it only couples to ordinary particles through gravitational strength interactions. Its decay width is

$$\Gamma \simeq O(\mu^6/m_{pl}^5), \tag{61}$$

which is much less than H (implying poor reheating) and leads to a reheat temperature of

$$T_{RH} \simeq O[(\Gamma m_{pl})^{1/2}], \tag{62a}$$

$$\simeq O(\mu^3/m_{pl}^2), \tag{62b}$$

$$\simeq 10 MeV. \tag{62c}$$

Such a reheat temperature safely returns the Universe to being radiation-dominated before primordial nucleosynthesis, and produces a smooth patch containing an enormous entropy—for $c_2 \simeq 10^{-8}$, $c_1 \simeq 1$, $S_{patch} \simeq (m_{pl}^3/\mu^2 T_{RH})e^{3N} \simeq 10^{35} exp(3 \times$

10^9), but does not reheat it to a high enough temperature for baryogenesis. Poor reheating is a problem which plagues almost all potentially viable models of inflation. Achieving $(\delta\rho/\rho)_H \lesssim 10^{-4}$ requires the scalar potential to be very flat, which necessarily means that ϕ is very weakly-coupled, and therefore $T_{RH}(\propto \Gamma^{1/2})$ tends to be very low.

(4) CERN SUSY/SUGR Models[79]

Early on members of the CERN theory group recognized that supersymmetry might be of use in protecting the very small couplings necessary in inflationary potentials from being overwhelmed by radiative corrections. They explored a variety of SUSY/SUGR models[79] (and dubbed their brand of inflation 'primordial inflation'). In the process, they encountered a difficulty which plagues almost all supersymmetric models of inflation based upon minimal supergravity theories.

It is usually assumed that at high temperatures the expectation value of an inflating field is at the minimum of its finite temperature effective potential (near $\phi = 0$); then as the Universe cools it becomes trapped there, and then eventually slowly evolves to the low temperature minimum (during which time inflation takes place). In SUSY models $< \phi >_T$ is not necessarily zero at high temperatures. In fact in essentially all of their models $< \phi >_T > 0$ and the high temperature minimum smoothly evolves into the low temperature minimum (as shown in Fig. 15).[80] As a result the Universe in fact would never have inflated!

There are two obvious remedies to this problem: (i) arrange the model so that $\langle\varphi\rangle_T \leq 0$ (as shown in Fig. 15), then ϕ necessarily gets trapped near $\phi = 0$; or (ii) assume that ϕ is never in thermal equilibrium before the phase transition so that ϕ is not constrained to be in the high temperature minimum of its finite temperature potential at high temperatures. Variants of the CERN models[79] based on these two remedies have been constructed by Ovrut and Steinhardt[81] and Holman, Ramond, and Ross[82].

8. LESSONS LEARNED–A PRESCRIPTION FOR SUCCESSFUL NEW INFLATION

The unsuccessful models discussed above have proven to be very useful in that they have allowed us to 'write a prescription' for the kind of potential that will successfully implement inflation[65]. The following prescription incorporates these lessons, together with other lessons which have been learned (sometimes painfully). As we will see all but the last of the prescribed features, that the potential be part of a sensible particle physics model, are relatively easy to arrange.

(1) The potential should have an interval which is sufficiently flat so that ϕ evolves slowly (relative to the expansion timescale H^{-1})—that is, flat enough so that a slow–rollover transition ensues. As we have seen, that means an interval

$$[\phi_s,\ \phi_e]$$

where

$$|V''| \leq 9H^2,$$
$$|V' m_{pl}/V| \leq (48\pi)^{1/2}.$$

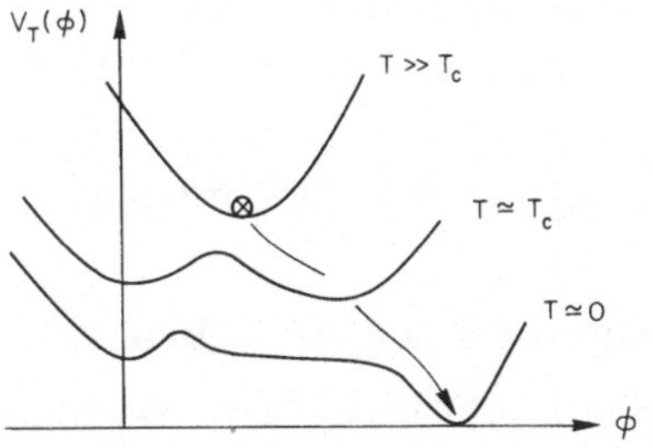

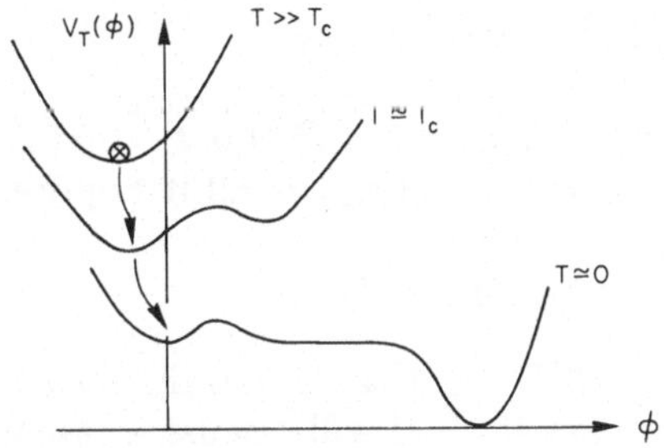

Figure 15: In SUSY/SUGR models $< \varphi >_T$ is not necessarily equal to zero. If $< \varphi >_T > 0$, then there is the danger that $< \varphi >_T$ smoothly evolves into the zero temperature minimum of the potential, thereby eliminating the possibility of inflation (upper figure). A sure way of preventing this is to design the potential so that $< \varphi >_T \leq 0$ (lower figure).

(2) The length of the interval where ϕ evolves slowly should be much greater than $H/2\pi$, the scale of the quantum fluctuations, so that the semi-classical approximation makes sense. [Put another way the interval should be long enough so that quantum fluctuations do not quickly drive ϕ across the interval.] Quantitatively, this calls for

$$|\phi_e - \phi_s| \gg (\dot{H}\Delta t)^{1/2}(H/2\pi)$$

where Δt is the time required for ϕ to evolve from $\phi = \phi_s$ to $\phi = \phi_e$. [More precisely, the semi-classical change in ϕ in a Hubble time, $\Delta\phi_{Hubble} \simeq -V' m_{pl}^2/8\pi V$, should be much greater than the increase in $< \phi^2 >_{QM}^{1/2} \simeq H/2\pi$, due to the addition of another quantum mode; see ref. 56.]

(3) In order to solve the flatness and homogeneity problems the time required for ϕ to roll from $\phi = \phi_s$ to $\phi = \phi_e$ should be greater than about 60 Hubble times

$$N \equiv \int_{\phi_s}^{\phi_e} H dt \simeq \int_{\phi_s}^{\phi_e} 3H^2 d\phi/(-V') \simeq 3H^2/V'' \geq 60$$

The precise formula for the minimum value of N is given in Eqn(39).

(4) The scalar field ϕ should be smooth in a sufficiently large patch (say size L) so that the energy density and pressure associated with the $(\nabla\phi)^2$ term is negligible:

$$1/2(\nabla\phi)^2 \simeq (\phi_0/L)^2 \ll V(\phi_0) \simeq M^4.$$

(Otherwise the $(\nabla\phi)^2$ term will dominate ρ and p, so that $R(t) \propto t$—that is, inflation does not occur.) Usually this condition is easy to satisfy, as all it requires is that

$$L \gg \phi_0/M^2 \simeq (\phi_0/m_{pl})H^{-1};$$

since ϕ_0 is usually $\ll m_{pl}$, $(\phi_0/m_{pl})H^{-1} \ll H^{-1}$—that is, ϕ only need be smooth on a patch comparable to the physics horizon H^{-1}. [I will discuss a case where it is not easy to satisfy—Linde's chaotic inflation.] Once inflation does begin, any initial inhomogeneities in ϕ are rapidly smoothed by the exponential expansion.

(5a) In order to insure a viable scenario of galaxy formation (and microwave anisotropies of an acceptable magnitude) the amplitude of the adiabatic density perturbations must be of order $10^{-5} - 10^{-4}$. [In a Universe dominated by weakly-interacting relic particles such as neutrinos or axions, $(\delta\rho/\rho)_{MD}$ must be a few $\times 10^{-5}$.] This in turn results in the constraint

$$few \times 10^{-5} \simeq (\delta\rho/\rho)_{MD} \simeq (2-3)(\delta\rho/\rho)_H \simeq (2-3)(H^2/\pi^{3/2}\dot{\phi})_{Galaxy},$$

$$(H^2/\dot{\phi})_{Galaxy} \simeq 10^{-4}$$

In general, this is by far the most difficult of the constraints (other than sensible particle physics) to satisfy and leads to the necessity of extremely flat potentials. I should add, if one has another means of producing the density perturbations necessary for galaxy formation (e.g., cosmic strings or isothermal perturbations), then it is sufficient to have

$$(H^2/\dot{\phi})_{Galaxy} \lesssim 10^{-4}$$

(5b) Isothermal perturbations produced during inflation, e.g., as discussed for the case of an axion-dominated Universe, also lead to microwave anisotropies and possibly structure formation. The smoothness of the microwave background dictates that

$$(\delta\rho/\rho)_{ISO} \lesssim 10^{-4}$$

while if they are to be relevant for structure formation

$$(\delta\rho/\rho)_{ISO} \simeq 10^{-5} - 10^{-4}$$

In the case of isothermal axion perturbations this is easy to arrange to have $(\delta\rho/\rho)_{ISO} \ll 10^{-4}$ unless the scale of PQ symmetry is larger than about $10^{18} GeV$.

(6a) Sufficiently high reheat temperature so that the Universe is radiation-dominated at the time of primordial nucleosynthesis ($t \simeq 10^{-2} - 10^{2} sec$, $T \simeq 10 MeV - 0.1 MeV$). Only in the case of poor reheating is T_{RH} likely to be anywhere as low as $10 MeV$, in which case $T_{RH} \simeq (\Gamma m_{pl})^{1/2}$ and the condition that T_{RH} be $\geq 10 MeV$ then implies

$$\Gamma \gtrsim 10^{-23} GeV \simeq (6.6 \times 10^{-2} sec)^{-1}$$

(6b) The more stringent condition on the reheat temperature is that it be sufficiently high for baryongenesis. If baryongenesis proceeds in the usual way[17], the out-of-equilibrium decay of a supermassive particle whose interactions violate B, C, P conservation, then T_{RH} must be greater than about 1/10 the mass of the particle whose out-of-equilibrium decays are responsible for producing the baryon asymmetry. Assuming that this particle couples to ordinary quarks and leptons, its mass must be greater than $10^{9} GeV$ or so to insure a sufficiently longlived proton, implying that the reheat temperature must be greater than about $10^{8} GeV$ (at the very least). On the other hand if the baryon asymmetry can be produced by the decays of the ϕ particles themselves, then

$$n_B/s \simeq (0.75)(T_{RH}/m_\phi)\epsilon$$

and a very low reheat temperature may be tolerable

$$T_{RH} \simeq 10^{-10}\epsilon^{-1} m_\phi$$

where as usual ϵ is the net baryon number produced per ϕ-decay.

(7) If ϕ is not a gauge singlet field, as in the case of the original Coleman-Weinberg SU(5) model, one must be careful that 'ϕ rolls in the correct direction'. It was shown that for the original Coleman-Weinberg SU(5) models ϕ might actually begin to roll toward the SU(4) $\times$ U(1) minimum of the potential even though the global minimum of the potential was the SU(3) $\times$ SU(2) $\times$ U(1) minimum[83]. This is because near $\phi = 0$ the SU(4) $\times$ U(1) direction is usually the direction of steepest descent. This is the so-called problem of 'competing phases'. As mentioned earlier, the extreme flatness required to obtain sufficiently small density perturbations probably precludes the possibility that ϕ is a gauge non-singlet, so the problem of competing phases does not usually arise. [Although in SUSY/SUGR models ϕ is often complex and one has to make sure that it rolls in the correct direction.]

(8) In addition to the scalar density perturbations discussed earlier, tensor or gravitational wave perturbations also arise (these correspond to tensor perturbations in

the metric $g_{\mu\nu}$)[84]. The amplitude of these perturbations is easy to estimate. The energy density in a given gravitational wave mode (characterized by wavelength λ) is

$$\rho_{GW} \simeq m_{pl}^2 h^2/\lambda^2 \tag{63}$$

where h is the dimensionless amplitude of the wave. As each gravitational wave mode crosses outside the horizon during inflation deSitter space produced fluctuations lead to

$$(\rho_{GW})_{\lambda \simeq H^{-1}} \simeq H^4, \text{ or } h \simeq H/m_{pl}. \tag{64}$$

While outside the horizon the dimensionless amplitude h remains constant (h behaves like a minimally coupled scalar field), and so each mode enters the horizon with a dimensionless amplitude

$$h \simeq H/m_{pl}. \tag{65}$$

Gravitational wave perturbations with wavelength of order the present horizon lead to a quadrupole anisotropy in the microwave temperature of amplitude h. The upper limit to the quadrupole anisotropy of the microwave background ($\delta T/T \lesssim 10^{-4}$) leads to the constraint

$$H/m_{pl} \leq 10^{-4},$$
$$M \leq O(10^{17} GeV).$$

In turn this leads to a constraint on the reheat temperature (using $g_* \simeq 10^3$)

$$T_{RH} \leq g_*^{-1/4} M \leq few \times 10^{16} GeV$$

(9) One has to be mindful of various particles which may be produced during the reheating process. Of particular concern are stable or very long-lived, NR particles (including other scalar fields which may be set into oscillation and thereafter behave like NR matter). Since $\rho_{NR}/\rho_R \propto R(t)$ and today $\rho_{NR}/\rho_R \simeq 3 \times 10^4$ or so one has to be careful that ρ_{NR}/ρ_R is sufficiently small at early times

$$\rho_{NR}/\rho_R \leq \begin{cases} 3 \times 10^4 & \text{today} \\ 10^{-8} & T = 1GeV \\ 10^{-18} & T = 10^{10} GeV \end{cases}$$

which is not always easy—just ask any experimentalist about suppressing some effect by 18 orders-of-magnitude!

Of particular concern in supersymmetric models are gravitinos which, if produced, can decay shortly after nucleosynthesis and photodissociate the light elements produced (particularly D and ^7Li)[85]. [In fact, the constraint that gravitinos not be overproduced during the reheating process leads to the very restrictive bound: $T_{RH} \leq 10^9 GeV$ or so.] In supersymmetric models where SUSY breaking is done ala Polonyi[86], the Polonyi field can be set into oscillation[87] and these oscillations

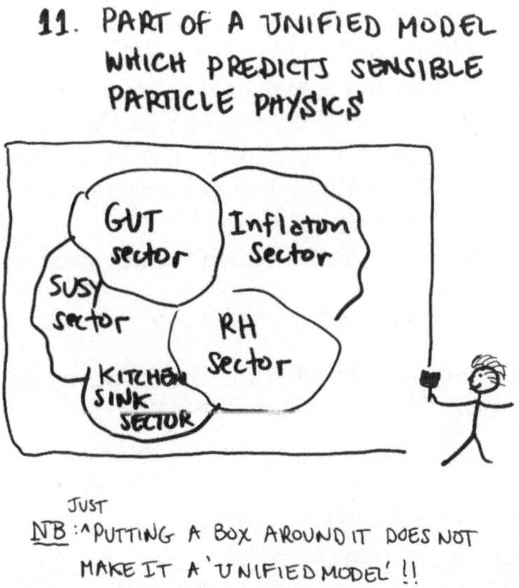
11. PART OF A UNIFIED MODEL
WHICH PREDICTS SENSIBLE
PARTICLE PHYSICS
GUT sector
Inflaton Sector
SUSY sector
RH Sector
KITCHEN SINK SECTOR
NB: JUST PUTTING A BOX AROUND IT DOES NOT
MAKE IT A 'UNIFIED MODEL' !!

Figure 16: Constraint (11) in 'The Prescription for Successful Inflation'.

which behave like NR matter can come to dominate the energy density of the Universe too early (leading to a Universe which if far too youthful when it cools to 3K) or even worse decay into dread gravitinos! In sum, one has to be mindful of the weakly-interacting, longlived particles which may be produced during reheating as they may eventually lead to an energy crisis.

(10) In SUSY/SUGR models where the scalar field responsible for inflation is in thermal equilibrium before the inflationary transition, one has to make sure that $<\varphi>_T$ does not smoothly evolve into the zero temperature minimum of the potential. A sure way of doing this is to arrange to have

$$\langle\varphi\rangle_T \leq 0$$

this is the so-called thermal constraint[80].

(11) Last (in my probably incomplete list) but certainly not least, the scalar potential necessary for successful inflation should be but a part of a 'sensible, perhaps even elegant, particle physics model' (see Fig. 16). We do not want cosmology to be the tail that wags the dog!

These conditions are spelled out in more detail in ref. 65. In general they lead to a potential which is 'short and squat' and has a dimensionless coupling of order 10^{-15} somewhere. In order that radiative corrections not spoil the flatness, it is all but mandatory that ϕ be a gauge singlet field which couples very weakly to other fields in the theory.

[Suppose that ϕ has a nice, flat potential which will successfully implement inflation and has a ϕ^4 term whose coefficient $\lambda \simeq O(10^{-15})$ (as is usually the case). Now suppose that ϕ couples to another scalar field ψ or to a fermion field f through terms like: $\lambda'\psi^2\phi^2$ and $h\bar{f}f\phi$. One-loop corrections to the $\lambda\phi^4$ term in the scalar potential arise due to the coupling of ϕ to ψ of f: $(\lambda'^2 ln + h^4 ln)\phi^4$. In order that they not spoil the flatness of $V(\phi)$ by these 1-loop corrections, the couplings λ' and h must be small: $\lambda' \lesssim ln^{-1/2}\lambda^{1/2}$; $h \lesssim ln^{-1/4}\lambda^{1/4}$.]

To give an idea of the kind of potential which we are seeking consider

$$V = V_0 - a\phi^2 - b\phi^3 + \lambda\phi^4$$

The constraints discussed above are satisfied for the following sets of parameters

$$SET\ 1 \begin{cases} \lambda \leq 4 \times 10^{-16} \\ \mathrm{b} \simeq 4 \times 10^7 \lambda^{3/2} m_{pl} \\ \mathrm{a} \leq H^2/40 \simeq 10^{28}\lambda^3 m_{pl}^2 \\ \sigma \simeq 3 \times 10^7 \lambda^{1/2} m_{pl} \\ \mathrm{M} \simeq V_0^{1/4} \simeq 3 \times 10^7 \lambda^{3/4} m_{pl} \\ \quad \simeq \lambda^{1/4}\sigma \end{cases}$$

$$SET\ 2 \begin{cases} \mathrm{V} = \lambda(\phi^2 - \sigma^2)^2 \quad (b = 0,\ a = 2\lambda\sigma^2,\ V_0 = \lambda\sigma^4) \\ \sigma/m_{pl} = 1/2,\ 2,\ 3,\ 10 \\ \lambda = 2 \times 10^{-44},\ 5 \times 10^{-20},\ 10^{-15},\ 2 \times 10^{-15},\ 3 \times 10^{-16} \\ \mathrm{M} \simeq \lambda^{1/4}\sigma \end{cases}$$

9. TWO SIMPLE MODELS THAT WORK

To date a handful of models that satisfy the prescription for successful inflation have been constructed[81,82,88–91,95–97]. Here, I will discuss two particularly simple and illustrative models. The first is an SU(5) GUT model proposed by Shafi and Vilenkin[89] and refined by Pi[90]. [Note, there is nothing special about SU(5); it could just as well be an E6 model.] I will discuss Pi's version of the model. In her model the inflating field $\vec{\phi}$ is a complex gauge singlet field whose potential is of the Coleman-Weinberg form[74]

$$V(\phi) = B[\phi^4 ln(\phi^2/\sigma^2) + (\sigma^4 - \phi^4)/2]/4 \tag{66}$$

where $\phi = |\vec{\phi}|$ and B arises due to 1-loop radiative corrections from other scalar fields in the theory and is set to be $O(10^{-14})$ in order to successfully implement inflation. (Note, in Eqn(66) I have only explicitly shown the part of the potential relevant for inflation.) Since the 1-loop corrections due to other fields in the model are of order

$(\lambda^2 ln)\phi^4$ (λ is a typical quartic coupling, e.g., $\lambda\phi^2\psi^2$) the dimensionless couplings of ϕ to other fields in the theory must be of order 10^{-7} or so. In her model, $\vec{\phi}$ is the field responsible for Peccei–Quinn symmetry breaking; the vacuum expectation value of $|\vec{\phi}|$ breaks the PQ symmetry and the argument of $\vec{\phi}$ is the axion degree of freedom. In addition, the vacuum expectation value of $|\vec{\phi}|$ induces SU(5) SSB as it leads to a negative mass-squared term for the 24-dimensional Higgs in the theory which breaks SU(5) down to SU(3) × SU(2) × U(1). In order to have the correct SU(5) breaking scale, the vacuum expectation value of $|\vec{\phi}|$ must be of order $10^{18}GeV$. In addition to the usual adiabatic density perturbations there are isothermal axion fluctuations of a similar magnitude[69]. The model reheats (barely) to a high enough temperature for baryogenesis. So far the model successfully implements inflation, albeit at the cost of a very small number ($B \simeq 10^{-14}$) whose origin is not explained and whose value is not stabilized (e.g., by supersymmetry).

The second model is a SUSY/SUGR model proposed by Holman, Ramond, and Ross[82] which is based on a very simple superpotential. They write the superpotential for the full theory as

$$W = I + S + G \tag{67}$$

where I, S, G pieces are the inflation, SUSY, and GUT sectors respectively. For the I piece of the superpotential they choose the very simple form

$$I = (\Delta^2/M)(\varphi - M)^2, \tag{68}$$

where $M = m_{pl}/(8\pi)^{1/2}$, Δ is an intermediate scale, and ϕ is the field responsible for inflation. Their potential has one free parameter: the mass scale Δ. This superpotential leads to the following scalar potential

$$\begin{aligned} V_I(\phi) &= exp(|\phi|^2/M^2)[|\partial I/\partial\phi + \phi^* I/M^2|^2 - 3|I|^2/M^2] \\ &= \Delta^4 exp(\phi^2/M^2)[\phi^6/M^6 - 4\phi^5/M^5 + 7\phi^4/M^4 - 4\phi^3/M^3 - \phi^2/M^2 + 1]. \end{aligned}$$

Expanding the exponential one obtains

$$V_I(\phi) = \Delta^4(1 - 4\phi^3/M^3 + 6.5\phi^4/M^4 - 8\phi^5/M^5 + ...), \tag{69a}$$

$$V_I' = \Delta^4(-12\phi^2/M^3 + 26\phi^3/M^4 - 40\phi^4/M^5 + ...) \tag{69b}$$

It is sufficient to keep just the first two terms in $V_I(\phi)$ to solve the equations of motion

$$\phi/M \simeq [12(N(\phi) + 1/3)]^{-1} \tag{70a}$$

$$H^2/\dot{\phi} \simeq (12\sqrt{3})^{-1}(\Delta/M)^2(\phi/M)^{-2} \simeq (12/\sqrt{3})(\Delta/M)^2 N^2, \tag{70b}$$

By choosing $\Delta/M \simeq 9 \times 10^{-5}$ density perturbations of an acceptable magnitude result (and about 2×10^6 e-folds of inflation!). Taking $\Delta/M \simeq 9 \times 10^{-5}$ corresponds to an intermediate scale in the theory of about $\Delta \simeq 2 \times 10^{14}GeV$—a very suggestive value.

The ϕ field couples to other fields in the theory only by gravitational strength interactions and

$$\Gamma \simeq m_{\varphi}^3/M^2 \simeq \Delta^6/M^5, \tag{71}$$

where $m_{\varphi}^2 \simeq 8e\Delta^4/M^2$.

The resulting reheat temperature is

$$T_{RH} \simeq (\Gamma m_{pl})^{1/2} \simeq (\Delta/M)^3 M \simeq 10^6 GeV. \tag{72}$$

The baryon asymmetry in this model is produced directly by ϕ-decays ($\phi \rightarrow H_3\bar{H}_3$; $H_3\bar{H}_3 \rightarrow q's\ l's$; H_3 = color triplet Higgs

$$\begin{aligned} n_B/s &\simeq (0.75\epsilon)T_{RH}/m_{\phi} \\ &\simeq 10^{-1}\epsilon(\Delta/M) \end{aligned}$$

Since $10^{-1}\Delta/M \simeq 10^{-5}$, a C, CP violation of about $\epsilon \simeq 10^{-5}$ is required to explain the observed baryon asymmetry of the Universe ($n_B/s \simeq 10^{-10}$).

Their model satisfies all the constraints for successful inflation except the thermal constraint. They argue that the thermal constraint is not relevant as the interactions of the ϕ field are too weak to put it into thermal equilibrium at early times. They therefore must take the initial value of ϕ ($\equiv \phi_0$) to be a free parameter and assume that in some regions of the Universe ϕ_0 is sufficiently far from the minimum so that inflation occurs ($\phi_0 \lesssim 10^{-3}M$). This model is somewhat *ad hoc* in that it contains a special sector of the theory whose sole purpose is inflation. Once again the model contains a small dimensionless coupling (the coefficient of the ϕ^4-term $\simeq 3 \times 10^{-16}$) or equivalently, a small mass ratio

$$(\Delta/M)^4 \simeq 10^{-16}$$

Since the model is supersymmetric that small number is stabilized against radiative corrections. Although the small ratio is not explained in their model, its value when expressed as a ratio of mass scales suggests that it might be related to one of the other small dimensionless numbers in particle physics (which also beg explanation)

$$\begin{aligned} (m_{GUT}/m_{pl}) &\simeq 10^{-4} \\ (m_W/m_{pl}) &\simeq 10^{-17} \\ g_e \simeq m_e/300GeV &\simeq 10^{-6} \end{aligned}$$

While neither of these models is particularly compelling and both have been somewhat contrived to successfully implement inflation, they are at the very least 'proof of existence' models which demonstrate that it is possible to construct a simple model which satisfies all the know constraints. Fair enough!

10. TOWARD THE INFLATIONARY PARADIGM

Guth's original model of inflation was based upon a strongly, first order phase transition associated with SSB of the GUT. The first models of new inflation were based upon Coleman-Weinberg potentials, which exhibit weakly-first order transitions. It now appears that the key feature needed for inflation is a very flat potential and that even potentials which lead to second order transitions (i.e., the $\phi = 0$ state is never metastable) will work just as well[92]. Most of the models for inflation now do not involve SSB, at least directly, they just involve the evolution of a scalar field which is initially displaced from the minimum of its potential. [There is a downside to this; in many models inflation is a sector of the theory all by itself.] Since the fields involved are very weakly coupled, thermal corrections can no longer be relied upon to set the initial value of ϕ. Inflation has become much more than just a scenario—it has become an early Universe paradigm!

On the horizon now are models which inflate, but are even more far removed from the original idea of a strongly-first order, GUT SSB phase transition; I'll discuss three of them here. Inflation—that is the rapid growth of our three spatial dimensions, appears to be a very generic phenomenon associated with early Universe microphysics.

(1) Chaotic Inflation

Linde[92] has proposed the idea that inflation might result from a scalar field with a very simple potential, say

$$V(\phi) = \lambda\phi^4$$

which due to 'chaotic initial conditions' (which thus far have not been well-defined) is displaced from the minimum of its potential—in this case $\phi = 0$ (see Fig. 17). With the initial condition $\phi = \phi_0$ this potential is very easy to analyze:

$$\begin{aligned} N(\phi_0) &\equiv \int_{\phi_0}^{0} H dt \simeq \pi(\phi_0/m_{pl})^2, \\ (\delta\rho/\rho)_H &\simeq (H^2/\dot{\phi}) \simeq (32/3)^{1/2}\lambda^{1/2}N(\phi)^{3/2}. \end{aligned} \tag{73}$$

In order to obtain density perturbations of the proper amplitude ($\delta\rho/\rho \simeq 10^{-4}$) λ must be very small

$$\lambda \simeq 10^{-14}$$

—as usual! In order to obtain sufficient inflation, the initial value of ϕ must be

$$\begin{aligned} N(\phi_0) &\simeq \pi(\phi_0/m_{pl})^2 \geq 60 \\ &\Rightarrow \phi_0 \geq 4.4 m_{pl} \end{aligned}$$

Both of these two conditions are rather typical of potentials which successfully implement inflation. However, when one talks about truly chaotic initial conditions one wonders if a large enough patch exists where ϕ is approximately constant. Remember the key constraint is that the gradient energy density be small compared to the potential energy

$$(\nabla\phi)^2/2 \ll \lambda\phi_0^4$$

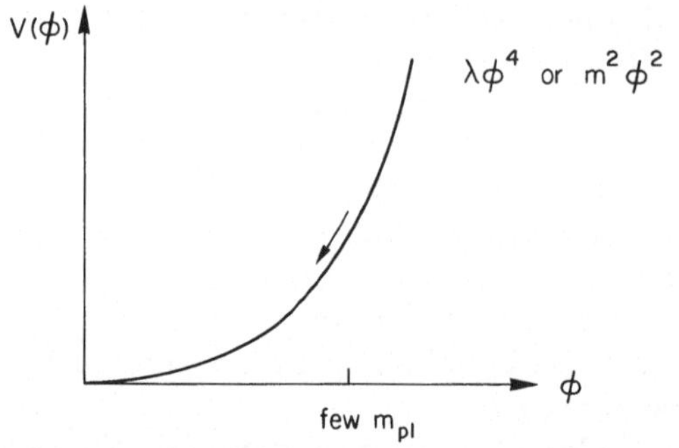

Figure 17: A potential for 'chaotic inflation'. In Linde's chaotic inflation, due to initial conditions, ϕ is displaced from the minimum of its potential ($\phi = 0$) and inflation occurs as it evolves to $\phi = 0$.

Labeling the typical dimension of the patch L, the above requirement translates to

$$L \gg \lambda^{-1/2}(m_{pl}/\phi_0)m_{pl}^{-1} \simeq 2(\phi_0/m_{pl})H^{-1}$$

which requires that L be rather large compared to the Hubble radius, therefore seeming to require rather special initial conditions. Still the simplicity of Linde's idea is very appealing.

[Note that the potential $V = \frac{1}{2}m^2\phi^2$ (corresponding to a massive scalar field) is also suitable for inflation. In this case

$$N(\phi_0) \simeq 2\pi(\phi_0/m_{pl})^2, \tag{74a}$$

$$(\delta\rho/\rho)_H \simeq H^2/\dot{\phi} \simeq 4(\pi/3)^{1/2}(m/m_{pl})N. \tag{74b}$$

Sufficient inflation requires: $\phi_0 \gtrsim 3m_{pl}$, and density perturbations of an acceptable magnitude requires: $(m/m_{pl}) \simeq 10^{-4}/(4N) \simeq 4\times10^{-7}$. This potential has been analyzed by I. Moss (private communication) and L. Jensen (private communication), and more recently by the authors of refs. 93.]

(2) Induced Gravity Inflation

Consider the Ginzburg-Landau theory of induced gravity based upon the effective Lagrangian[94]

$$\mathcal{L} = -\epsilon\phi^2 R/2 - \partial_\mu\phi\partial^\mu\phi/2 - \lambda(\phi^2 - v^2)^2/8, \tag{75}$$

where ϵ, λ are dimensionless couplings, R is the Ricci scalar, and $v \equiv \epsilon^{-1/2}(8\pi G)^{-1/2}$. The equation of motion for ϕ is

$$\ddot{\phi} + 3H\dot{\phi} + \dot{\phi}^2/\phi + [V' - 4V/\phi]/(1+6\epsilon) = 0 \tag{76a}$$

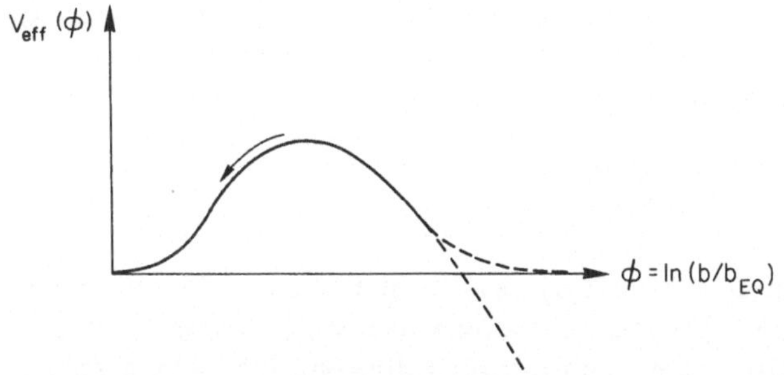

Figure 18: In theories with additional spatial dimensions there must be an effective potential associated with the size of the extra dimensions (shown here schematically). One might expect that very early on ($t \leq 10^{-43}$ sec) the size of the extra dimensions is displaced from its equilibrium value ($\equiv b_{eq}$), due to finite temperature corrections, initial conditions, or whatever. It is speculated that inflation might occur as the size of the extra dimensions evolves to its equilibrium value, thereby solving both the usual cosmological puzzles and the puzzle of why the extra dimensions are so small compared to our three familiar spatial dimensions.

supplemented by

$$H^2[1 + (2\dot{\phi}/\phi)/H] = (3\epsilon\phi^2)^{-1}[\dot{\phi}^2/2 + V(\phi)] \tag{76b}$$

Successful inflationary scenarios can be constructed for $\phi_0 \ll v$ and for $\phi_0 \gg v$(ϕ_0 = the initial value of ϕ), so long as $\epsilon \leq 10^{-2}$ and $\lambda \simeq O(10^{-12} - 10^{-16})$[95,96]. The small dimensionless coupling constant required in the scalar potential is by now a very familiar condition.

(3) The Compactification Transition

Ever increasing numbers of physicists are pursuing the idea that unification of the forces may require additional spatial dimensions (or as the optimist would say, unification of the forces is evidence for extra dimensions!), e.g., Kaluza-Klein theories, supergravity theories, and most recently, superstring theories. We know experimentally that these extra dimensions must be very small ($\ll 10^{-17}cm$) and indeed in most theories the extra dimensions form a compact manifold of characteristic size $10^{-34}cm$ or so. If our space-time is truly more than four dimensional, then we have yet another problem to add to our list of puzzling cosmological facts—the extreme smallness of the extra spatial dimensions, some $62 \simeq log(10^{28}cm/10^{-34}cm)$ or so orders of magnitude smaller than the more familiar three spatial dimensions. The possible use of inflation to explain this largeness problem has not escaped the attention of researchers in this field.

In these theories there is a natural candidate for the 'inflating field' (which is also automatically a gauge singlet)—the radius of the extra dimensions. If there are extra dimensions there must be some dynamics which determine their present, equilibrium size ($\equiv b_{eq}$), and in principle one should be able to construct an effective potential associated with the size of the extra dimensions

$$V_{eff} = V(\phi),$$
$$\phi = ln(b/b_{eq}).$$

(see Fig. 18). [The substitution $\phi = ln(b/b_{eq})$ results in the usual kinetic term for ϕ when the higher dimensional Einstein equations are written down.] If the extra dimensions are displaced from their equilibrium value—an idea which seems not at all unreasonable since very early on ($t < 10^{-43} sec$) one might expect all the dimensions to be on equal footing, then while they are evolving to their equilibrium value ($\phi = 0$) the Universe will be endowed with a large potential energy (and may very well inflate), thereby explaining the largeness of our three spatial dimensions as well as the usual cosmological puzzles. Inflationary models involving the compactification transition have already been investigated and the results are encouraging[97].

11. LOOSE ENDS

Inflation offers the possibility of making the present state of the Universe (on scales as large as our Hubble radius) insensitive to the initial data for the Universe. Since we have little hope of ever knowing what the initial data were this is a very attractive proposition. It has by no means yet achieved that lofty goal. There are a number of loose ends (and perhaps even a loose thread which may unravel the entire tapestry). I will briefly mention a few of them here.

(1) 'Who is ϕ?'

Inflationary models exist in which the scalar field ϕ: effects SSB of the GUT[89,90] effects SSB of SUSY[81], induces Newton's constant (in a Landau-Ginzburg model of induced gravity)[95,96], is $\sim ln(r_X/r_{XEQ})$ (where r_X is the radius of compactified extra dimensions) in theories with extra dimensions which become compactified[97], is $\propto$ (scalar curvature $)^{1/2}$ in R^2 theories of gravity,[98] is just some 'random' scalar field[92], or is merely in the theory to effect inflation.[79,82] Given the number of different kinds of inflationary scenarios which exist, it seems as though inflation is generic to early Universe microphysics, occurring whenever a weakly-coupled scalar field finds itself displaced from the minimum of its potential. Clearly, a key question at this point is just how 'the inflation sector' of the theory fits into the Big Picture!

(2) What Determines the Initial Value of ϕ?

One thing is certain, and that is that ϕ must be very weakly-coupled, as quantified by its small dimensionless coupling constant. Because of this fact, it is almost certain that ϕ was not initially in thermal contact with the rest of the Universe and so the initial value of $\phi(\equiv \phi_0)$ is unlikely to be determined by thermal considerations (in the earliest models of new inflation, ϕ_0 was determined by thermal considerations, however these models resulted in density perturbations of an unacceptably large amplitude). At present ϕ_0 must be taken as initial data. Some have argued that ϕ_0 might be determined in an anthropic-like way, as regions of the Universe where ϕ_0 is sufficiently far displaced from equilibrium will undergo inflation and eventually occupy most of the physical volume of the Universe. Perhaps the wave-function of the Universe approach will shed some light on the initial distribution of the scalar field ϕ. Or it could be that due to 'as-of-yet unknown dynamics' ϕ was indeed in thermal equilibrium at a very early epoch. It goes without saying that it is crucial that ϕ be initially displaced from its minimum.

(3) Validity of the Semi-Classical Equations of Motion for ϕ

While it may seem perfectly plausible that ϕ evolves according to its semi-classical equations of motion, the validity of this assumption has troubled inflationists from the 'dawn of new inflation'. While a full quantum field theory treatment of inflation is very difficult and has not been effected, a number of specific issues have been addressed. Several authors have studied the role of inhomogeneities in ϕ, and have found that for the very weakly-coupled fields one is dealing with, mode coupling is not important and the individual modes are quickly smoothed by the exponential expansion of their physical wavelengths.[99] I already mentioned the necessity of having ϕ smooth over a sufficiently large region so that the gradient terms in the stress energy do not dominate.

The effect of quantum fluctuations on the evolution of ϕ has been studied in some detail by Guth and Pi[91], Fischler et al.[98], Linde[75], Vilenkin and Ford[76], Semenoff and Weiss[101], and Evans and McCarthy[102]. The basic conclusion that one draws from the work of these authors is that the use of the semi-classical equations of motion is valid so long as $\phi_{cl} \gg \Delta\phi_{QM} \simeq N^{1/2}H/2\pi$, which is almost always satisfied for the very flat potentials of interest to inflationists (at least for the last 50 or so e-folds which affect our present Hubble volume). [More precisely, the semi-classical change in ϕ in a Hubble time, $\Delta\phi_{Hubble} \simeq -V'/3H^2 \simeq -V'm_{pl}^2/(8\pi V)$, should be much greater than the increase in $< \phi^2 >_{QM}^{1/2}, \simeq H/2\pi$, due to the addition of another quantum mode; see Bardeen et al.[56]] At present the validity of the semi-classical equations of motion seems to be reasonably well established.

(4) No Hair Conjectures

While inflation has been touted from the very beginning as making the present state of the Universe insensitive to the initial spacetime geometry, not much has been done to justify this claim until very recently. As I mentioned earlier, inflation is nearly always analyzed in the context of a flat, FRW cosmological model, making such a claim somewhat dubious. However, it has now been shown that all of the homogeneous models (with the exception of the highly-closed models) undergo inflation, isotropize and remain isotropic to the present epoch providing that the model would have inflated the requisite 60 or so e-folds in the absence of anisotropy.[103]

The proof of this result involves three parts. First, Wald[104] demonstrated that all homogeneous models with a positive cosmological term asymptotically approach deSitter (less the aforementioned highly-closed models which recollapse before the cosmological term becomes relevant). Wald's result follows because all forms of 'anisotropy energy density' decrease with increasing proper volume element, whereas the cosmological term remains constant, and so eventually triumphs. Of course, inflationary models do not, in the strictest sense, have a cosmological term, rather they have a positive vacuum energy as long as the scalar field is displaced from the minimum of its potential. Thus the dynamics of the scalar field comes into play: does ϕ stay displaced from the minimum of its potential long enough so that the vacuum energy comes to dominate? Due to the presence of anisotropy the expansion rate is *greater* than if there were only vacuum energy density, and so the friction felt by ϕ as it trys to roll (the $3H\dot{\phi}$ term) is greater and it takes ϕ longer to evolve to its minimum than without anisotropy. For this reason the Universe does become vacuum dominated before the vacuum energy disappears, and in fact the Universe inflates slightly longer in the presence of anisotropy (one or two e-folds)[105]. Finally, is the anisotropy reduced sufficiently so that the Universe today is still nearly isotropic? As it turns out, the requisite 60 or so e-folds needed to solve the other conundrums reduces the growing modes of anisotropy sufficiently to render them small today.

Allowing for inhomogeneous initial spacetimes makes matters much more difficult. Jensen and Stein-Schabes[106] and Starobinskii[107] have proven the analogue of Wald's theorem for spacetimes which are negatively-curved. Jensen and Stein-Schabes[106] have gone on to conjecture that spacetimes which have sufficiently large regions of negative curvature will undergo inflation, resulting in a Universe today which although not globally homogeneous, at least contains smooth volumes as large as our current Hubble volume.

Does this improve the situation that Collins and Hawking[13] discussed in 1973? While the work of Jensen and Stein-Schabes[106] seems to indicate that many inhomogeneous spacetimes undergo inflation and even leads one to speculate that the measure of the set of initial spacetimes which eventually inflate is non-zero, it is not possible to draw a definite conclusion without first defining a measure on the space of initial data. In fact, as Penrose[25] pointed out there is at least one way of defining the measure such that this is not the case. Consider the set of all Cauchy data at the present epoch; intuitively it is clear that those spacetime slices which are highly irregular are the rule, and those which are smooth in regions much larger than our current Hubble volume are the exception. Defining the measure today, it seems very reasonable that the smooth spacetime slices are a set of measure zero. Now evolve the spacetimes back to some initial epoch (for example $t = 10^{-43} sec$). Using the seemingly very reasonable measure defined today and the mapping back to 'initial' spacetimes, one could argue that the set of initial data which inflate is still of measure zero. While I believe that this argument is technically correct, I also believe that it is silly. First, upon close examination of all of those initial spacetimes which led to spacetimes today without smooth regions as large as our present Hubble volume, one would presumably find that the scalar field responsible for inflation would be very close to the minimum of its potential (in order that they not inflate)–not a very generic initial condition. Secondly, if one adopts the point-of-view of an evolving Universe which has an 'initial epoch' (and not everyone does), then there is a preferred epoch at which one would define a measure–the 'initial epoch,' and at that epoch I believe any reasonably defined measure would lead to the set of initial spacetimes which inflate being of non-zero measure.

Although it is not possible *yet* to claim rigorously that inflation has resolved the problem of the seemingly special initial data required to reproduce the Universe we see today (at least within our Hubble volume), I think that any fairminded person would admit that it has improved the situation dramatically. Extrapolating from the solid results that exist, it seems to me that starting with a general inhomogeneous spacetime, there will exist regions which undergo inflation and which today are much larger than our present Hubble volume, thereby accounting for the smooth region we find ourselves in. From a more global perspective, one might expect that on scales $\gg H^{-1}$ the Universe would be highly irregular. [The evolution of a model universe which is isotropic and homogeneous except for one spherically-symmetric region of false vacuum (where $\phi \neq \sigma$) has been studied by the authors of ref. 108. The results are interesting in that they begin to address the problem of general initial conditions. The vacuum-dominated bubble becomes causally detached from the rest of the spacetime, becoming a 'child Universe' spawned by inflation.]

(5) The Present Vanishingly Small Value of the Cosmological Constant

Inflation has shed no light on this difficult and very fundamental puzzle (nor has anything else for that matter!). In fact, since inflation runs on vacuum energy so to speak, the fate of inflation hinges upon the resolution of this puzzle. For example, suppose there were a grand principle that dictated that the vacuum energy of the Universe is always zero, or that there were an axion-like mechanism which operated and ensured that any cosmological constant rapidly relaxed to zero; either would be a disaster to inflation, shorting out its source of power–vacuum energy. [Another possibility which has received a great deal of attention recently is the possibility that deSitter space might be quantum mechanically unstable[109]–of course, if its lifetime were at least 60 some e-folds that would not necessarily adversely affect inflation.]

12. INFLATION CONFRONTS OBSERVATION

No matter how appealing a theory may be, it must meet and pass the test of experimental verification. Experiment and/or observation is the final arbiter. One of the few blemishes on early Universe physics is the lack, thus far, of experimental/observational tests of the many beautiful and exciting predictions. That situation is beginning to change as the field starts to mature. Inflation is one of the early Universe theories which is becoming amenable to verification or falsification. Inflation makes the following very definite predictions (postdictions?):

(1) $\Omega = 1.0$ (more precisely, $R_{curv} = R(t)|k|^{-1/2} = H^{-1}/|\Omega - 1|^{1/2} \gg H^{-1}$)

(2) Harrison-Zel'dovich spectrum of constant curvature perturbations (and possibly isocurvature perturbations as well) and tensor mode gravitational wave mode perturbations

The prediction of $\Omega = 1.0$ together with the primordial nucleosynthesis constraint on the baryonic contribution, $0.014 \lesssim \Omega_B h^2 \lesssim 0.035 \lesssim 0.15$ (ref. 6), suggests that most of the matter in the Universe must be nonbaryonic. The simplest and most plausible possibility is that it exists in the form of relic WIMPs (*W*eakly-*I*nteracting *M*assive *P*articles, e.g., axions, photinos, neutrinos; for a review, see ref. 110). Going a step further, these two original predictions then lead to testable consequences:

(3) $H_0 t_0 = 2/3$ (providing that the bulk of the matter in the Universe today is in the form of NR particles)

The observational data on both H_0 and t_0 are far from being definitive: $H_0 \simeq 40 - 100 km sec^{-1} Mpc^{-1}$ and $t_0 \simeq 12 - 20 Gyr$, implying only that $H_0 t_0 \simeq 0.5 - 2.0$.

(4) $\Omega = 1.0$

All of the dynamical observations suggest that the fraction of critical density contributed by matter which is clumped on scales $\lesssim 10 - 30 Mpc$ is only about: $\Omega_{\lesssim 30} \simeq 0.2 \pm 0.1$ (± 0.1 is not meant to be a formal error estimate, but indicates the spread in the observations) (see refs. 8). If inflation is not to be falsified, that leaves but two options: (1) the observations are somehow misleading or wrong; or (2) there exists a component of energy density which is smoothly distributed on scales $\lesssim 10 - 30 Mpc$ (and therefore would not be reflected in the dynamical determinations). Candidates for the smooth component include: relic, light neutrinos, which by virtue of the large length scale ($\lambda_\nu \simeq 13 h^{-2} Mpc$) on which neutrino perturbations are damped by freestreaming, would likely still be smooth on these scales; relic relativistic particles produced by the recent decay of an unstable WIMP species;[111] a relic cosmological term;[112] 'failed galaxies,' referring to a population of galaxies which have the same mix of dark matter to baryons, but are more smoothly distributed and are too faint to observe (at least thus far);[113] relic population of light strings–either fast moving non-intercommuting strings or a tangled network of non-Abelian strings.[114] All of these smooth component scenarios have testable consequences[115]—their predictions for $H_0 t_0$ differ from 2/3; the growth of perturbations is different; the evolution of the cosmic scale factor $R(t)$ is different from the matter-dominated model and various kinematic tests (magnitude-redshift, angular size-redshift, lookback time-redshift, proper volume element-redshift, etc.) can in principle differentiate between them.

(5) Microwave Fluctuations

Both the scalar and tensor metric perturbations lead to fluctuations in the CMBR on large angular scales ($\gg 1°$). On such large scales causal microphysical processes (such as reionization) cannot have erased the primordial fluctuations, and so if ever present, they must still be there. The scalar perturbations (if they have anything to do with structure formation) must be of amplitude $\gtrsim few \times 10^{-6}$, which is within a factor of 10 or less of the current upper limits on these scales.

(6) Two Detailed Stories of Structure Formation

The simplest possibility, namely that most of the mass density is in relic WIMPs ($\Omega_{WIMP} = 1.0 - \Omega_B \simeq 0.9$) leads to two very detailed scenarios of structure formation: hot dark matter (the case where the dark matter is neutrinos) and cold dark matter (essentially any other WIMP as dark matter). At present, the numerical simulations of these scenarios are sufficiently definite that it is possible to falsify them–and in fact, both of these simplest scenarios have difficulties (see the recent review by White[116]). In the hot dark matter case it is forming galaxies early enough. The large-scale structure which evolves in this case (voids, superclusters, froth) qualitatively agrees with what is observed; however, in order to get agreement with the galaxy-galaxy correlation function, galaxies must form very recently (redshifts $\lesssim 1$) in contradiction to all the galaxies (redshifts as large as 3.2) and QSO's (redshifts as large as 4.0) which are seen at redshifts $\gtrsim 1$.

With cold dark matter the simulations can nicely reproduce galaxy clustering, most of the observed properties of galaxies (masses and densities, rotation curves, etc).[117] However the simulations do not seem to be able to produce sufficient large-scale structure. In particular, they fail to account for the amplitude of the cluster-cluster correlation function (by a factor of about 3), large amplitude, large-scale peculiar velocities, and voids. [In fairness I should mention that our knowledge of large-scale structure of the Universe is still very fragmentary, with the first moderate sized ($\sim 10^4$), 3-dimensional surveys having just recently been completed.] In order to account for $\Omega = 1.0$, galaxy formation must be biased (i.e., only density-averaged peaks greater than some threshold, typically $2 - 3\sigma$, are assumed to evolve into galaxies which we see today, the more typical 1σ peaks resulting in 'failed galaxies' for some reason or another; see ref. 113).

[The situation with respect to large scale structure is becoming more interesting every moment. Several groups have now reported large-amplitude ($600 - 1000 km sec^{-1}$) peculiar velocities on large scales ($\sim 50h^{-1}Mpc$) (Burstein et al.;[118] Collins et al.[119]). Such large peculiar velocities are very difficult, if not impossible, to reconcile with either hot or cold dark matter (or even smooth component models) and the Zel'dovich spectrum (see ref. 120). If these data hold up they may pose an almost insurmountable obstacle to any scenario with the Zel'dovich spectrum of density perturbations. The frothy structure observed in the galaxy distribution by de Lapparent et al.[121], galaxies distributed on the surfaces on large ($\sim 30h^{-1}Mpc$), empty bubbles, although somewhat more qualitative, also seems difficult to reconcile with cold dark matter.]

There are a number of observations/experiments which can and will be done in the next few years and which should really put the inflationary scenario to the test. They include improved sensitivity measurements of the CMBR anisotropy. The microwave background anisotropies predicted in the hot dark matter scenario are very close to the observational upper limits on angular scales of both 5 or so arcminutes and $\gtrsim$ few degrees.[12] With cold dark matter, the predictions are a factor of $3 - 10$ away from the observational limits (for the isocurvature spectrum, the quadrupole upper limit may actually rule out this possibility; see, Efstathiou and Bond[122]). An improvement in sensitivity to microwave anisotropies of the order of $3 - 10$ could either begin to confirm one of the scenarios or rule them both out, and is definitely within the realm of experimental reality (Wilkinson in ref. 5).

The relic WIMP hypothesis for the dark matter can also be tested. While it was once almost universally believed that all WIMP dark matter candidates were, in spite of their large abundance, essentially impossible to detect because of the feebleness of their interactions, a number of clever ideas have recently been suggested (and are being experimentally implemented) for detecting axions,[123] photinos, sneutrinos, heavy neutrinos, etc.[124] Results and/or limits will be forth coming soon. With the coming online of the Tevatron at Fermilab, the SLC at SLAC, and hopefully the SSC it is possible that one of the candidates may be directly produced in the lab. Experiments to detect neutrino masses in the eV mass range also continue.

A geometric measurement of the curvature of the Universe (which uses the dependence of the comoving volume element as a function of redshift) has recently been made by Loh and Spillar.[125] Their preliminary results indicate $\Omega = 0.9^{+0.7}_{-0.5}$ (95% confidence) (for a matter-dominated model). This technique appears to have great cosmological leverage and looks very promising (especially the value!)—far more promising than the traditional approach of determining the density of the Universe through the deceleration parameter q_0.

Another area with great potential for improvement is 3d surveys of the distribution of galaxies. The largest redshift surveys at present contain only a few 1000 galaxies, yet have been very tantalizing, indicating evidence of voids and froth-like structure to the galaxy distribution.[121] The large, automated surveys which are likely to be done in the next decade could very well lead to a quantum leap in our understanding of the large scale features of the Universe and help to provide hints as to how they evolved.

The peculiar velocity field of the Universe is potentially a very valuable and direct probe of the density field of the Universe:

$$|\delta v_k| = |\dot{\delta}_k/k| \quad (= (\lambda H/2\pi)\delta_k \ for\ \Omega = 1) \tag{77}$$

$$(\delta v/c)_\lambda \simeq (\lambda/10^4 h^{-1} Mpc)(\delta\rho/\rho)_\lambda \tag{78}$$

where δ_k and δv_k are the $k-th$ Fourier components of $\delta\rho/\rho$ and $\delta v/c$ respectively. The very recent measurements which indicate large amplitude peculiar velocities on scales of $\sim 50h^{-1}Mpc$ are surprising in that they indicate substantial power on these scales, and are problematic to almost every scenario of structure formation. Should they be confirmed they will provide a very acute test of structure formation in inflationary models.

Of course, theorists are very accommodating and have already started suggesting alternatives to the simplest scenarios for structure formation. As I mentioned earlier, scenarios with a smooth component to the energy density have been put forward to solve the Ω problem. Cosmic strings present a radically different approach to structure formation with their non-gaussian spectrum of density fluctuations (for further discussion see refs. 131). [It is interesting to note that cosmic strings of the right 'weight' ($G\mu \simeq 10^{-6}$ or so, where μ is the string tension) seem to be somewhat incompatible with inflation, as they must necessarily be produced after inflation and require reheating to a temperature $\gtrsim \mu^{1/2} \simeq 10^{16} GeV$ which seems difficult.] Somewhat immodestly I mention a proposal Silk and I recently made: 'double inflation'.[126] While the Harrison-Zel'dovich spectrum is a beautiful prediction both because of its geometric simplicity and its definiteness, it may well be in conflict with observation because it does not seem to allow enough power on large scales to account for the recent observations of froth and large amplitude peculiar velocities. In the variant we have proposed there are two (or more) episodes of inflation, with the final episode lasting only about 40 e-folds or so, so that the amplitudes of perturbations on large scales are set by the first episode and those on small scales by the second episode. That there might be multiple episodes of inflation seems quite plausible given the number of different microphysical scenarios which result in inflation. Arranging the most recent episode to last for only 40 or so e-folds so that some of the scales within our present Hubble volume crossed outside the horizon during an earlier episode of inflation is a more formidable task—but not an impossible or implausible one! If this can be arranged then it is possible to have very large amplitude perturbations on small scales (of order 10^{-1}) and larger than usual amplitude perturbations on large scales (nearly saturating the large scale microwave limits), thereby providing enough power for the large scale structure which the recent redshift surveys and peculiar velocity measurements indicate. The large amplitude perturbations on small scales allow for very early galaxy formation (and reionization of the Universe, thereby erasing the CMBR fluctuations on small angular scales). If the second episode of inflation proceeds via the nucleation of bubbles, they might directly explain the froth-like structure recently reported by de Lapparent et al.[121]

13. EPILOGUE

Despite the absence of a compelling model which successfully implements the inflationary paradigm, inflation remains a very attractive means of accounting for a number of very fundamental cosmological facts by microphysics that we have some understanding of: namely, scalar field dynamics at sub-planck energies. The lack of a compelling model at present must be viewed in the light of the fact that at present we have no compelling, detailed model for the '*Theory of Everything*' and the fact that despite vigorous scrutiny there has yet to be a *No-Go Theorem* for inflation unearthed. It is my belief that the undoing of inflation (if it should come) will involve observations and not theory. At the very least The Inflationary Paradigm is still worthy of further consideration–and I hope that I have convinced you of that fact!

Due to space/time limitations my review of inflation has necessarily been incomplete, for which I apologize. I refer the interested reader to the more complete reviews by Linde[127]; by Abbott and Pi[128]; by Steinhardt[129]; by Brandenberger[130]; by Bonometto and Masiero[138]; and by Blau and Guth.[132] My prescription for successfully implementing inflation borrows heavily from the paper written by Steinhardt and myself.[65] This work was supported in part by the DoE (at Chicago) and by my Alfred P. Sloan Fellowship.

REFERENCES

1. S. Djorgovski, H. Spinrad, P. McCarthy, and M. Strauss, *Astrophys.J.* **299**, L1 (1985).
2. B.A. Peterson *et al.*, *Astrophys.J.* **260**, L27 (1982); D. Koo, *Astron.J.* **90**, 418 (1985); C. Hazard, R-G. McMahon, and W.L.W. Sargent, *Nature* **322**, 38 (1986); *Nature* **322**, 40 (1986); S.J. Warren, et al., *Nature* **325**, 131 (1987).
3. P.J.E. Peebles, *Physical Cosmology* (Princeton Univ. Press, 1971); S. Weinberg, *Gravitation and Cosmology* (Wiley, NY, 1972), chapter 15; *Physical Cosmology*, eds. J. Audouze, R. Balian, and D.N. Schramm (North-Holland, Amsterdam, 1980).
4. J. Peterson, P. Richards, and T. Timusk, *Phys.Rev.Lett.* **55**, 332 (1985); G.F. Smoot et al., *Astrophys.J.* **291**, L23 (1985); D. Meyer and M. Jura, *Astrophys.J.* **276**, L1 (1984); D. Woody and P. Richards, *Astrophys.J.* **248**, 18 (1981).
5. D. Wilkinson, in *Inner Space/Outer Space*, eds. E. Kolb et al. (Univ. of Chicago Press, Chicago, 1986).
6. J. Yang, M.S. Turner, G. Steigman, D.N. Schramm, and K. Olive, *Astrophys.J.* **281**, 493 (1984); A. Boesgaard and G. Steigman, *Ann.Rev.Astron.Astrophys.* **23**, 319 (1985).
7. J. Huchra, and A. Sandage and G. Tammann, in *Inner Space/Outer Space*, eds. E. Kolb et al. (Univ. of Chicago Press, Chicago, 1986); J. Gunn and B. Oke, *Astrophys.J.* **195**, 255 (1975); J. Kristian, A. Sandage, and J. Westphal, *Astrophys.* **221**, 383 (1978); R. Buta and G. deVaucouleurs, *Astrophys.J.* **266**, 1 (1983); W.D. Arnett, D. Branch, and J.C. Wheeler, *Nature* **314**, 337 (1985).
8. S. Faber and J. Gallagher, *Ann.Rev.Astron.Astrophys.* **17**, 135 (1979); V. Trimble, *Ann.Rev.Astron.Astrophys.* **25**, in press (1987).
9. I. Iben, Jr., *Ann.Rev.Astron.Astrophys.* **12**, 215 (1974); A. Sandage, *Astrophys.J.* **252**, 553 (1982); D.N. Schramm, in *Highlights of Astronomy*, ed. R. West (Reidel, Dordrecht, 1983), vol.6; I. Iben and A. Renzini, *Phys.Rep.* **105**, 329 (1984).

9a. C.B. Collins and M.J. Perry, *Phys.Rev.Lett.* **34**, 1353 (1975).

10. G.F.R. Ellis, *Ann.NY.Acad.Sci.* **336**, 130 (1980).
11. R. Sachs and A. Wolfe, *Astrophys.J.* **147**, 73 (1967).
12. N. Vittorio and J. Silk, *Astrophys.J.* **285**, L39 (1984); J.R. Bond and G. Efstathiou, *Astrophys.J.* **285**, L44 (1984); J. Silk, in *Inner Space/Outer Space*, eds. E.W. Kolb et al. (University of Chicago Press, Chicago, 1986).
13. C.B. Collins and S. Hawking, *Astrophys.J.* **180**, 317 (1973).
14. See, e.g., P.J.E. Peebles, *Large-Scale Structure of the Universe* (Princeton Univ. Press, Princeton, 1980); G. Efstathiou and J. Silk, *Fund.Cosmic Phys.* **9**, 1 (1983); S.D.M. White, in *Inner Space/Outer Space*, eds. E. Kolb et al. (Univ. of Chicago Press, Chicago, 1986).

14a. R.H. Dicke and P.J.E. Peebles, in *General Relativity: An Einstein Centenary Survey*, eds. S.W. Hawking and W. Israel (Cambridge University Press, Cambridge, 1979).

15. G. Steigman, *Ann.Rev.Astron.Astrophys.* **14**, 339 (1976).
16. G.G. Ross, *Grand Unified Theories* (Benjamin/Cummings, Menlo Park, 1984).
17. E.W. Kolb and M.S. Turner, *Ann.Rev.Nucl.Part.Sci.* **33**, 645 (1983).
18. G. 't Hooft, *Nucl.Phys.* **B79**, 276 (1974); A.M. Polyakov, *JETP Lett.* **20**, 194 (1974).
19. T.W.B. Kibble, *J.Phys.* **A9**, 1387 (1976); J. Preskill, *Phys.Rev.Lett.* **43**, 1365 (1979); Ya. B. Zel'dovich and M. Yu. Khlopov, *Phys.Lett.* **79B**, 239 (1978).
20. J. Preskill, *Ann.Rev.Nucl.Part.Sci.* **34**, 461 (1984); M.S. Turner, in *Monopole '83*, ed. J. Stone (Plenum Press, NY, 1984).
21. A. Guth and S-H. H. Tye, *Phys.Rev.Lett.* **44**, 631 (1980).
22. A. Guth, *Phys.Rev.* **D23**, 347 (1981).
23. S. Bludman and M. Ruderman, *Phys.Rev.Lett.* **38**, 255 (1977); Ya. B. Zel'dovich, *Sov.Phys.Uspekhi* **11**, 381 (1968).
24. C.W. Misner, *Astrophys. J.* **151**, 431 (1968); in *Magic Without Magic*, ed. J. Klauder (Freeman, San Francisco, 1972); R. Matzner and C.W. Misner, *Astrophys. J.* **171**, 415 (1972).
25. R. Penrose, in *General Relativity: An Einstein Centenary Survey*, eds. S.W. Hawking and W. Israel (Cambridge University Press, Cambridge, 1979).
26. B. deWitt, *Phys.Rev.* **90**, 357 (1953).
27. L. Parker, *Nature* **261**, 20 (1976).
28. Ya.B. Zel'dovich, *JETP Lett.* **12**, 307 (1970).
29. A.A. Starobinskii, *Phys.Lett.* **91B**, 99 (1980).
30. P. Anderson, *Phys.Rev.* **D28**, 271 (1983).
31. J.M Hartle and B.-L Hu, *Phys.Rev.* **D20**, 1772 (1979).
32. M.V. Fischetti, J. Hartle, and B.-L. Hu, *Phys.Rev.* **D20**, 1757 (1979).
33. J.M. Hartle and S.W. Hawking, *Phys.Rev.* **D28**, 2960 (1983).
34. B.J. Carr and M.J. Rees, *Nature* **278**, 605 (1979); J.D. Barrow and F. Tipler, *The Anthropic Cosmological Principle* (Oxford University Press, Oxford, 1986).
35. D. Lindley, Fermilab preprint (1985).
36. A. Guth and E. Weinberg, *Nucl.Phys.* **B212**, 321 (1983); S. Hawking, I. Moss, and J. Stewart, *Phys.Rev.* **D26**, 2681 (1982).
37. A. Linde, *Phys.Lett.* **108B**, 389 (1982).
38. A. Albrecht and P.J. Steinhardt, *Phys.Rev.Lett.* **48**, 1220 (1982).
39. S. Coleman and F. de Luccia, *Phys.Rev.* **D21**, 3305 (1980); S. Coleman, *Phys.Rev.* **D15**, 2929 (1977); C.G. Callan and S. Coleman, *Phys.Rev.* **D16**, 1762 (1977).
40. S.W. Hawking and I.G. Moss, *Phys.Lett.* **110B**, 35 (1982); L. Jensen and P.J. Steinhardt, *Nucl.Phys.* **239B**, 176 (1984); K. Lee and E.J. Weinberg, *Nucl.Phys.* **B267**, 181 (1986).
41. A. Albrecht, P.J. Steinhardt, M.S. Turner, and F. Wilczek, *Phys.Rev.Lett.* **48**, 1437 (1982).
42. L. Abbott, E. Farhi, and M. Wise, *Phys.Lett.* **117B**, 29 (1982).
43. A. Dolgov and A. Linde, *Phys.Lett.* **116B**, 329 (1982).
44. P.J. Steinhardt and M.S. Turner, *Phys.Rev.* **D29**, 2162 (1984).

45. J.R. Gott, *Nature* **295**, 304 (1982); J.R. Gott and T.S. Statler, *Phys.Lett.* **136B**, 157 (1984).
46a. M.S. Turner, *Phys.Lett.* **115B**, 95 (1982); J. Preskill, in *The Very Early Universe*, eds. G. Gibbons, S. Hawking, and S. Siklos (Cambridge Univ. Press, Cambridge, 1983); G. Lazarides, Q. Shafi, and W.P. Trower, *Phys.Rev.Lett.* **49**, 1756 (1982).
46b. G. Lazarides and Q. Shafi, in *The Very Early Universe, ibid*; K. Olive and D. Seckel, in *Monopole '83*, ed. J. Stone (Plenum Press, NY, 1984).
47. M.S. Turner, *Phys.Rev.* **D28**, 1243 (1983).
48. J.D. Barrow and M.S. Turner, *Nature* **292**, 35 (1981); J.D. Barrow and D.H. Sonoda, *Phys.Rep.* **139**, 1 (1986).
49. D. Nanopoulos, K. Olive, and M. Srednicki, *Phys.Lett.* **127B**, 30 (1983).
50. R.J. Scherrer and M.S. Turner, *Phys.Rev.* **D31**, 681 (1985).
51. T. Bunch and P.C.W. Davies, *Proc.Roy.Soc.London* **A360**, 117 (1978); also see, G. Gibbons and S. Hawking, *Phys.Rev.* **D15**, 2738 (1977).
52. L. Abbott and M. Wise, *Nucl.Phys.* **B244**, 541 (1984).
53. S. Hawking, *Phys.Lett.* **115B**, 295 (1982).
54. A.A. Starobinskii, *Phys.Lett.* **117B**, 175 (1982).
55. A. Guth and S.-Y. Pi, *Phys.Rev.Lett.* **49**, 1110 (1982).
56. J. Bardeen, P. Steinhardt, and M.S. Turner, *Phys.Rev.* **D28**, 679 (1983).
57. R. Brandenberger, R. Kahn, and W. Press, *Phys.Rev.* **D28**, 1809 (1983); W. Fischler, B. Ratra, and L. Susskind, *Nucl.Phys.* **B259**, 730 (1985); R. Brandenberger, *Rev.Mod.Phys.* **57**, 1 (1985).
58. E.M. Lifshitz and I.M. Khalatnikov, *Adv.Phys.* **12**, 185 (1963); E.M. Lifshitz, *Zh.Ek.Teor.Fiz.* **16**, 587 (1946).
59. W. Press and E.T. Vishniac, *Astrophys.J.* **239**, 1 (1980).
60. J.M. Bardeen, *Phys.Rev.* **D22**, 1882 (1980).
61. W. Press, in *Cosmology and Particles*, eds. J. Audouze et al. (Editions Frontieres, Gif-sur-Yvette, 1981), p. 137.
62. J.A. Frieman and M.S. Turner, *Phys.Rev.* **D30**, 265 (1984); R. Brandenberger and R. Kahn, *Phys.Rev.* **D29**, 2172 (1984).
63. E. Harrison, *Phys.Rev.* **D1**, 2726 (1970).
64. Ya.B. Zel'dovich, *Mon.Not.Roy.Astron.Soc.* **160**, 1p (1972).
65. P.J. Steinhardt and M.S. Turner, *Phys.Rev.* **D29**, 2162 (1984).
66. M.S. Turner and E.T. Vishniac, in preparation (1985).
67. J. Bond and A. Szalay, *Astrophys.J.* **276**, 443 (1983); P.J.E. Peebles, *Astrophys.J.* **263**, L1 (1982); J. Bond, A. Szalay, and M.S. Turner, *Phys.Rev.Lett.* **48**, 1036 (1982).
68. M. Axenides, R. Brandenberger, and M.S. Turner, *Phys.Lett.* **120B**, 178 (1983); P.J. Steinhardt and M.S. Turner, *Phys.Lett.* **129B**, 51 (1983).
69. D. Seckel and M.S. Turner, *Phys.Rev.* **D32**, 3178 (1985).
70. A.D. Linde, *JETP Lett.* **40**, 1333 (1984); *Phys.Lett.* **158B**, 375 (1985).
71. J. Preskill, M. Wise, and F. Wilczek, *Phys.Lett.* **120B**, 127 (1983); L. Abbott and P. Sikivie, *Phys.Lett.* **120B**, 133 (1983); M. Dine and W. Fischler,

Phys.Lett. **120B**, 137 (1983).
72. M.S. Turner, F. Wilczek, and A. Zee, *Phys.Lett.* **125B**, 35, 519(E) (1983); J. Ipser and P. Sikivie, *Phys.Rev.Lett.* **50**, 925 (1983).
73. M.S. Turner, *Phys.Rev.* **D33**, 889 (1986).
74. S. Coleman and E. Weinberg, *Phys.Rev.* **D7**, 1888 (1973).
75. A.D. Linde, *Phys.Lett.* **116B**, 335 (1982).
76. A. Vilenkin and L. Ford, *Phys.Rev.* **D26**, 1231 (1982).
77. S. Dimopoulos and S. Raby, *Nucl.Phys.* **B219**, 479 (1983).
78. A. Albrecht, S. Dimopoulos, W. Fischler, E. Kolb, S. Raby, and P. Steinhardt, *Nucl.Phys.* **B229**, 528 (1983).
79. J. Ellis, D. Nanopoulos, K. Olive, and K. Tamvakis, *Phys.Lett.* **118B**, 335 (1982); D.V. Nanopoulos, K. Olive, M. Srednicki, and K. Tamvakis, *Phys.Lett.* **123B**, 41 (1983); J. Ellis, K. Enqvist, D. Nanopoulos, K. Olive, and M. Srednicki, *Phys.Lett.* **152B**, 175 (1985); K. Enqvist and D. Nanopoulos, *Phys.Lett* **142B**, 349 (1984); C. Kounnas and M. Quiros, *Phys.Lett.* **151B**, 189 (1985); for a review of the CERN SUSY/SUGR models see D.V. Nanopoulos, *Comments on Astrophysics* **X**, 219 (1986).
80. B. Ovrut and P.J. Steinhardt, *Phys.Lett.* **133B**, 161 (1983); L. Jensen and K. Olive, *Nucl.Phys.* **B263**, 731 (1986).
81. B. Ovrut and P.J. Steinhardt, *Phys.Rev.Lett.* **53**, 732 (1984); *Phys.Lett.* **147B**, 263 (1984).
82. R. Holman, P. Ramond, and G.G. Ross, *Phys.Lett.* **137B**, 343 (1984); C.D. Coughlan, R. Holman, P. Ramond, and G.G. Ross, *Phys.Lett.* **140B**, 44 (1984); **158B**, 47 (1985).
83. J. Breit, S. Gupta, and A. Zaks, *Phys.Rev.Lett.* **51**, 1007 (1983).
84. A.A. Starobinsky, *JETP Lett.* **30**, 682 (1979); V. Rubakov, M. Sazhin, and A. Veryaskin, *Phys.Lett.* **115B**, 189 (1982); R. Fabbri and M. Pollock, *ibid* **125B**, 445 (1983); L. Abbott and M. Wise, *Nucl.Phys.* **B244**, 541 (1984).
85. J. Ellis, J. Kim, and D. Nanopoulos, *Phys.Lett.* **145B**, 181 (1984); L.L. Krauss, *Nucl.Phys.* **B227**, 556 (1983); M.Yu. Khlopov and A.D. Linde, *Phys.Lett.* **138B**, 265 (1984).
86. J. Polonyi, Budapest preprint KFKI 1977-93, unpublished (1977).
87. C. Coughlan, W. Fischler, E. Kolb, S. Raby, and G.G. Ross, *Phys.Lett.* **131B**, 54 (1983).
88. S. Gupta and H.R. Quinn, *Phys.Rev.* **D29**, 2791 (1984).
89. Q. Shafi and A. Vilenkin, *Phys.Rev.Lett.* **52**, 691 (1984).
90. S.-Y. Pi, *Phys.Rev.Lett.* **52**, 1725 (1984).
91. A. Guth and S.-Y. Pi, *Phys.Rev.* **D32**, 1899 (1985).
92. A.D. Linde, *Phys.Lett.* **129B**, 177 (1983).
93. V.A. Belinsky, L.P. Grischuk, I.M. Khalatnikov, and Ya.B. Zel'dovich, *Phys.Lett.* **155B**, 232 (1985); T. Piran and R.M. Williams, *ibid* **163B**, 331 (1985).
94. S. Adler, *Rev.Mod.Phys.* **54**, 729 (1982); L. Smolin, *Nucl.Phys.* **B160**, 253 (1979); A. Zee, *Phys.Rev.Lett.* **42**, 417 (1979).

95. F. Accetta, D. Zoller, and M.S. Turner, *Phys.Rev.* **D31**, 3046 (1985).
96. B.L. Spokoiny, *Phys.Lett.* **147B**, 39 (1984).
97. Q. Shafi and C. Wetterich, *Phys.Lett.* **129B**, 387 (1983); **152B**, 51 (1985).
98. A.A. Starobinskii, *Phys.Lett.* **91B**, 99 (1986); M.B. Mijic, M.S. Morris, and W.-M. Suen, *Phys.Rev.* **D34**, 2934 (1986).
99. A. Albrecht and R. Brandenberger, *Phys.Rev.* **D31**, 1225 (1985); *ibid* **D32**, 1280 (1985).
100. W. Fischler, B. Ratra, and L. Susskind, *Nucl.Phys.* **B259**, 730 (1985).
101. G. Semenoff and N. Weiss, *Phys.Rev.* **D31**, 689 (1985).
102. M. Evans and J. McCarthy, *Phys.Rev.* **D31**, 1799 (1985).
103. M.S. Turner and L. Widrow, *Phys.Rev.Lett.* **57**, 2237 (1986); L. Jensen and J. Stein-Schabes, *Phys.Rev.* **D34**, 931 (1986).
104. R.M. Wald, *Phys.Rev.* **D28**, 2118 (1983).
105. G. Steigman and M.S. Turner, *Phys.Lett.* **128B**, 295 (1983).
106. L. Jensen and J. Stein-Schabes, *Phys.Rev.*, in press (1987).
107. A.A. Starobinskii, *JETP Lett.* **37**, 66 (1983).
108. S.K. Blau, E.I. Guendelman, and A.H. Guth, *Phys.Rev.D*, in press (1987); K. Sato, M. Sasaki, H. Kodama, and K. Maeda, *Phys.Lett.* **108B**, 103 (1982); *Prog.Theor.Phys.* **65**, 1443 (1981); *ibid* **68**, 1979 (1982); *ibid* **66**, 2287 (1981).
109. N. Myhrvad, *Phys.Lett.* **132B**, 308 (1983); E. Mottola, *Phys.Rev.* **D31**, 754(1985); **D33**, 2136; L. Parker, *Phys.Rev.Lett.* **50**, 1009 (1983); L. Ford, *Phys. Rev.* **D31**, 710 (1985); P. Anderson, *Phys.Rev.* **D32**, 1302 (1985); J. Traschen and C.T. Hill, *Phys. Rev.* **D33**, 3519 (1986).
110. M.S. Turner, in *Dark Matter in the Universe*, eds. J Kormendy and G. Knapp (Reidel, Dordrecht, 1987), p. 445.
111. M.S. Turner, G. Steigman, and L.L. Krauss, *Phys.Rev.Lett.* **52**, 2090 (1984); D.A. Dicus, E.W. Kolb, and V. Teplitz, *Phys.Rev.Lett.* **39**, 168 (1977); K.A. Olive, D. Seckel, and E.T. Vishniac, *Astrophys.J.* **292**, 1 (1985).
112. P.J.E. Peebles, *Astrophys.J.* **284**, 439 (1984); M.S. Turner, et al. in ref. 111.
113. N. Kaiser, *Astrophys.J.* **273**, L17 (1983); J.M. Bardeen, J. Bond, N. Kaiser, and A. Szalay, *Astrophys.J.* **304**, 15 (1986).
114. A. Vilenkin, *Phys.Rev.Lett.* **53**, 1016 (1984).
115. J.C. Charlton and M.S. Turner, *Astrophys.J.* **313**, 495 (1987).
116. S.D.M. White, in *Inner Space/Outer Space*, eds. E.W. Kolb et al. (University of Chicago Press, Chicago, 1986).
117. G. Blumenthal, S. Faber, J. Primack, and M. Rees, *Nature* **311**, 517 (1984); M. Davis, G. Efstathiou, C. Frank, and S.D.M. White, *Astrophys.J.* **292**, 371 (1985).
118. D. Burstein, et al., in *Galaxy Distances and Deviations from Universal Expansion*, eds. B. Madore and R. Tully (Reidel, Dordrecht, 1987), p. 123.
119. C.A. Collins, et al., *Nature* **320**, 506 (1986).
120. N. Vittorio and M.S. Turner, *Astrophys.J.* **316**, in press (1987).
121. V. deLapparent, M. Geller, and J. Huchra, *Astrophys.J.* **302**, L1 (1986).
122. G. Efstathiou and J.R. Bond, *Mon.Not.r.Astron.Soc.* **218**, 103 (1986).

123. P. Sikivie, *Phys.Rev.Lett.* **51**, 1415 (1983).
124. M. Goodman and E. Witten, *Phys.Rev.* **D31**, 3059 (1985).
125. E. Loh and E. Spillar, *Astrophys.J.* **307**, L1 (1986); *Phys.Rev.Lett.* **57**, 2865.
126. J. Silk and M.S. Turner, *Phys.Rev.* **D35**, 419 (1986); M.S. Turner, et al., *Astrophys.J.*, in press (1987).
127. A. Linde, *Rep.Prog.Phys.* **47**, 925 (1984); *Comments on Astrophysics* **10**, 229 (1985); *Prog.Theo.Phys. (Suppl.)* **85**, 279 (1985).
128. *The Inflationary Universe*, eds. L. Abbott and S.-Y. Pi (World Publishing, Singapore, 1986).
129. P.J. Steinhardt, *Comments on Nucl.Part.Phys.* **12**, 273 (1984).
130. R. Brandenberger, *Rev.Mod.Phys.* **57**, 1 (1985).
131. A. Vilenkin, *Phys.Rep.* **121**, 263 (1985); A. Albrecht and N. Turok, *Phys.Rev.Lett.* **54**, 1868 (1985); N. Turok, *Phys.Rev.Lett.* **55**, 1801 (1985); R. Scherrer, *Astrophys.J.*, in press (1987); A. Melott and R. Scherrer, *Nature*, in press (1987).
132. S.K.Blau and A.H. Guth, in *300 Years of Gravitation*, eds. S.W. Hawking and W. Israel (Cambridge University Press, Cambridge, 1987).
133. S. Bonometto and A. Masiero, *La Rivista del Nuovo Cimento* **9**, 3 (1986).

THE PROBLEM OF ORIGIN OF THE PRIMORDIAL COSMOLOGICAL PERTURBATIONS

V.N. Lukash and I.D. Novikov
Space Research Institute, Academy of Sciences of the USSR,
Profsoyuznaja 84/32, Moscow 117810, USSR

ABSTRACT. The theory of the parametric amplification of potential perturbations in the isotropic Universe is reviewed. The problem of production of the primordial cosmological perturbations due to the parametric effect is considered. The related fluctuations of the relic background radiation are analysed. The spottiness structure of the large-scale background anisotropy is discussed.

INTRODUCTION

The formation of the structure of the Universe is one of the most important problems of cosmology.

In these notes we shall describe some of the modern approaches to the solution of this problem.

Now the majority of the specialists believe that the observed large-scale structure of the Universe (galaxies and their clusters) is the result of the development of the primordial perturbations of the uniform matter distribution which were in the very Early Universe. It is so-called adiabatic density perturbations, or potential perturbations.

There are some other rather exotic possibilities of the explanation of the origin of the large-scale structure of the Universe. But we shall not discuss them here.

The most difficult part of the problem under consideration is the investigation of the creation of the small primordial perturbations in the early Universe. We want to know the physical processes which resulted in their origin, the amplitudes and spectrum of these perturbations.

Now we believe that the mistery of the creation of the primordial perturbations is related with the physics at very high energies.

At the beginning allow us to remind very briefly and qualitatively the main difficulties of that problem which were in the classical cosmology before the new development of the high energy physics.

P. Galeotti and D.N. Schramm (eds.), Gauge Theory and the Early Universe, 71–98.

The theory of the small fluctuations in the Friedmann universe was created by Lifshitz in 1946.

According to this theory there are 3 types of perturbations of the uniform isotropical model:

1) The density perturbations (potential perturbations), 2) the vortex perturbations and 3) gravitational waves. We are interested now in the first type of perturbations - the potential perturbations because we believe that they are the source of the galaxy formation. Let us consider perturbations of this type in the hot Universe.

It is usual to expand density perturbations $\delta\rho$ in a Fourier expansion:

$$\frac{\delta\rho}{\rho} = (2\pi)^{-3/2} \int \delta_{\vec{k}}\, e^{i\vec{k}\vec{x}}\, d^3\vec{k} \quad ,$$

$$\lambda_{ph} = \frac{2\pi}{k} a(t) \text{ - the physical wavelength,}$$

k - wave number ($k=|\vec{k}|$)

Sometimes one uses the value:

$$\left(\frac{\delta\rho}{\rho}\right)^2_k = (2\pi)^{-3} \, |\delta_{\vec{k}}|^2 k^3$$

In the Hot Universe when the equation of state is $p=\varepsilon/3$, the solution for $(\delta\rho/\rho)_k$ is

$$\left(\frac{\delta\rho}{\rho}\right)_k = C_1 \left|-\cos\varkappa + 2\left(\frac{\sin\varkappa}{\varkappa} + \frac{\cos\varkappa - 1}{\varkappa^2}\right)\right| +$$

$$+ C_2 \left|\sin\varkappa + 2\left(\frac{\cos\varkappa}{\varkappa} - \frac{\sin\varkappa}{\varkappa^2}\right)\right|,$$

$$C_1 = C_1(k),\ C_2 = C_2(k),$$

$$\varkappa = \frac{k}{\sqrt{3}} (t/t_o)^{1/2} \quad \text{- the ratio of the scale } \ell$$

which the sound travels during the cosmological time t to the perturbations scale λ_{ph}. The sound spead is

$$v_s = c/\sqrt{3} \simeq c,$$

so $\varkappa \simeq ct/\lambda_{ph}$. The expression with the coefficient C_1 is so-called the growing mode, and with the coefficient C_2 is the decaying mode. In order of magnitude the perturbations of the gravitational field (metric perturbations) are:

$$h_1 \simeq C_1 \left(\frac{1-\cos\varkappa}{\varkappa^2}\right) \quad \text{for the growing mode,}$$

$$h_2 \simeq C_2 \frac{\sin\varkappa}{\varkappa^2} \quad \text{for the decaying mode.}$$

The most important feature of these perturbations is the following. Neither growing mode nor decaying mode increases catastrophically: both of them are described by sin and cos, so if C_1 and C_2 are less then unity (and it must be so because in the other case h_1 and h_2 would be > 1 at small $\varkappa$) both these modes become sound waves with constant (independent of time) amplitudes C_1 and C_2.

First of all, this result means that the Hot Universe is absolutely stable against gravitational instability. If the initial perturbations were small they will be small forever.

These sound waves with small amplitudes existed in the Hot Universe up to the moment of a few hundred years after the beginning of the expansion. At this epoch Universe became cold enough; the equation of state $p = \frac{1}{3}\rho c^2 = \frac{1}{3}\varepsilon$ is not correct more, universe is not hot. Now the clumps of the medium in the sound waves begin to grow because of the real gravitational instability (at least some components of the medium undergo this instability) and this process develops causing the fragmentation of the medium into separate bodies.

We shall not discuss these late processes of galaxy formation here. It is a separate problem.

The matter is that for the formation of the galaxies we need definite amplitude of sound waves $\sim 10^{-4} - 10^{-6}$ in the linear scale which encompasses the number of barions big enough for a galaxy formation. So, C_1 or/and C_2 must be in this scale of the order of $10^{-6} - 10^{-4}$.

It is a very serious demand on the initial perturbations. Indeed, when t is small, $\varkappa \ll 1$, we have

$$h_1 \simeq C_1/2 \ll 1,$$

$$h_2 \simeq C_2/\varkappa \ll 1$$

From these expressions we can see that C_2 must be extremely small and could not be of order of $10^{-6} - 10^{-4}$.

$$(1)\ h_1 \simeq C_1 \simeq 10^{-4}$$

$$(2)\ C_2 \ll C_1.$$

Both of these conclusions look very strange.

Indeed

I. As we shall see, in any natural assumption about the process of the origin of the perturbations C_1 must be equal C_2 and (2) could not be correct.

II. And the second: if we believe that initial perturbations arised very early (for example at $t \simeq t_{Planck} \simeq 10^{-43}$) the C_1 and C_2 have to be dozens orders of magniture less than 10^{-4}.

The last conclusion can be seen from the following example.

Let us suppose that the time of the origin of the fluctuations is the Planck time $t_o = t_{pl}$, and let us denote $k = 1$ for λ_{pl}. In the scale of galaxies $k_{gal} \simeq 10^{-26}$. Now let us suppose (as an example) that the spectrum of $\delta\rho/\rho$ fluctuation at that moment had a thermal shape with maximum at $\lambda = \lambda_{pl}$; then the amplitude of the perturbation would be proportional to $k^{3/2}$, and in the scale of galaxies the amplitude is $K_{gal}^{3/2} \simeq 10^{-40}$. Thus C_1 have to be $\sim 10^{-40}$ and it is the 35 orders of magnitude less than we need.

Our conclusions are the following:

The classical cosmology of the Hot Universe has great difficulties in the explanation of the primordial seed perturbations of matter. Fluctuations in the classical Hot Universe need:

1) $C_1 \gg C_2$ - not "natural".

2) $C_1 \simeq 10^{-4}$ much greater than "natural" value $C_{gal} \simeq 10^{-40}$.

Now let us come to the modern cosmology and analyse the problem of generation of density perturbations.

The cosmological model including the period of the inflation (when the scale factor of the expansion is $a(t) \sim e^{Ht}$) at the beginning of the Universe is the most popular now. The inflationary universe scenario was born as a natural consequence of the application of the Grand Unified Gauge Theories (GUTs) to the very early stage of the Universe.

Unfortunately the different versions of these scenarioes have some difficulties (for example extreme fine tuning of the parameters of the theory for a sufficient inflation to be achieved). This is true for the versions: "new inflationary" universe and "Newer Inflation". On the other hand some models of the inflationary universe were proposed, in which the cause of inflation was not GUT's. It was for example Starobinskij"s theory (1980) of primordial inflation at the Planckian time or the Linde's (1983) chaotic inflation scenario.

The inflationary universe scenario itself solved many fundamental problems of cosmology and it is too fascinating to be abandoned. Thus it looks natural to separate the inflation from GUTs and from any specific scheme and analyse the different possibilities. And the first of all we want to analyse all consequences of the change of the rate of the expansion of the universe from power law to the exponential one and otherwise. There are the following ideas about the physical phenomena responsible for the perturbations production.

1) Quantum-gravitational one loop effects (Starobinsky 1980; Hartle and Horowitz 1981; Mukhanov and Chibisov 1981).

2) Vacuum phase transitions (Kirzhnits 1972; Guth 1981; Guth and Pi 1982; Hawking 1982; Bardeen et al. 1983).

3) Cosmic strings, walls and monopoles (Kibble 1976, Zeldovich 1980, Vilenkin 1981, 1984).

4) Parametric amplification (production of new phonons) (Lukash 1980; Kompaneets et al. 1982; Lukash and Novikov 1983, 1985).

The last effect dominates on scales greater than the cosmological horizon $\lambda > ct$.

In this paper we shall discuss the last possibility. In this lecture we follow our paper Lukash and Novikov (1985) (For the Newtonian aspects of the problem of the evolution of the small perturbations see Bonnor (1956, 1957)).

Parametric amplification means production of gravitational potential inhomogeneities from the nonstationary gravitational background of the expanding Universe (Lukash 1980a). The essence of this effect may easily be demonstrated when considering a massless scalar field with minimal coupling:

$$L = \frac{1}{2} \varphi_{,i} \varphi^{,i}. \tag{1}$$

The equation of motion of the φ-field in the isotropic Universe is as follows:

$$\frac{(a^3 \dot{\varphi})^{\dot{}}}{a3} - \frac{\Delta \varphi}{a2} = 0, \tag{2}$$

where $ds^2 = dt^2 - a^2(t)dl^2$ is the Friedmann metric (1922, 1924), $(\dot{}) = \partial/\partial t$, Δ is the Laplacian operator in the space dl. Let us use the conformally-static variables:

$$\eta = \int \frac{dt}{a}, \quad \bar{\varphi} = a\varphi . \tag{3}$$

Then Eq (2) turns to

$$\Box \bar{\varphi} = U \bar{\varphi} , \quad ds^2 = a^2 d\bar{s}^2 , \tag{4}$$

where $\Box = \partial^2/\partial\eta^2 - \Delta$ is the d'Alambertian operator in the metric $d\bar{s}^2 = d\eta^2 - dl^2$, $U = a''/a$ is the effective potential of the φ-field in metric (2), $(') = \partial/\partial\eta$. It is $U \neq 0$ that causes the parametric amplification. (The adjective "parametric" rather than "superadiabatic" is used since the process develops in time and Eq (4) is typical for the usual parametric effect; φ-field characteristic frequencies excited by $U \sim \eta^{-2}$, are $\omega \lesssim \eta^{-1}$).

If the scale factor is linear in the conformal time, then the effective potential vanishes,

$$a \sim \eta , \quad U = 0 \tag{5}$$

and φ-field appears to be conformally invariant:

$$\varphi \rightarrow \bar{\varphi} , \quad \text{when} \quad s \rightarrow \bar{s} . \tag{6}$$

There is no interaction between the φ-field and the background metric S in this case, i.e. the number of the φ-field quanta does not vary with time (adiabatic invariant is conserved). For any other expansion rate than of Eq (5) $U \neq 0$ and φ-field quanta are produced. This may well be seen when the scattering problem is discussed.

Let a be linear in η when $\eta < \eta_1$ and $\eta > \eta_2$, so that non-vanishing U exists only in the interval $\eta_1 < \eta < \eta_2$. Then, depending on the φ-field phase at $\eta < \eta_1$, the resulting amplitude of φ when $\eta > \eta_2$ may appear to be smaller or greater than the initial one. However, after averaging over initial phases one would always get an increase in the field energy. Due to ergodicity, the creation of new quanta is always there when the initial state is random in phase (such as vacuum state, for instance).

A similar amplification behaviour is inherent in gravitational waves (Grishchuk 1974) and in potential (density) perturbations (Lukash 1980a). The parametric effect could be the cause of the primordial cosmological perturbations which were necessary for the formation of galaxies. Quaside Sitter stage may well be used as an intermediate stage (with nonvanishing U) in the very early Universe. The post-recombination growth of the primordial perturbations gave rise to the present large-scale structure of the Universe. At the early stages of the expansion primordial inhomogeneities perturbed the relic background radiation. Measurements of this background anisotropy provide an independent test of the parametric amplification theory. Gravitational waves which were produced at the same time as potential perturbations, might as well be used to test the theory. But detection of the cosmological gravitational waves is still far beyond the possibilities of modern experimental techniques, whereas the search for relic microwave background fluctuations is quite within up-to-date experiment powers.

We should note here, that there exist some other possible mechanisms of producing density perturbations in the Early Universe. The quantum decay of false vacuum may serve as an example (The Very Early Universe 1983). We shall not consider these mechanisms in the paper. The perturbations we are dealing with are inevitable in a given model of the Universe. All other mechanisms can only add more perturbations to the lower limit we are speaking of. Further on, the units $8\pi G = c = \hbar = 1$ are used.

1. POTENTIAL MOTIONS OF PERFECT FLUID AND THE FIELD THEORY

Perfect fluid is described by the energy-momentum tensor

$$T_{ik} = (\varepsilon + p)\, u_i u_k - p\, g_{ik} \tag{1.1}$$

with the pressure p being an arbitrary function of the matter density ε. We can present this relationship in terms of the parameter w,

$$w = \exp \int \frac{dp}{\varepsilon + p} \ . \tag{1.2}$$

Then the equation of matter state takes the following form:

$$p = p(w), \qquad \varepsilon = w\frac{dp}{dw} - p \quad . \tag{1.3}$$

Note, that the scalar w can multiply an arbitrary constant, while p and ε are invariant under this scale-transformation.

The hydrodynamic equations $T^k_i{}_{;k} = 0$ (semicolon is a covariant derivative in the manifold metric g_{ik}) when projected on the 4-velocity of the matter u^i and on the orthogonal directions $p^{ik} = g^{ik} - u^i u^k$, yield the conservation law for the pressure derivative

$$\left(\frac{dp}{dw}u^i\right)_{;i} = 0 \tag{1.4}$$

and the Euler equations

$$u^k(p_{i,k} - p_{k,i}) = 0, \quad p_i = w\, u_i . \tag{1.5}$$

The hydrodynamic particular examples of the perfect fluid are a barotropic matter (with a constant specific entropy) and an equilibrium matter (with zero chemical potential). In these cases, scalars w and dp/dw are identical to the specific enthalpy and the particle density, and to the temperature and the entropy of the matter, respectively.

Potential motions of a perfect fluid are described in the most general form by a single real scalar $\varphi = \varphi(x^i)$ that allows for determining all the other matter quantities:

$$p_i = \varphi_{,i} \; , \quad w = (\varphi_{,i}\,\varphi^{,i})^{1/2} . \tag{1.6}$$

Eqs (1.5) are satisfied identically, and the continuity equation (1.4) turns to the second-order hyperbolic wave-equation for the φ-field:

$$\left(\frac{dp}{wdw}\varphi^{,i}\right)_{;i} = 0 . \tag{1.7}$$

A Lagrangian density of the field is obviously an arbitrary function of the kinetic term $\varphi_{,i}\,\varphi^{,i}$:

$$L(\varphi) = p(w), \quad w = (\varphi_{,i}\,\varphi^{,i})^{1/2} \tag{1.8}$$

So, the theory of perfect fluid potential motions is mathematically equivalent to the real scalar field theory with the Lagrangian depending only on the kinetic term. Identity of these two theories is ensured by Eq (1.8). We can discard the perfect fluid and operate by the field theory notions:

$$T_{ik} = \frac{2\,\partial\sqrt{-g}\,p(w)}{\sqrt{-g}\,\partial g^{ik}} = \frac{dp}{wdw}\varphi_{,i}\,\varphi_{,k} - pg_{ik}, \tag{1.9}$$

$$\varepsilon = T_{ik}\,u^i u^k ,$$

where $g = \det g_{ik}$, u^i is the unity vector orthogonal to hypersurfaces φ = const (see (1.6), 1.5)).

Note, that φ-field from Eq (1.8) may be of any physical nature; it is not necessarily the collective field of the perfect fluid motions. The simplest example is given by Eq (1). Surely, the physical sense of φ-field depends on the model considered. As for generic hydrodynamic media, the analytical analogy between the potential streams and the field theory of an ensemble of real scalars can be extended (Lukash 1983). Here we restrict ourselves to gravitating fields of type (1.8) to clarify in the simplest way the effect of parametric amplification. Besides, the matter states (1.8) have important applications in the Early Universe physics and thus they may be employed in the problem of origin of the cosmological primordial perturbations.

Here, it is relevant to stress how parametric generation of primordial inhomogeneities is related with other theories that single out one or another physical phenomenon responsible for the perturbations production - for example, quantum-gravitational one-loop effects (Starobinsky 1980 a, b, Hartle & Horowitz 1981, Mukhanov & Chibisov 1981), vacuum phase transitions (Kirshnits 1972, Guth 1981, Guth & Pi 1982, Hawking 1982, Linde 1982, 1983, Starobinsky 1982, Bardeen et al. 1983), cosmic strings, walls and monopoles (Kibble 1976, Zeldovich 1980, Vilenkin 1981, 1984), possible anisotropy of the Early Universe, dissipative processes, particle creation, etc. (e.g., Parker 1968, Hartle 1980, Hogan 1980, Barrow & Turner 1981). The parametric effect is always present in any theory, since the Lagrangian of the fields in question must depend on their first derivatives hence it contains terms corresponding to interaction of the fields with background metric (e.g., $\varphi_{,i}\, g^{ik}$, $\varphi_{,i}\, \varphi_{,k}\, g^{ik}$ etc.). Owing to nonstationarity of the background, these terms ensure parametric creation of fields' fluctuations (and related metric fluctuations). Thus, the parametric effect produces the minimum, inevitable in any theory, level of cosmological perturbations whereas the physical mechanisms listed above give birth to additional perturbations (see Introduction). The choice of the Lagrangian of the simplest scalar field in the form (1.8) allows for investigating the parametric effect proper, without the influence of other mechanisms of perturbations production. (For example, Higgs theories of vacuum phase transitions introduce additional dependence of the Lagrangian on the field itself, $L(\varphi) = \frac{1}{2}\varphi_{,i}\,\varphi^{,i} + V(\varphi)$, which is the source of additional statistical fluctuations. The latter have nothing to do with the nonstationarity of the background metric, they are created in the process of the phase transition when the φ-field changes. In particular, statistical fluctuations vanish and only parametric fluctuations remain at $V(\varphi) = 0$; this corresponds to $p = \varepsilon = \frac{1}{2} w^2$).

Sections 2, 3, 4 survey the Lagrangian theory of potential perturbations and the parametric amplification theory in the Friedmann models (Lukash 1980 a, b, 1983). Applications to the Early Universe and models for the parametric production of the primordial perturbations are given in Sections 5, 6 (Kompaneets et al. 1982, Lukash & Novikov 1983 a, b). Section 7 shows how primordial perturbations parameters may be observed by the microwave background fluctuations. Sections 8, 9 deal with the results of calculations of the correlation and multipole characteristics of relic radiation anisotropy for different spectra of primordial perturbations and different assumptions as to the nature of the missing mass (Lukash et al. 1984). Section 10 discusses the spottiness structure of the large-scale background anisotropy (Novikov 1968, Lukash 1977, 1982, Bisnovatyi-Kogan et al. 1980).

Before coming to the theory of potential perturbations we would like to emphasize in what respect the theory differs from other studies devoted to potential perturbations in the isotropic Universe (Hawking 1966, Harrison 1967, Sachs & Wolfe 1967, Field & Sheptley 1968, Bardeen 1980, Press & Vishniac 1980, Brandenberger et al. 1983). In each paper developing the Lifshitz theory (1946) gauge-invariant investigation methods, independent of perturbed reference systems, have been elaborated. Following this approach, Bardeen (1980) constructed a number of gauge-invariant functions, that characterize joint perturbations of metric and matter and gave their physical interpretation. Contrary to this the theory of parametric effect, based on Lagrangian approach, operates with a single gauge-invariant function which describes potential perturbations in terms of the field theory. All the other perturbations of metric and matter may be derived as unambiguous functions of this field variable. (There are few such field variables for multicomponent media).

2. THE LAGRANGIAN THEORY OF GRAVITATIONAL POTENTIAL PERTURBATIONS

Potential motions (1.7) represent a complex non-linear pattern. Self-gravity of the φ-field affects the geometry of space-time and creates gravitational waves of a certain polarization which, in turn, influence the dynamics of the φ-field. (For a perfect fluid it means that Eq (1.6) holds, i.e. the gravitational waves do not bring about a vortical velocity component).

These processes are described by the following action function:

$$W[\varphi, g^{ik}] = \int (p - \tfrac{1}{2} R) \sqrt{-g}\, d^4x \ , \tag{2.1}$$

where $R = R^i_i$, R_{ik} is the Ricci tensor. Evolution equations for the φ-field (see (1.7)) and for the coupled gravitational waves

$$R_{ik} - \tfrac{1}{2} g_{ik} R = T_{ik} \tag{2.2}$$

are obtained while varying $\delta W = 0$ over φ (with fixed g^{ik}) and over g^{ik} (with fixed φ) respectively.

One of the basic problems is the analysis of small perturbations of some exact solutions of type (1.6). In this case functions φ and g^{ik} are the sums of the known ones that determine a background solution and small functions that are the subject of analysis:

$$\varphi = \varphi^{(o)} + wv, \quad g^{ik} = g^{ik(o)} - h^{ik} \ . \tag{2.3}$$

The small functions v and h_{ik} are scale-invariant (in contrast with w and φ, see Eqs (1.3), (1.6)) and gauge-dependent. The gauge transformation is given by

$$\tilde{v} = v + u^i \xi_i , \quad \tilde{h}_{ik} = h_{ik} + 2 \xi_{(i;k)}, \tag{2.4}$$

where ξ_i - small functions, index parentheses mean symmetrization.

The Lagrangian of the perturbation fields is a result of the expansion of the integrand (2.1) up to the second order in v and h_{ik} (the first order vanishes due to the background equations):

$$W [v, h_{ik}] = W^{(2)} [\varphi^{(o)} + wv, g^{ik(o)} - h^{ik}] =$$
$$= \int L \sqrt{- g^{(o)}} d^4x,$$
$$L = \frac{\varepsilon + p}{2} (v_i v^i - 2v_i \psi^i_k u^k + \delta^2 (\beta^{-2} - 1)) +$$
$$+ \frac{\varepsilon - p}{8} (h_{ik}h^{ik} - \tfrac{1}{2}h^2) + 1/8 \; \psi_{ik;l} \; h^{ik;l} -$$
$$- 1/4 \; \psi_{ik;l} \; \psi^{il;k} , \tag{2.5}$$

where $v_i = w^{-1}(wv)_{,i}$, $\psi^k_i = h^k_i - \tfrac{1}{2} h \; \delta^k_i$, $h = h^i_i$,

$$\delta = \frac{\delta w}{w} = v_i u^i - \tfrac{1}{2} h_{ik} u^i u^k , \quad \beta^{-2} = \frac{d^2 p}{dw^2} \left(\frac{dp}{wdw}\right)^{-1} = \frac{d \varepsilon}{dp}$$

(All the manipulations with Latin indices are carried out in the background metric $g^{(o)}$). While varying (2.5) two linear second-order hyperbolic equations for two functions v and h_{ik} are obtained. The system reminds two couples oscillators. Potential and gravitational fields are coupled through the background shear deformation.

The problem of small perturbations is topical for two reasons
i) Most important solutions with some symmetry group are particular cases of (1.6), e.g., simplest anisotropic cosmologies, Friedmann models (in particular, some Early Universe models), spherically-symmetric collapse, etc.
ii) The problem of production and evolution of matter density primordial perturbations in the isotropic Universe is a key for explaining the origin of the large-scale Universe structure and the related anisotropy of the relic background radiation.

The only solution with zero anisotropic (shear) deformation of the spatial expansion is Friedmann's model. Gravitational waves propagate freely in this background and do not interact with potential perturbations. Einstein equations for metric potential perturbations h_{ik} turn out to be of elliptic type (as Poisson equation). So, the functions h_{ik} are unambiguously specified by the v-potential and the resulting system for the perturbations is reduced to one wave-equation for a real scalar field $q = q(x^i)$. Quantities v and h_{ik} are found as unambiguous functions of q.

Next Sections deal with potential perturbations to be formalized as a test Hamiltonian scalar field on the classical Friedmann background.

3. GAUGE-INVARIANT THEORY OF POTENTIAL PERTURBATIONS IN THE ISOTROPIC UNIVERSE

A real 4-scalar q which utterly describes the field of potential perturbations is fixed by the requirement that the field Lagrangian is a function only of the field derivatives $q_{,i}$ (i.e. it does not depend on the q-field itself, see Eq (1.8)). The $\dot{q}$-scalar is obviously a gauge-invariant analog of v, i.e. q is a gauge-invariant superposition of the v-potention and some gravitational potentials of h_{ik}. Let's find this relation for the isotropic Universe.

In the homogeneous reference system (t, x^{α}) the background is as follows:

$$ds^2 = dt^2 - a^2 \gamma_{\alpha\beta}\, dx^{\alpha}\, dx^{\beta}, \quad u_i = \delta^{o}_{i}, \tag{3.1}$$

where $a = a(t)$ is the scale factor, $\gamma_{\alpha\beta} = \gamma_{\alpha\beta}(x^{\gamma})$ is the space metric tensor, the constant spatial curvature being $-h^2 = 0, \mp 1$ for the flat, closed and open models. (Introduction h instead of frequently used $\varkappa = -h^2$ allows for $h > 0$ in the open Universe, see Section 10).

In a perturbed reference system

$$v = V + C + u^i D_{,i},$$
$$\tfrac{1}{2} h_{ik} = \Phi\, e_{ik} + (Cu_{(i})_{;k)} + D_{;ik}, \tag{3.2}$$

where $e_{ik} = 2u_i u_k - g_{ik}$ is the Euclidean tensor (background index $(^{o})$ is omitted), C and D are arbitrary functions due to the gauge freedom (see Eq (2.4)). The q-function is

$$q = 2\sqrt{3}\,(HV + \Phi), \tag{3.3}$$

where $H = 1/3\, u^{k}_{;k} = \frac{\dot{a}}{a} = \sqrt{\frac{\varepsilon}{3} + (\frac{h}{a})^2}$ is the "Hubble constant". Gauge invariant gravitational potential Φ is related to the q-field through the Poisson type equation:

$$a^{-2}\Delta\Phi = \sqrt{3}\, H\, \xi^2 q_{,i} u^i,$$
$$\delta\varepsilon = H(2\sqrt{3}\,\xi^2 q_{,i}u^i - 3(\varepsilon + p)v) \tag{3.4}$$

where $\Delta = \gamma^{\alpha\beta}\dfrac{\partial^2}{\partial x^{\alpha}\partial x^{\beta}}, \quad \xi^2 = (2\beta)^{-2}\gamma,$

$$\gamma = \frac{w}{3H^2}\frac{dp}{dw} = \frac{\varepsilon + p}{3H^2}.$$

The inverse transformations to Eq (3.3) are

$$V = \frac{\sqrt{3}}{2}\left(\frac{1}{3H} q - \frac{1}{2a}\int a \gamma q \, dt\right),$$

$$\Phi = \frac{\sqrt{3}\, H}{4a}\int a \gamma q \, dt \tag{3.5}$$

The Lagrangian is calculated after substituting (3.5) into Eq (2.5) and omitting additive terms of type $S^{i}_{;i}$:

$$L(q) = \tfrac{1}{2} D^{ik} q_{,i} q_{,k} , \quad D_{ik} = \xi^2 (u_i u_k + \beta^2 p_{ik}) \tag{3.6}$$

where $p_{ik} = g_{ik} - u_i u_k$ is the projection tensor. The wave-equation for the q-field

$$(D^{ik} q_{,i})_{;k} = 0 \tag{3.7}$$

can be got directly from Eqs (3.4), (3.5) as well.

The effect of amplification of potential perturbations in the expanding Universe and the Hamiltonian formalism are based on Eqs (3.6), (3.7) (Lukash 1980a). Let's rewrite Eq (3.7) in the conformally - static coordinates (η, x^{α}), $\eta = \int \frac{dt}{a}$ (compare Eqs (1)-(4)):

$$\Box_{\beta} \bar{q} = U \bar{q} , \tag{3.8}$$

where $\bar{q} = a \xi q$, $\Box_{\beta} = \frac{\partial^2}{\partial \eta^2} - \beta^2 \Delta$, $U = \frac{(a \xi)''}{a \xi}$ is the effective potential reflecting the nonstationarity of the Universe. The β-function is the velocity of the potential perturbation field. (For the perfect fluid case in a short-wave approximation β is the sound velocity). The equation of motion of the perturbations of the gravitating φ-field differs from that of the nongravitating field by the nonzero right-hand of Eq (3.8). It provides for the parametric generation of long-wave potential perturbations in the course of the cosmological expansion (see Section 1). The difference between the U-function (3.8) and the purely gravitational U-potential from Eq (4) is due to $\xi(\eta)$. This comes from the Special Relativity effect of the pressure spatial gradient in the co-moving coordinates (see Introduction). It accelerates the q-wave because of negative time derivative of the background pressure $\dot{p} < 0$. (This effect is similar to the increase in amplitude of an acoustic wave propagating to the upper layers of the atmosphere). In the nonrelativistic limit $U/a^2 \simeq 4\pi G \varepsilon/3$ in accordance with the Jeans's instability formular. Note, that for scales greater than the horizon all pressure gradient effects are negligible. q-field is mainly metric potential perturbations in this limit hence the nature of the parametric amplification of potential and gravitational-wave perturbations is the same.

Although the parametric effect has a classical nature the quantization is needed to find some quantitative characteristics for spontaneous creation.

4. QUANTUM THEORY OF q-FIELD

Let's denote by $\sigma = \sigma(x^i)$ the canonically conjugate to q gauge-invariant scalar:

$$\sigma = \frac{\partial L(q)}{\partial q_{,i}} u_i = \xi^2 q_{,i} u^i \tag{4.1}$$

The scalars q and σ describe invariant changes of the field potential (see Eq (3.2)) and density perturbation (see Eq (3.4)).

Canonical quantization of potential perturbations in the expanding Universe is based on the Lagrangian L(q) and on the simultaneous commutation relation for the canonically conjugate operators q and σ

$$[q(t,x^\alpha)\ \sigma(t,y^\alpha)] = i\delta_{xy}, \tag{4.2}$$

where $[ab] = ab - ba$, δ_{xy} is the δ-function in the metric space (3.1). (For the nongravitating static perfect fluid this relation is analogous to the commutation between the operators of velocity potential and density perturbation, see Lifshitz & Pitaevskij 1978).

Let's define the field Hamiltonian and the Cauchy hypersurface t = const as

$$H = \frac{\bar{H}}{a} = a^3 \int \mathcal{E}\, dV,$$

$$\mathcal{E} = \frac{1}{2a^4}(\bar{q}'^2 + \beta^2 \bar{q}_{,\alpha}\bar{q}'^{\alpha} - U\bar{q}^2) \tag{4.3}$$

where $dV = \sqrt{\det \gamma_{\alpha\beta}}\, d^3x$ is the space measure, Greek indices are manipulated by $\gamma^{\alpha\beta}$. $\mathcal{E}$ is the local energy density of the field q (see Eq (3.8)). In the nonrelativistic limit

$$\mathcal{E} \simeq \varepsilon \frac{\bar{v}^2}{2} + \frac{(\delta p)^2}{2\varepsilon} \tag{4.4}$$

where $\bar{v} = (\frac{v_{,\alpha}}{a})$ is the potential 3-velocity. (For the perfect fluid Eq (4.4) describes the local energy of sound waves).

It is the locality of potential perturbations that differs them principally from gravitational waves. One can calculate the energy density $\mathcal{E}$ or the energy momentum tensor τ_{ik} (see Eq (1.9)) of potential perturbations even for scales greater than the horizon. (Note, that by definition (4.3), $\mathcal{E} \neq \tau_{ik} u^i u^k$; for details see Lukash 1980 b, 1983).

To find most favourable parameters for the best amplification amplitude on cluster's scales it is enough to consider the flat model: $h = 0$, $\gamma_{\alpha\beta} = \delta_{\alpha\beta}$, $\vec{x} = (x^{\alpha})$, see Eq (3.1).

Secondary quantization of q-field over plane waves yields:

$$q = \int_{-\infty}^{+\infty} d^3 \vec{k}\, (a_{\vec{k}}\, q_{\vec{k}} + a^{+}_{\vec{k}}\, q^{+}_{\vec{k}}) \quad , \tag{4.5}$$

where $(q_{\vec{k}}, q_{\vec{k}'}) = [a_{\vec{k}}\, a^{+}_{\vec{k}'}] = \delta(\vec{k} - \vec{k}')$,

$$(q_{\vec{k}}\ ,\ q^{*}_{\vec{k}'}) = [a_{\vec{k}}\, a_{\vec{k}'}] = 0.$$

$$(a,b) = 1 \int_{\Sigma} d\Sigma_i\, D^{ik}\, (a^{*} b_{,k} - a^{*}_{,k} b)$$

is the scalar product in Hilbert's space ($d\Sigma_i$ is the invariant measure on the Cauchy hypersurface Σ); $a_{\vec{k}}$ and $a^{+}_{\vec{k}}$ are the annihilation and creation operators of the field quanta. We shall call these particles phonons (or scalarons) to stress the fact that q-field is coupled with density perturbations (see Eq (3.4)).

In the homogeneous coordinates (3.1)

$$q_{\vec{k}} = q_{\vec{k}}\,(x^{i}) = \frac{\nu_k}{(2\pi)^{3/2}\, \gamma\, a}\, e^{i\vec{k}\vec{x}} \tag{4.6}$$

where $\nu_k = \nu_k(\eta)$ satisfies the equations

$$\nu''_k + (\beta^2 k^2 - U)\, \nu_k = 0, \quad \nu_k\, \nu^{*\prime}_k - \nu^{*}_k\, \nu'_k = i\ ,$$

$$k = |\vec{k}|\ .$$

An essential aspect is that in a hot Universe with $p = \varepsilon/3$ ($a \sim \eta$) the q-field is conformally coupled:

$$U = 0\ , \qquad \bar{H} = \text{const} \tag{4.7}$$

The integral of motion (4.7) of the q-field exists because phonons are neither created nor annihilated in the process of the cosmological expansion with $a \sim \eta$. When $U \neq 0$, phonons interact with the background metric (see Eq (3.8)) and their number is not conserved. This provides spontaneous and induced phonon creation.

To solve the question quantitatively two items have to be considered: the initial seed fluctuations and the intermediate stage $p \neq \varepsilon/3$. Before considering these items, we need operators for the "growing" and "decaying" modes at the stage when Eqs (4.7) hold:

$$C_g = \left(\frac{\omega}{2}\right)^{1/2} \frac{a_{\vec{k}} - a^{+}_{-\vec{k}}}{i\,A}\ , \quad C_d = \left(\frac{\omega}{2}\right)^{1/2} \frac{a_{\vec{k}} + a^{+}_{-\vec{k}}}{A}$$

where $\nu_k = (2\omega)^{-1/2} \exp(-i\,\omega\,\tau)$ is the phonon representation, $a = A\tau$, $\tau = \eta + \text{const}$, $\omega = k/\sqrt{3}$, $A = \text{const}$.

Eqs (4.5) and (4.8) yield

$$q = (2\pi)^{-3/2} \int_{-\infty}^{+\infty} d^3\vec{k} e^{I\vec{k}\vec{x}} \left(C_g \frac{\sin \omega\tau}{\omega\tau} + C_d \frac{\cos \omega\tau}{\omega\tau}\right),$$

$$\bar{H} = \frac{A^2}{2} \int_{-\infty}^{+\infty} d^3\vec{k} (|C_g|^2 + |C_d|^2), \qquad (4.9)$$

$$\bar{H}_{Reg} = \int_{-\infty}^{+\infty} d^3\vec{k}\, \omega\, N_{\vec{k}} \quad ,$$

where $|C|^2 = C^+C = CC^+$, $N_{\vec{k}} = a^+_{\vec{k}} a_k$ is the operator of the number of phonons with momentum $\vec{k}/a$. In the last Eq (4.9) the contribution from point-zero oscillations of the q-field is subtracted. Eigenvalues of the mean energy density operator:

$$E = \frac{H}{a^3 V} \quad , \quad V = \int d^3\vec{x} \quad , \qquad (4.10)$$

are as follows at the stage (4.7):

$$(2\pi a)^{-3} \int d^3\vec{k}\ \varepsilon_k\, (n_{\vec{k}} + ½) \ ,$$

where $\varepsilon_k = \omega/a$ and $n_{\vec{k}}$ are the phonon energy and the occupation number.

5. EQUIPARTITION HYPOTHESIS

The most tempting way of getting primordial perturbations is from initial quantum point-zero metric fluctuations. But it is not the only possibility: initial fluctuations might be of another type - statistical thermal etc. The equipartition hypothesis displays general features of initial fluctuations.

We propose that at moment t_o close to the Plank time the energy of fluctuations should be equally distributed over all modes of perturbations (Kompaneets et al. 1982, Lukash & Novikov 1983a). This means that the initial phases of the fluctuations were random. It is exactly the situation which holds for quantum or thermal fluctuations, so they are special cases of the equipartition hypothesis. (For another approach see Zeldovich & Novikov 1969).

Further on we assume that at $t = t_o + 0$ relativistic particles predominated, so the equation of state was $p = \varepsilon/3$ and the temperature $T = T_p\ (t_p/t\ \sqrt{N})^{½}$, where N is a number of massless degrees of freedom (for three families of leptons and quarks $N \sim 10^2$).

For the case of potential perturbations the hypothesis is as follows: the density matrix ρ of the q-field depends either (weak variant) on the phonon number operators $N_{\vec{k}}$ at the stage $t = t_o + 0$, or (strong variant) on the Hamiltonian H_o at $t = t_o + 0$ (see (4.8)).

The weak variant follows from the requirement that the density matrix does not depend on the initial phases of the field of potential perturbations: $\rho = \mathrm{inv}(a_{\vec{k}} \to a_{\vec{k}} \cdot \exp i \Phi_{\vec{k}})$. It leads to the equipartition of a given scale fluctuation energy over both physical modes:

$$\langle |c_g|^2 \rangle = \langle |c_d|^2 \rangle \tag{5.1}$$

where $\langle \ldots \rangle = \mathrm{Trace}(\rho \ldots)$, $\mathrm{Trace}\ \rho = 1$.

For scales much greater that the horizon ($\omega \eta_0 \ll 1$) it would be natural to strengthen the assumption about initial phase randomness with the condition of statistical independence (over scales) of the fluctuation field amplitudes. In this case the density matrix is factorized over representation (4.8); we call such states statistical.

According to the strong variant the density matrix is a function of the linear superposition of the $N_{\vec{k}}$ - operators identical to the Hamiltonian (4.9) (rather than ρ is a function of $N_{\vec{k}}$ themselves as in the weak variant). Note, that the statistical distributions of the strong variant are identical to the normal ones:

$$\rho \sim \exp\left(- \frac{\bar{H}_{o\ Reg}}{\sigma_0} \right) \tag{5.2}$$

$$\langle a^+_{\vec{k}}\, a_{\vec{k}'} \rangle = n_k\ \delta(\vec{k} - \vec{k}'), \quad n_k = (e^{\omega/\sigma_0} - 1)^{-1},$$

where σ_0 is a constant. For $\sigma_0 = 0$, $(aT)_0$ the states (5.2) are vacuum and thermal respectively.

The strong variant is not as natural as the weak one. It follows from the idea of an initially stationary state of the Universe with tunnelling into the Friedmann state. According to it the initial state of the expanding Universe was characterized by a single parameter - the total energy of all the fields - and this energy was distributed over all the internal degrees of freedom in the most probable way, i.e. the distribution corresponded to the largest number of microscopic configurations. This assumption will hold if the strengths of all the interactions are unified. The notion of the space-time extent (and of the energy) arises here after the spontaneous breaking of the initial supersymmetry by tunnelling to a state that was characterized by the local Lorentz group (gravitational field) and the fields unifying all the other types of interactions. (Massive particles and the interactions of our present epoch are produced by subsequent symmetry breaking of the unified fields). After this tunnelling transition the density matrix of any free field (q-field or gravitational waves) for $t > t_0$ is likely to depend only on the Hamiltonian $H_0 = H(t_0)$ of this field at the initial moment $t = t_0 + 0$.

It is not difficult to show that on large scales the field distribution is of the Hibbs from if, in the simplest model of tunnelling transition, the initial field state before transition corresponded to a vacuum of high simmetry. Indeed, if we neglect the effects of interactions associated with the details of the transition, then relative to the final vacuum with the broken symmetry the field state will obey the Gauss distribution with dispersion equal to the energy difference of the initial and final vacuums (Kompaneets & Lukash 1981, Sciama et al. 1981). This means that for large scales the same energy ($\sim T_o$) will account for each interval of momenta $d^3\vec{k}$. Note that the thermal state of the Universe at $t = t_o + 0$ was, in principle, prepared in this model at the very beginning since the initial state (vacuum with the restored symmetry) had been already stationary with Gauss's law of distribution.

6. PARAMETRIC EFFECT IN THE EARLY UNIVERSE

The equipartition hypothesis determining initial states of the field of potential fluctuations allows for further investigation of the amplification effect and discussion models of the primordial perturbations origin.

Let us consider a cosmological model q with a change of the expansion rate:

$$\eta_0 < \eta < \eta_1 : a = \eta = \tau = \sqrt{2t} \qquad (p = \varepsilon/3), \quad (6.1a)$$
$$\eta_1 < \eta < \eta_2 : a = a(\eta) \qquad (p = p(\varepsilon)), \quad (6.1b)$$
$$\eta_2 < \eta : a = A\tau \qquad (p = \varepsilon/3). \quad (6.1c)$$

The nature of this change at the b-stage is immaterial: the cause may be massive short-lived particles, de Sitter stages, phase transitions etc. (see e.g., Polnarev & Khlopov 1980, Starobinsky 1980a, Guth 1981, Linde 1982, 1983). At the $p = \varepsilon/3$ stages, the "growing" and "decaying" modes of the q-field are determined by the phase $\omega\tau$, where

$$\tau = \frac{a}{a'} = \eta - \int_{\eta_1} \frac{aa''}{a'^2}\, d\eta = \eta_0 + \tfrac{1}{2} \int_{\eta_0} \left(1 + \frac{3p}{\varepsilon}\right) d\eta . \quad (6.2)$$

We choose the following normalization of the wave-vector: k = 1 corresponds to the Planck length at the Planck time. A = const.

Let us $a_{\vec{k}}$ and $c_{\vec{k}}$ be the phonon representations at the stages "a" and "c" respectively. Then

$$c_{\vec{k}} = \alpha_k a_{\vec{k}} + \beta_k^* a^+_{-\vec{k}} , \quad |\alpha_k|^2 - |\beta_k|^2 = 1 \quad (6.3)$$

where α_k and β_k are Bogolubov's coefficients. The density matrix (see Section 5) is independent of time in the Heisenberg representation. Eqs (4.9), (4.10) yield for the mean occupation numbers

$$\langle a^+_{\vec{k}} a_{\vec{k}'} \rangle = n_{ka}\, \delta(\vec{k}' - \vec{k}'), \quad \langle c^+_{\vec{k}} c_{\vec{k}'} \rangle = n_{kc}\, \delta(\vec{k} - \vec{k}')$$

$$n_{kc} = n_{ka} + |\beta_k|^2 (2n_{ka} + 1) \quad (6.4)$$

and for the relative energy of the field of potential perturbations at the a- and c-stages:

$$\frac{1}{\varepsilon} E_{Reg} = \frac{1}{24\pi^3} \int d^3\vec{k}\, \omega \cdot \begin{cases} n_{ka} \\ A^{-2} n_{kc} \end{cases} \tag{6.5}$$

n_{kc} is larger than n_{ka} because of induced and spontaneous creation. The factor A^{-2} takes into account the magnitude of phonon cooling at the intermediate b-stage.

The exact solution of Eq (3.8) in the asymptotic region $\omega\eta \ll 1$

$$q = C_1(\vec{x}) + C_2(\vec{x}) \int \frac{d\eta}{(\xi a)^2} \tag{6.6}$$

affords an opportunity to derive the amplification coefficient for large scales independently of the rate of cosmological expansion at the intermediate stage (Kompaneets & Lukash 1981, Kompaneets et al. 1982):

$$2B_k A^{-1} = -\frac{i\varkappa}{\omega} - 1 + A^{-2} + O(\omega\eta_1), \tag{6.7}$$

where $\varkappa = \dfrac{1}{\eta_1} - \dfrac{1}{A^2 \tau_2} - \displaystyle\int_{\eta_1}^{\eta_2} d\eta/(\xi a)^2 .$

The spectrum of q-field at the c-stage is given for $A \gg 1$ and $\omega\eta_1 < 1$ (it exponentially dies for $\omega\eta_1 > 1$):

$$\langle q^2 \rangle = \int_0^\infty q_k^2 \frac{dk}{k}, \quad q_k = (3/4)^{1/4} \frac{\varkappa k}{\pi} (n_{ka}+\tfrac{1}{2})^{1/2} \frac{\sin \omega\tau}{\omega\tau}. \tag{6.8}$$

So, the amplification coefficient is equal to $\varkappa/\omega$.

A very important point is that after amplification the mutual correlation of phases of the created perturbations corresponds to the "growing" mode (compare Eq (5.1)):

$$C_{gc} = C_{ga} + \frac{\varkappa}{\omega} C_{da}, \quad C_{dc} = A^{-2} C_{da}, \tag{6.9}$$

$$\langle |C_{gc}|^2 \rangle \gg \langle |C_{dc}|^2 \rangle \quad \text{for} \quad \omega\eta_1 \ll 1 .$$

So, from initial random perturbations the "growing" mode is created. If initially only the "growing" mode were present, then no amplification would take place at all since the evolution of the "growing" mode does not depend on the rate of the expansion (metric fluctuations are constant for large scales). The possibility of amplification is directly connected with presence of the "decaying" mode at the beginning. The equipartition of energy means that the metric perturbation amplitude on large scales is almost all in the "decaying" mode: $q_d \sim c/\omega\eta \gg q_g \sim c$. To clarify the physical meaning of the amplification effect, let us consider a situation in which the duration of the intermediate stage ($p \neq \varepsilon/3$) is rather short. Then the perturbation amplitude q will not change significantly after this stage but now it will be approximately the same in every mode ($q_d \sim q_g \sim c/\omega\eta_1$), so we have a large increase in the "growing" mode. Analysis shows that for extended intermediate stages in cosmology a standard situation is the same: at the epoch close to t_1, when first $p \neq \varepsilon/3$ and a parametric barrier (non-zero effective potential) arises, the amplitude of the "growing" mode increases and becomes of the order of the "decaying" mode amplitude at $t = t_1$, and then remains constant as far as metric perturbations on large scales are concerned. The amplification coefficient $\varkappa/\omega$ is equal in order of magnitude to the ratio of the perturbation scale to the horizon at $\sim t_1$. The amplification effect does not practically depend on the duration of the intermediate stage and on the rate of expansion at this stage.

Knowing that at the Planck time the amplitude of the initial fluctuations is about unity on the horizon scale, we obtain that, for the effect to work, the first change in the expansion rate must occur not later than in $\sim 10^4$ Planck times, which corresponds to $M \sim 10^{17}$ Gev. This figure does not yet conflict with the estimate of the amount of gravitational waves produces in the change, that will influence the large-scale isotropy of the relict radiation.

Now the question about scales. It is necessary to have a scale that initially (at t_1) was $\sim 10^4$ times larger than the Planck length to be the scale of clusters by now. To stretch the scale factor we need the total duration of de Sitter stages to be about several tens of cosmological times.

Let us model the simplest inflationary stage in the Early Universe by a high Λ-term which transforms to heat after some moment t_2. This allows to remain within the "perfect fluid" framework and hence to apply the developed theory (see Eq (1.3)). At $t < t_2$ the q-field of potential perturbations is coupled with relativistic particles ($p = \varepsilon/3$) since $\delta\Lambda = 0$. So, the background model is the following:

$$
\begin{array}{ll} t_o < t < t_2 : \\ t_2 < t \quad : \end{array}
\quad a = \begin{cases} M^{-1}\,\mathrm{Sh}^{1/2}\, t/t_1 \\ (2A(t-\Delta))^{1/2} \end{cases}, \quad \varepsilon = \begin{cases} 3M^4\,\mathrm{cth}^2\, t/t_1 \\ 3A^2 a^{-4} \end{cases} \tag{6.10}
$$

where $M = \eta_1^{-1} = (2t_1)^{-1/2}$, $\Delta = t_2 - t_1$, $A = \mathrm{Sh}\, t_2/t_1 \simeq \exp \Delta/t_1 \gg 1$.

The Λ-term $3M^4$ predominates the $t_1 < t < t_2$. The effective potential U(t) is localized near $t \sim t_2$,

$$
t < t_2: \quad U = \frac{(a\xi)''}{a\xi} = -8M^2\,\mathrm{Sh}\,\frac{t}{t_1}\,\mathrm{ch}^{-2}\,\frac{t}{t_1} \quad , \tag{6.11}
$$

and $\varkappa \sim \eta_1^{-1}$ for $k\eta_1 < 1$ (Kompaneets et al. 1982, Lukash & Novikov 1983a), i.e. the amplification coefficient has a standard form and it does not depend on the inflation duration Δ. (The Δ-parameter determines the scales). Fig 1 shows the resulting spectra of the created perturbations. Note, that the q-perturbations die exponentially at smaller scales ($k\eta_1 > 1$). However, if some test φ-field of type (1) (e.g., Higgs' field) dominates the q-field at the de Sitter stage, then the effective potential U(t) has the following form (compare (6.11))

$$t < t_2: \quad U = \frac{a''}{a} = 2M^2 \, \mathrm{Sh} \, \frac{t}{t_1} \, , \tag{6.12}$$

and the resulting spectrum of the created perturbations is flat for $k\eta_1 > 1$ (Lukash 1980 a, b, Lukash & Novikov 1983 b).

Figure 1. Spectra of the primordial cosmological perturbations q_k ($k\tau < 1$), created from initially :
solid lines: (a) thermal ($n_{ka} \simeq k^{-1}$), (b) quantum ($n_{ka}=0$) potential fluctuations of metric, $q_k \simeq \pi^{-1} kM(n_{ka} + \frac{1}{2})^{1/2}$, $k < M$;
dashed line: quantum fluctuations of a test scalar field at the de Sitter stage

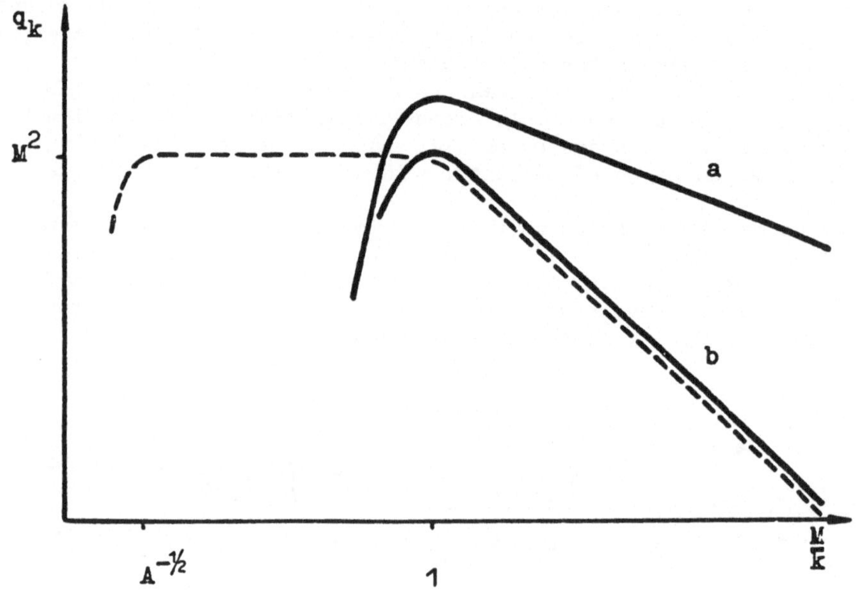

The physical difference between q- and φ-fields results in different spectra of the primordial perturbations at $k\eta_1 > 1$. The former exists since very beginning as a perturbation of the background (Friedmann) fields while the latter can appear as an external (test) field in the unperturbed background (6.10) and only at $t > t_2$ it transforms to the metric and density perturbations. The spectrum shape $q_k \sim k\eta_1^{-1}$ (for $k\eta_1 < 1$) is universal in any theory since it originates due to the parametric amplification of longwave quantum fluctuations.

Before coming to observational tests, some formulae are given to make clear the connection between the gauge invariant spectrum q_k and that in synchronous reference system frequently used in literature. The synchronous gauge $h_{ik}u^k = 0$ means the special choice of the C- and D-functions of Eq (3.2). For the longwave "growing" mode ($\omega\tau < 1$, see (6.8)):

$$h_{\alpha\beta} = \frac{1}{\sqrt{3}} q \, \delta_{\alpha\beta} , \quad \frac{\delta\varepsilon}{\varepsilon} = \frac{(\omega\tau)^2}{2\sqrt{3}} q , \quad q = q(\vec{x}) \qquad (6.13)$$

7. WHY IS IT NECESSARY TO MEASURE $\Delta T/T$-ANISOTROPY?

The investigation of possible anisotropy of the microwave relic background is very important for two reasons.

i) Observations of the background anisotropy are a clue to the Early Universe physics. They provide direct information about the primordial cosmological perturbations in the Universe and consequently, about the processes which caused them.

ii) The features of angular correlations in the background temperature fluctuations over large scales tell us about the overall space curvature and, thus, about the total matter density in the Universe including the missing mass. To gain this sort of information in any other way is quite difficult.

In the simplest model of parametric production of the primordial perturbations, the characteristic scale of the changing slope of the spectrum (see Fig 1) depends on the inflation duration in the Early Universe and it is an arbitrary parameter in the given model. The new theories of vacuum phase transitions (see Section 1) predict both the flat spectrum of the primordial perturbations and zero spatial curvature (h = 0) in the Universe. So, calculations of the anisotropy of the relic radiation and their confrontation with the observational data provide a unique opportunity to test theories of the very early stages of the Universe expansion.

By no means belittling the importance of searching for small-scale $\Delta T/T$ -fluctuations which correspond to linear dimensions encompassing the masses typical of galaxy clusters, we have always tried to draw special attention to the measurements of the $\Delta T/T$-anisotropy of large angular scales (from $\theta \sim 5°$ and up to the dipole component). This anisotropy bears direct information as to the primordial metric fluctuations and is not coupled to the details of recombination dynamics. The spottiness effect in the large-scale $\Delta T/T$ correlations make it possible to evaluate the total matter density $\Omega = \varepsilon/3H^2$ (including the invisible mass). Detection of the $\Delta T/T$-variations (say, a quadrupole moment) - or even setting a reliable upper limit for them - would enable most important conclusions on the spectrum of the primordial perturbations.

Here the results of calculations of the $\Delta T/T$-anisotropy in the whole range of angular scales for different cosmological models (Lukash et al. 1984) are presented.

8. THE PHYSICAL CAUSES OF $\Delta T/T$-FORMATION

When we observe the relic radiation we see the wall of the last scattering of the photons by the ionized matter. It corresponds to the moment of the hydrogen recombination in the past, when the optical depth by Thompson scattering is unity. For the largest angular scales $\theta \gtrsim 5°$, corresponding to the horizon linear scale as it is seen on the recombination sphere, the fluctuations of $\Delta T/T$ are caused by the perturbations of the gravitational field (Sachs & Wolfe 1967) in combination with the overall space curvature (Novikov 1968, Lukash 1977, 1982, 1983, Bisnovatyi-Kogan et al. 1980, Lukash & Novikov 1984).

On scales $\theta \lesssim 11'$, corresponding to the angular resolution of the recombination width taken by the cosmic plasma to transform from the opaque to transparent state, the dominant effect is the Doppler shift experienced by the relic photons having been scattered off the moving condensations (Peebles & Yu 1970, Zeldovich & Sunyaev 1970, Doroshkevich et al. 1978, Shandarin et al. 1983). At $11' < \theta < 5°$ this process is augmented by the density perturbation effect accounting for the fact that just before the recombination plasma density fluctuations were accompanied by radiation temperature variations (Silk 1968). On smaller scales this effect vanishes because the radiation has time to escape density enhancements, and the dominant contribution becomes that due to the Doppler shift since translucent condensation boundaries move freely. For scales $\theta \lesssim 11'$ one should take into account the mutual compensation effect when light rays pass successively through many condensations and rarefactions along the recombination width, here we use the numerical simulation results (Zabotin & Naselskij 1982, 1983).

For scales $\theta > 11'$ we can treat the recombination as an instantaneous process and explicitly calculate the background anisotropy:

$$\frac{\Delta T}{T}(\vec{e}) = -\tfrac{1}{2} e^{\alpha} e^{\beta} \int_{\eta_{rec}}^{\eta_0} h'_{\alpha\beta}\, d\eta + (v' + e^{\alpha} v_{,\alpha})\Big|_{\eta=\eta_{rec}} \tag{8.1}$$

where $h_{\alpha\beta}$, $\delta_b = 3v'$ and $u^i_b = (1, -v_{,\alpha}/a)$ are metric, barion density and 4-velocity perturbations in the synchronous co-moving to relic particles reference system, functions $h_{\alpha\beta}$ and v are taken on the light-cone, $\vec{e} = (e^{\alpha})$ is the unit vector along the line of sight. Three terms on the right-hand-side of Eq (8.1) are obviously responsible for the corresponding effects mentioned above.

9. $\Delta T/T$ - CALCULATIONS AND CONFRONTATION WITH OBSERVATIONS

The amplitude of temperature fluctuations is a function of: the field of primordial cosmological perturbations $q = q(\vec{x})$ (see Eq (6.13)); the dynamics of the Early Universe that relates the primordial perturbations with the temperature fluctuations and includes missing mass parameters, the hydrogen recombination dynamics, reheating and other factors; the parameter $\Omega = (1+3h^2/\varepsilon a^2)^{-1}$ that is the total matter density of the present Universe in the critical density units (see (3.1)). Observations give the map of the sky radiation temperature $\Delta T/T(\vec{e})$.

To compare these two functions $\Delta T/T(\vec{e})$ the correlation analysis is employed. This possibility is based on the statistical independence (randomness) of the primordial perturbation amplitudes on different scales (see Section 5). It shows that the observed sky $\Delta T/T$-pattern is the result of a random superposition of the independent perturbation amplitudes. In particular the root-mean-square temperature fluctuation

$$\frac{\Delta T}{T}(\theta) = \left\langle \left(\frac{T(\vec{e}) - T(\vec{e}\,')}{T}\right)^2 \right\rangle^{1/2} \tag{9.1}$$

obviously depends on the angle between the observation directions, $\cos\theta = \vec{e}\,\vec{e}\,'$. Ergodicity theorem allows identifying (9.1) with the observed $\Delta T/T$-function where the averaging is taken over all directions on the celestial sphere with the fixed angle θ, for $\theta > \theta_o$ (the beam-width of the antenna to be used). This function is directly detected in the small-scale experiments ($\theta \sim 4'$-$50'$, see Parijskij et al. 1977, Partridge 1980, Uson & Wilkinson 1984). It can also be derived from the sky $\Delta T/T$-pattern obtained in the large-scale experiments ($\theta > 5°$, see Fabbri et al. 1980, Fixen et al. 1983, Lubin et al. 1983, Strukov et al. 1984). In addition to $\Delta T/T(\theta)$, we construct the expected amplitudes $(\Delta T/T)_l = \langle a_{lm}^2 \rangle^{1/2}$ of multipole components of the large-scale anisotropy for the spherical harmonic expansion:

$$\frac{\Delta T}{T}(\vec{e}) = \sum_{l,m} a_{lm} \Psi_{lm}(\vec{e}) \; . \tag{9.2}$$

Let us now come to the investigated models. Three variants of missing mass are considered: (a) barions, (b) massive thermodynamic neutrinos ($m_\nu \simeq$ 6-30 eV) and (c) very heavy relic weak interacting particles ($m_r >$ 100 eV) such as axions, primordial black holes, monopoles, gravitino, etc. The physical nature of the superheavy particles is not important for $\Delta T/T$-calculations. It matters only that they become nonrelativistic rather early before the recombination epoch. We have also considered three types of the primordial spectra:

$$q_k \sim k^n, \qquad n - \begin{cases} 0 & \text{(flat spectrum)}, \\ 1 & \text{(white noise)}, \\ 0,1 & \text{(see fig. 1)}. \end{cases} \tag{9.3}$$

The characteristic scale of the slope change ($0 \rightarrow 1$) in the third spectrum is taken within 20 to 600 Mpc. The first structure of the Universe ($\delta\varepsilon/\varepsilon \simeq 1$) emerges at red-shift z_s. The following results are given for $z_s = 3$.

Fig. 2 displays the $\Delta T/T(\theta)$-function (9.2), dipole moment is subtracted. Arrows - observational data. Data for θ = 90° correspond to the upper limit for the quadrupole anisotropy (Strukov et al. 1984). Data for $\theta \sim 10'$ correspond to $\Delta T/T\ (\theta)$ (Parijskij et al. 1977). For the curve (c) objects with masses $\sim 10^{14} M_\odot$ form at z = 3. The expected amplitudes of dipole and quadrupole moments (9.2) are summarized in the table. Two values of the quadrupole anisotropy in c-model correspond to two different normalizations - the one taken for the curve (c) and the other taken by Peebles (1982) (objects with masses $\sim 10^{12} M_\odot$ form at z = 3).

Figure 2. The anisotropy of the relic radiation $\Delta T/T\ (\theta)$
solid lines: Ω = 1, n = 0, missing mass: (a) barions, (b) massive neutrinos, (c) superheavy particles;
dashed line: variant (c) for Ω = 0.5

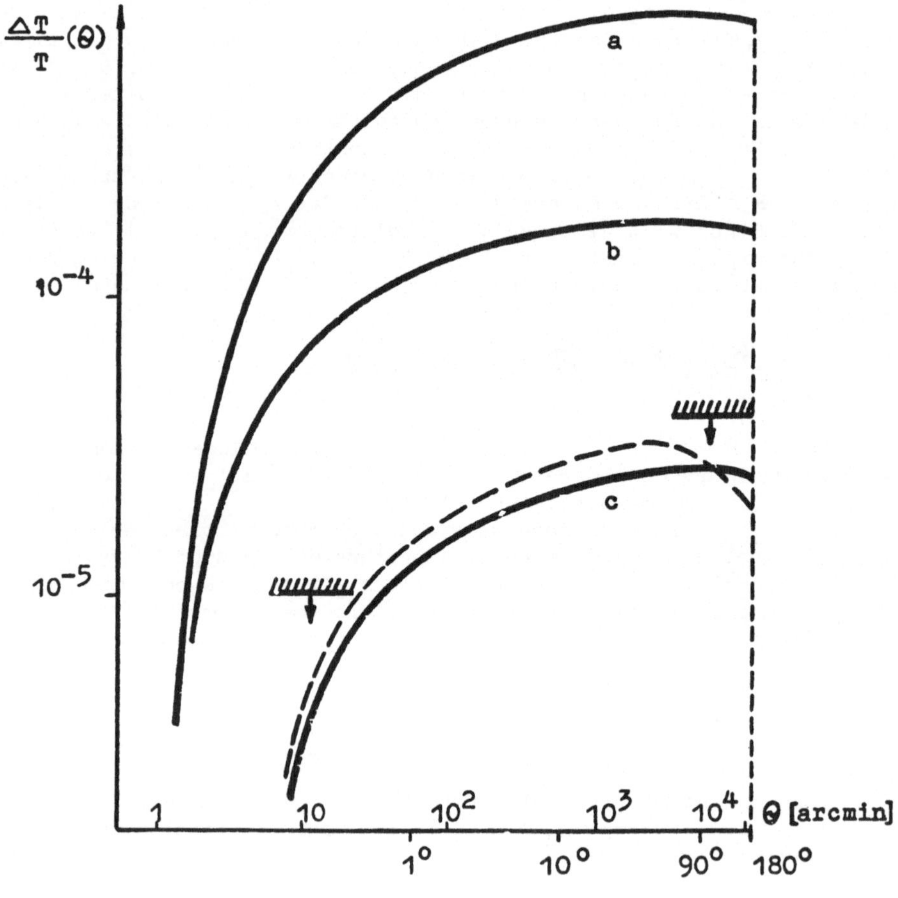

Curves (a) and (b) show that the models with the flat spectrum and usual barions and thermodynamic stable neutrinos as the missing mass are in contradiction with both observational limits. (The model with superheavy free particles is in agreement with the existing observations). The dipole moments are calculated for the rather long waves that evolve linearly up to now. For confrontation with the observations, one will have to take into account the kinematics of the local group of galaxy clusters in order to subtract the contribution of the nonlinear modes to the observed dipole anisotropy.

Table. The dipole (l = 1) and quadrupole (l = 2) ($\Delta T/T)_l$-anisotropies of the relic radiation in the massive neutrino (b) and superheavy particle (c) dominated universes

	l	n = 0	n = -1	n = 1
b	1	$1.5\ 10^{-2}$	$2.2\ 10^{-2}$	$1.5\ 10^{-2}$
	2	$3\ 10^{-5}$	$9\ 10^{-4}$	$2\ 10^{-5}$
c	1	$1.8\ 10^{-3}$	$2.8\ 10^{-2}$	$1.8\ 10^{-3}$
	2	$2\ 10^{-5}$	$6\ 10^{-5}$	10^{-6}
		$3\ 10^{-6}$		

10. THE SPOTTINESS STRUCTURE OF SKY $\Delta T/T$-PATTERN

We consider only those states of the primordial perturbation field q(x) which are equally excited at every point of the space (see Section 5). They can be represented as random superpositions of plane waves (see (4.5), (6.4)). In the Lobachevski space (3 - space of the open Universe) the plane wave analogs are the following (Lukash 1977):

$$f^{\pm}_{k,\vec{e}} = \left(\sqrt{1 + (h\vec{x})^2} \pm h\vec{e}\vec{x} \right)^{\pm i\frac{k}{h}} , \tag{10.1}$$

where $\vec{x}$ labels the locally Cartesian coordinates, h^{-1} is the space curvature radius (see (3.1)). For $h \to 0$, functions (10.1) turn to usual plane waves in the Euclidean space ($\vec{k} = k.\vec{e}$). For $k \to 0$ they become Bianchi V spatially homogeneous modes. In general, every uniformly limited function $|q(\vec{x})| <$ const in the Lobachevski space can be Fourier-expanded over the nonhomogeneous plane waves (10.1).

A single density perturbation plane wave in the flat Friedmann model produces the sky $\Delta T/T$-pattern with the large-scale shape presented as a quadrupole dependence on the angle θ between the wave-vector and the line of sight ($\Delta T/T \sim \sin^2\theta$). The large-scale

$\Delta T/T$-distribution produced by a single plane wave (10.1) in the open model has the form of a "spot":

$$\frac{\Delta T}{T} \sim \left(\frac{2\,\varkappa \tan\theta/2}{1+(\varkappa_0 \tan\theta/2)^2}\right)^2 \qquad (10.2)$$

where $\varkappa_0 = \exp(-h\chi)$, $\chi = \eta_0 - \eta_{rec}$ is the geodetic 3-distance of the observer from the recombination sphere. The spot shape (10.2) is independent of the wavevector modulus k and identical with that of Bianchi V model (Novikov 1968). The $\Delta T/T$-anisotropy fully develops inside the annulus of the angular radius $\theta_0 = 2\tan^{-1}\varkappa_0$ and the width $\sim \theta_0$. For $\Omega = 1$ (h = 0), the spot (10.2) reduces to a quadrupole dependence. For small Ω, $\varkappa_0 = \Omega/4$ and the angular size of the spot is equal to Ω.

The observed sky $\Delta T/T$-pattern is a random superposition of the uncorrelated spots. It means that for $\Omega < 1$ a statistical distribution of $\frac{\Delta T}{T}(\theta)$ with an angular correlation scale $\sim\Omega$ takes place (see fig. 2). The correlation function also depends on the primordial perturbation spectrum q_k. So, the large-scale $\frac{\Delta T}{T}(\theta)$-distribution allows for determining both the spectrum q_k and the overall matter density Ω.

CONCLUSION

Primordial cosmological perturbations are the key problem of the modern cosmology. It is the node where the theories of the Early Universe based on the physics of elementary particles in intense gravity field link with the modern experiment. The present experimental capabilities make it possible already today to test what is directly implied by these theories, the predictions that concern the spectrum and other parameters of the primordial perturbations. This can be done using angular temperature variations of the relic radiation over the celestial sphere. (The observed large-scale structure of the Universe allows only approximate estimates of the primordial perturbation parameters). The limitations known up to date on the quadrupole anisotropy of $\Delta T/T$ forbid all standard models with a flat spectrum of primordial perturbations, $\Omega = 1$ and massive stable neutrinos as missing mass. The detection of $\Delta T/T$ on scales from 5° and up to 180° would enable one to draw the conclusions on the primordial spectrum $\delta\varepsilon/\varepsilon$ and the total matter density $\Omega = \varepsilon/\varepsilon_{cr}$.

REFERENCES

Bardeen, J.M. (1980). Phys. Rev. D22, 1882.
Bardeen, J., Steinhardt, P. & Turner, M. (1983). Phys. Rev. D28, 679.
Barrow, J.D. & Turner, M.S. (1981). Nature 292, 38.
Bisnovatyi-Kogan, G.S., Lukash, V.N. & Novikov, I.D. (1980). In Variability in Stars and Galaxies (IAU/EPS), Institut d'Astrophysique, Liege, G.1.1.

Bonnor, W.B. (1956). Zeits. f. Astroph. 39, 143.
Bonnor, W.B. (1957). MNRAS 117, 104.
Brandenberger, R., Kahn, R. & Press, W.H. (1983). Phys.Rev. D28, 1809.
Doroshkevich, A.G., Zeldovich, Ya.B. & Sunyaev, R.A. (1978). Astron. Zh. 55, 913.
Fabbri, R., Guidi, J., Melchiorri, F. & Natale V. (1980). Phys. Rev. Lett. 44, 1563.
Field, G.B. & Shepley, L.C. (1968). Astroph. Spa. Sci. 1, 309.
Fixen, D., Cheng, E. & Wilkinson, D. (1983). Phys. Rev. Let. 50, 620.
Friedmann, A. (1922). Z. Phys. 10, 377.
Friedmann, A. (1924). Z. Phys. 21, 326.
Grishchuk, L.P. (1974). JETP 67, 825.
Guth, A.H. (1981). Phys. Rev. D23, 347.
Harrison, E.R. (1967). Rev. Mod. Phys. 39, 862.
Hartle, J.B. (1980). Phys. Rev. D22, 2091.
Hartle, J.B. & Horowitz, G.T. (1981). Phys. Rev. D24, 257.
Hawking, S.W. (1966). Astrophys. J. 145, 544.
Hogan, C.J. (1980). Nature 286, 360.
Kibble, T.W. (1976). J. Phys. A9, 1387. Phys. Rep. 67, 183.
Kirzhnits, D.A. (1972). JETP Lett. 15, 745.
Kompaneets, D.A. & Lukash, V.N. (1981). Astron. Zh. 58, 482.
Kompaneets, D.A., Lukash, V.N. & Novikov, I.D. (1982). Astron. Zh. 59, 424. Quantum Gravity, ed. M.A. Markov, V.A. Berezin, V.P. Frolov, Nuclear Res. Inst., Moscow, 82.
Lifshitz, E.M. (1946). JETP 16, 587.
Lifshitz, E.M. & Pitaevskij, L.P. (1978). Statistical Physics, part II, § 24, Nauka, Moscow.
Linde, A.D. (1982). Phys. Lett. 108B, 389. 114B, 431. 116B, 340.
Linde, A.D. (1983). JETP Lett. 38, 149. Phys. Lett. 129B, 177.
Lubin, P., Epstein, G. & Smoot, G. (1983). Phys. Rev. Lett. 50, 616.
Lukash, V.N. (1977). In Contr. papers 8-Int. Conf. GRG, Waterloo, 237.
Lukash, V.N. (1980a). JETP Lett. 31, 631. JETP 79, 1601.
Lukash, V.N. (1980b). Generation of Sound Waves in the Isotropic Universe. Preprint Space Res. Inst. IIp -559, Moscow.
Lukash, V.N. (1982). In Early Evolution of the Universe and its Present Structure, ed. G.O. Abell, G. Chincarini, D. Reidel Publ. Comp., 149.
Lukash, V.N. (1983). Dynamics of the Early Universe and Origin of its Structure. Doctoral Thesis, Space Res. Inst., Moscow.
Lukash, V.N., Naselskij, P.D. & Novikov, I.D. (1984). Quantum Gravity, ed. M.A. Markov, V.A. Beresin, F.P. Frolov, Nuclear Res. Inst., Moscow.
Lukash, V.N. & Novikov, I.D. (1983a). In The Very Early Universe, ed. G.W. Gibbons, S.W. Hawking, S. Siklos, Cambridge. Univ.Press, 311.
Lukash, V.N. & Novikov, I.D. (1983b). In Contr. papers 10-Int. Conf. GRG, ed. B. Bertotti, F. de Felice, A. Pascolini, Consiglio Nazionale delle Ricerche, Roma, 844.
Lukash, V.N. & Novikov, I.D. (1984). The Effect of Spottiness in Large-Scale Structure of the Microwave Background. Preprint Space Res. Inst. IIp -954, Moscow; Nature 316, 46, 1985.
Lukash, V.N. & Novikov, I.D. (1985). In Galaxies, Auxisymmetric

Systems and relativity, Essays to W.B. Bonnor's 65th Birthday, ed. M.A.H. MacCallum, Cambridge Univ. Press, 23.
Mukhanov, V.F. & Chibisov, G.V. (1981). JETP Lett. 33, 549.
Novikov, I.D. (1968). Astron. Zh 45, 538.
Parijskij, Yu.N., Petrov, Z.N. & Chernov, A.N. (1977). Astron. Zh. Lett. 3, 483.
Parker, L. (1968). Phys. Rev. Lett. 21, 562.
Partridge, R.B. (1980). Phys. Scripta, 21, 624.
Peebles, P.J.E. (1982). Astroph. J. Lett. 263, L1.
Peebles, P.J.E. & Yu, I.T. (1970). Astroph. J. 162, 815.
Polnarev, A.G. & Khlopov, M.Yu. (1980). Phys. Lett. 97B, 383.
Press, W.H. & Vishniac, E.T. (1980). Astrophys. J. 239, 1.
Sachs, R.K. & Wolfe, A.M. (1967). Astrophys. J. 147, 73.
Sciama, D.W., Candelas, P. & Deutch, D. (1981). Advances in Phys. 30, 367.
Shandarin, S.F., Doroshkevich, A.G. & Zeldovich, Ya.B. (1983). Usp. Fiz. Nauk 139, 83.
Silk, J. (1968). Astroph. J. 151, 459.
Starobinsky, A.A. (1980a). Phys. Lett. 91B, 99.
Starobinsky, A.A. (1980b). JETP Lett. 34, 460.
Starobinsky, A.A. (1982). Phys. Lett. 117B, 175.
Strukov, I.A., Sagdeev, R.Z., Kardashev, N.S., Skulachev, D. & Eysmont, N. (1984). In Advances in Space Research, Pergamon Press, COSPAR.
The Very Early Universe (1983). Ed. G.W. Gibbons, S.W. Hawking, S. Siklos, Cambridge Univ. Press.
Uson, J.M. & Wilkinson, D.T. (1984). Astrophys. J. Lett. 277, L1.
Vilenkin, A. (1981). Phys. Rev. Lett. 46, 1169.
Vilenkin, A. (1984). Cosmic Strings and Domain Walls. Preprint Tufts Univ., Medford, MA 02155.
Zabotin, N.A. & Naselskij, P.D. (1982). Astron. Zh. 59, 447.
Zabotin, N.A. & Naselskij, P.D. (1982). Astron. Zh. 60, 467.
Zeldovich, Ya.B. (1980). MNRAS 192, 663.
Zeldovich, Ya.B. & Novikov, I.D. (1969). Astron. Zh. 46, 960.
Zeldovich, Ya.B. & Sunyaev, R.A. (1970). Astroph. Spa. Sci. 6, 358.

GALAXY FORMATION IN COLD DARK MATTER DOMINATED UNIVERSES: OBSERVATIONAL TESTS

Nicola Vittorio
Istituto Astronomico, Università di Roma, La Sapienza

I.Introduction

There are three observational facts which strongly support the hot Big- Bang cosmology: the universal Hubble expansion, the existence of the cosmic microwave background (CMB), and the agreement between the predictions of the primordial nucleosynthesis theories and the observed abundance of light elements. The cosmological Friedmann models are parametrized by the density parameter Ω_0. This measures the present cosmological density in units of the critical density: $\rho_{crit} \equiv 3H_0^2/[8\pi G] = 2 \cdot 10^{-29}\ h^2\ g\ cm^{-3}$; H_0 is the Hubble constant and $h = H_0\ /[100\ km\ s^{-1}\ Mpc^{-1}]$. If Ω_0 is less than or equal to unity (open or flat model respectively), the universe will expand forever; if Ω_0 is greater than unity, the universe will recollapse eventually.

Despite of their successes, these models present several puzzles (i.e., horizon, flatness, etc.), which are naturally resolved if the universe underwent through an early accelerated expansion phase. The inflationary scenario (see Turner this volume) provides the physical framework for justifying such an accelerated expansion and predicts that the universe must be today indistinguishible from the flat (Einstein- de Sitter) model, quite independently of the initial conditions. In order to produce through primordial nucleosynthesis the observed amount of light elements the baryons can not exceed the 20% of the critical density (i.e., $\Omega_b \lesssim 0.2$; Yang et al., 1984). Then, for Ω_0 to be unity, most of the mass in the universe must be non- baryonic. There is a long

P. Galeotti and D. N. Schramm (eds.), Gauge Theory and the Early Universe, 99–118.

list of candidate particles (e.g. massive neutrinos, axions, gravitinos, photinos, etc.) that might be left over from the Big-Bang and whose predicted abundance is such that they could be the dark, non-baryonic component able to close the universe (see Turner, 1986 for a recent review).

Particle physicists strongly argue in favour of a flat universe. However, no astronomical observations support $\Omega_0 = 1$; rather, $\Omega_{obs} \sim 0.20$ (see, e.g., Faber and Gallagher, 1979). The disagreement between the theoretical prejudice for a high density universe ($\Omega_0 = 1$) and the observational evidence for a low density one ($\Omega_{obs} < 1$) is commonly indicated as the Ω_0 problem. In order to resolve it, more speculative scenarios have been proposed where the 80% of the critical density is in the form of a smoothly-distributed, undetectable component. Suggestions for this component include "failed" galaxies (Kaiser, 1986; Bardeen et al., 1986), relativistic particles produced by the recent decay of unstable particles (Dicus, Kolb, and Teplits, 1977; Turner, Steigman, Strauss, 1984; Gelmini, Schramm, and Valle, 1984; Olive, Seckel, and Visnhiac, 1985), a relic cosmological term (Turner, Steigman, Strauss, 1984; Peebles, 1984), or even fast moving strings (Vilenkin, 1984).

We have at our disposal at least three different observational tests, for discriminating among different scenarios: the angular structure of the cosmic microwave background, the large scale peculiar velocity field, and the large scale matter distribution.

Here we want to consider inflationary cosmological models dominated by "cold" dark matter (CDM) and to discuss their ability in explaining some recent observations. Cold dark matter is constituted by weakly interacting particles which have negligible velocity dispersion also at early times, either because they are very massive (~ 1 GeV) or because they are created in this state, as the axions. We refer to Neil Turok (this volume) for discussions on the alternative approach to the galaxy formation problem based on the cosmic string model . The plan of the paper is the following. In Sect.II, we will review cosmological models and gravitational instability theories. In Sect.III we will describe the mechanismis which may induce angular anisotropies in the CMB. In Sec.IV we will discuss the large scale peculiar velocity field, both from the observational and the theoretical point of view. Finally, in Sec.V, discussion and conclusions.

II. Cosmological models and perturbation theory

a) The background universe

The background cosmological model is described by the Friedmann equation. This gives the expansion rate, H(t), as a function of the cosmic scale factor a(t), normalized to unity at the present time t_0:

$$H^2 \equiv (\frac{\dot{a}}{a})^2 = \frac{8\pi G}{3}[\rho_{NR}(a) + \rho_{ER}(a)] - \frac{kc^2}{a^2} + \frac{1}{3}Lc^2 \tag{1}$$

where k is the curvature constant and L is a relic cosmological constant. The quantities ρ_{ER} and ρ_{NR} are the density of the relativistic (e.g., photons , massless neutrinos, etc.) and non-relativistic (e.g., baryons, non-relativistic dark matter, etc.) component. We have:

$$\rho_{NR}(a) = \frac{1.8 \cdot 10^{-29}}{a^3}\Omega_{NR}h^2 g\ cm^{-3} \tag{2a}$$

$$\rho_{ER}(a) = \frac{4.5 \cdot 10^{-34}}{a^4}[1 + N_\nu \frac{7}{8}(\frac{4}{11})^{4/3}](\frac{T_{\gamma 0}}{2.7})^4 g\ cm^{-3} \tag{2b}$$

Here N_ν is the number of massless neutrinos, and T_{γ_0} is the present CMB temperature. The redshift $1 + z_{EQ} \equiv a_{EQ}^{-1} = 42000\Omega_0 h^2/[1 + N_\nu \frac{7}{8}(\frac{4}{11})^{4/3}](2.7/T_{\gamma_0})^4$ defines the transition from the radiation dominated era ($\rho_{ER} > \rho_{NR}$) to the matter dominated era ($\rho_{NR} > \rho_{ER}$). The particle horizon at the so-called matter- radiation equality (a = a_{EQ}) defines a characteristic scale $L_{EQ} = 13/[\Omega_0 h^2] Mpc$. This scale is important, as we will see, in the framework of the gravitational instability theory. For redshifts smaller than $1 + z_c \equiv a_c{}^{-1} \sim \Omega_0^{-1}$ the curvature term dominates in the rhs of Eq.(1). The expansion becomes kinematic : the matter is so rarefied that it is unable to decelerate the cosmological expansion. Eq.(1) has the following solution:

$$t^{1/2}; \qquad a < a_{EQ} \tag{3a}$$

$$a(t) \propto t^{2/3}; \qquad a_{EQ} < a < a_c \tag{3b}$$

$$t\ ; \qquad a_c < a \tag{3c}$$

and, when the L term dominates,

$$a(t) \propto \exp\{\sqrt{L/3}\ ct\}\ . \tag{3d}$$

In the early phases of the cosmological expansion, Compton scattering of photons by free electrons ensures enough coupling between matter and radiation that they behave as a single relativistic fluid with a sound speed $c_s \approx c/\sqrt{3}$. At redshift $z_{rec} \sim 1000$, the temperature is too low for maintaining a high level of hydrogen ionization. Matter and radiation decouple and the photons mean free path increases by many orders of magnitude.

We will focus hereafter on a 4- component universe , in which radiation, baryons, cold dark matter, and massless neutrinos are present. We fixed $N_\nu = 3$, $\Omega_b = 0.03$, in agreement with the observations of luminous matter, and $\Omega_0 \leq 1$. We will also consider low density ($\Omega_0 < 1$) cosmological models which are flat because of a non zero relic cosmological constant $\Lambda \equiv Lc^2/(3H_0^2) = 1 - \Omega_0$.

b) Perturbation theory

The observed large scale structure is belivied to be the result of the gravitational amplification of initially small, statistical fluctuations of the density field. We have to assume that the universe was initially slightly inhomogeneous, since a universe exactly homogeneous is gravitationally "stable" and remains homogeneous.

The phenomenology of the gravitational instability presents some differences, depending upon the nature (collisional or collisionless) of the medium in which it develops. In a collisional self- gravitating medium (e.g., baryons + radiation before recombination), the fate of sound waves depends upon the concorrent rival action of the pressure, which forces the wave to propagate with the sound speed at a constant amplitude, and gravity which wants the compressional part of the wave to collapse, and then to growth in amplitude. In a collisionless self- gravitating medium (e.g., weakly interacting massive particles), gravity still wants the denser regions to collapse; this tendency is however counterbalanced by the fact that particles can move unimpeded from the denser regions. The perturbation profile is no more conserved: there is either a growth in amplitude or a damping of the density wave, at a rate depending upon the random velocity dispersion of the collisionless particles.

We can state the gravitational instability criterium by requiring that in a free fall time, $t_{ff} \sim (4\pi G\rho)^{-1/2}$, the denser regions (of scale λ) still remain denser than the background, i.e. $t_{ff} < t_c$. Here $t_c = \lambda/v$ is the characteristic time either for oscillation of the acustic wave, or for damping in a collisionless medium. The quantity v is either the sound velocity (for a collisional fluid) or the dispersion velocity (for a collisionless medium). The condition $t_{ff} \sim t_c$ defines the Jeans length, λ_J, or, equivalently, the Jeans wavenumber,$k_J \equiv 2\pi/\lambda_J = \sqrt{4\pi G\rho/v^2}$. Gravity amplifies perturbations of wavenumber $0 < k < k_J$. It is clear that the gravitational instability is so to speak maximized in a cold ($v \to 0$) medium: in fact, in this case, $k \to \infty$. However, the growth of fluctuations is very slow, also in this favourable case: the universe expands on a time scale comparable with the perturbation free fall time.

The growth of structure in the universe can be viewed as an initial data problem. Firstival, we have to specify the nature of the fluctuations. We will focus on adiabatic fluctuations: all the different cosmological components have initially the same $\delta n/n$ [n is a number density]. The perturbations are called adiabatic because the specific entropy per baryon, $\propto n_\gamma/n_b$, is spatially constant ($n_{\gamma,b}$ are the photon and baryon number density respectively) . This is also predicted in baryogenesys scenarios for the origin of the baryon asymmetry (Kolb and Turner, 1983). Secondly, we have to specify the relative amplitude of fluctuations on different length scales. This is done by considering a perturbed density field $\rho(\vec{x},t) = <\rho(\vec{x},t)> [1 + \Delta(\vec{x},t)]$, by Fourier analyzing the density fluctuations, and by specifying the spectral index n for a scale free density fluctuation spectrum: $|\delta_k|^2 = A\ k^n$. The amplitude A of the density fluctuations is fixed a posteriori as discussed in the next subsection. We will restrict ourselves to the so called Harrison-Zel'dovich power spectrum $|\delta_k|^2 = A\ k$, n=1. This spectrum is predicted in the inflationary scenario (Hawking, 1982; Starobinski, 1982; Guth and Pi, 1982; Bardeen, Steinhardt, and Turner, 1983) and has the advantage that no divegences are introduced in the metric perturbation neither on small nor on large scale. Finally, it will be also assumed that the density fluctuations are gaussian distributed:

$$p(\delta)d\delta = \frac{1}{\sqrt{2\pi}\Sigma(R,t)} e^{-\delta^2/2\Sigma^2(R,t)} d\delta \tag{4}$$

Here δ is the density contrast averaged, e.g., inside a sharp edged sphere of radius R, randomly placed in the universe and the quantity $\Sigma(R,t)$ is its rms value.

For the Harrison- Zel'dovich power spectrum, $\Sigma(R, t_H) \sim const$, $t_H = R/c$: perturbations of different sizes have the same amplitude at the horizon crossing. Let us consider two perturbation of size $R_1 < L_{EQ}$ and $R_2 > L_{EQ}$ rispectively in a CDM dominated universe. The first one crosses into the horizon when the universe was radiation dominated. In this regime, the growth of fluctuations is strongly inhibited (Mezsaros, 1976) until the matter radiation equality epoch. The second one crosses into the horizon when the universe is already matter dominated and it undergoes continuous growth. The period of interrupted growth of those perturbations which enter the horizon before matter domination flattens the initial density fluctuation spectrum below the characteristic scale L_{EQ}. Once the universe is matter dominated, density fluctuations on scales relevant for the large scale structure are amplified irrespectively of their scale, the shape of the spectrum is frozen, and the overall growth factor D(t) obeys the equation:

$$\frac{1}{a^2}\frac{d}{dt}a^2\frac{d}{dt}D = 4\pi G\rho D \tag{5}$$

The solution of this equation in the matter dominated era is:

$$D(t) \propto \begin{cases} t^{2/3}; & a \ll a_c \\ const; & a_c \ll a \end{cases} \tag{6}$$

Baryonic density fluctuations which enter the horizon before matter radiation decoupling undergo acoustic oscillations. In fact, in this period baryons and radiation behave as a single relativistic fluid. After recombination the baryons are able to answer to the gravitational field of the CDM perturbations and, at a redshift ~ 100, the amplitude of the baryonic and CDM density fluctuations are equal (see Fig.1). So, the density field at the onset of non-linearity is well described by the CDM density fluctuation spectrum. The power spectrum, as deduced from numerical calculation, is well fitted by the following analytical formula:

$$|\delta(k, t_0)|^2 = Ak(1 + \alpha k + \beta k^{1.5} + \gamma k^2)^{-2} \tag{7}$$

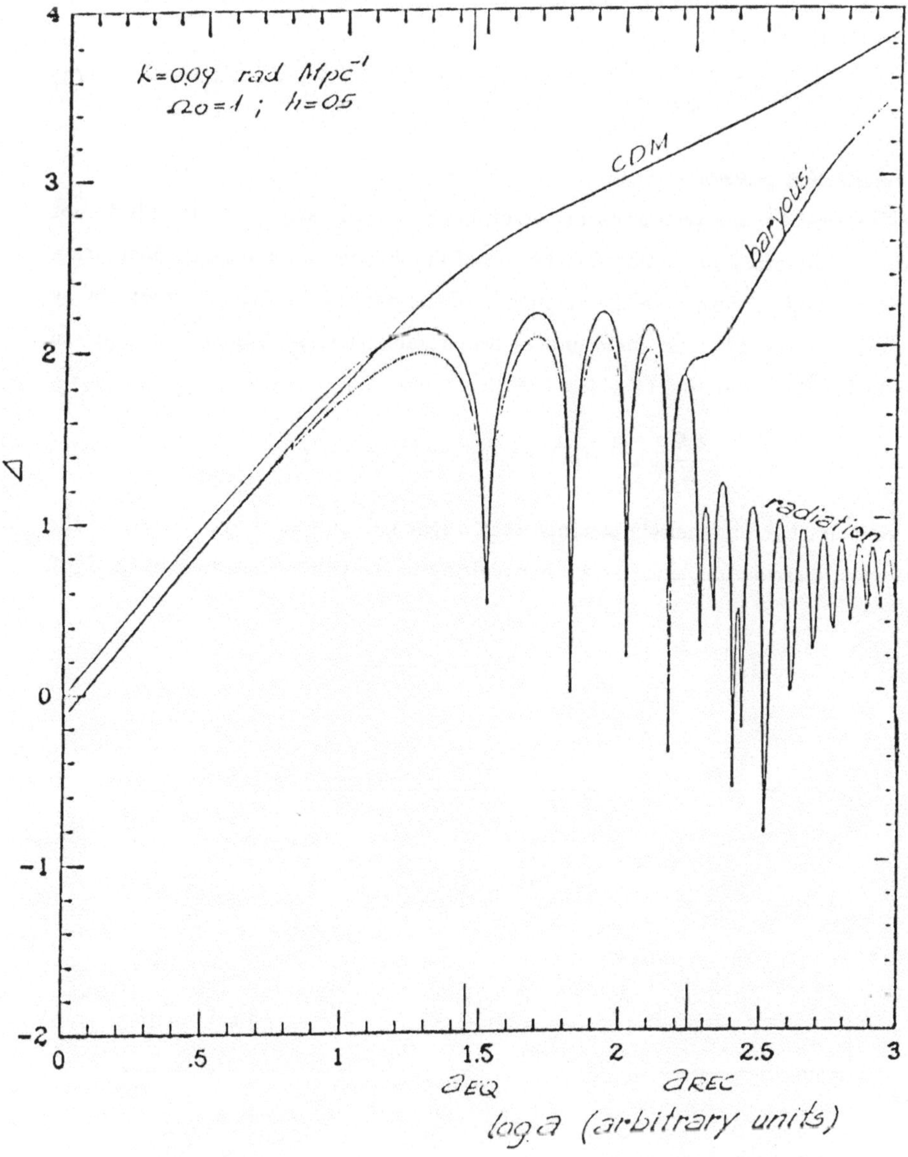
k=0.09 rad Mpc^{-1}
Ω_0=1 ; h=0.5
CDM
baryons
radiation
Δ
4
3
2
1
0
-1
-2
0
.5
1
1.5
2
2.5
3
a_{EQ}
a_{REC}
log a (arbitrary units)

with $\alpha = 1.7(\Omega h^2)^{-1}$ Mpc, $\beta = 9.0(\Omega h^2)^{-1.5}$ Mpc$^{1.5}$; and $\gamma = 1.0(\Omega\, h^2)^{-2}$ Mpc^2 (*e.g.*, Davis *et al.* 1985). In terms of the power spectrum (7), we also have (Peebles, 1980):

$$\Sigma^2(R, t_0) = \frac{1}{2\pi^2} \int_0^\infty k^2\, dk |\delta(k, t_0)|^2 [3\frac{sin\ kR - kR\ cos\ kR}{(kR)^3}]^2 \tag{8}$$

c) Normalization procedure

The theories for the generation of perturbations in the early universe still do not provide a satisfactory prediction for the initial amplitude of the density fluctuation field. It is usual to normalize the amplitude of the density fluctuation spectrum by fitting the variance $\delta N/N$ in the counts of bright galaxies inside a spherical region of radius R. It is then required (Peebles, 1982):

$$\frac{\delta N}{N}(8h^{-1}Mpc) = \Sigma(8h^{-1}Mpc, t_0) = 1 \tag{9}$$

Here it has been implicitely assumed that bright galaxies are a good tracer of the overall mass distribution, since we are normalizing to the galaxy distribution. In Fig.2

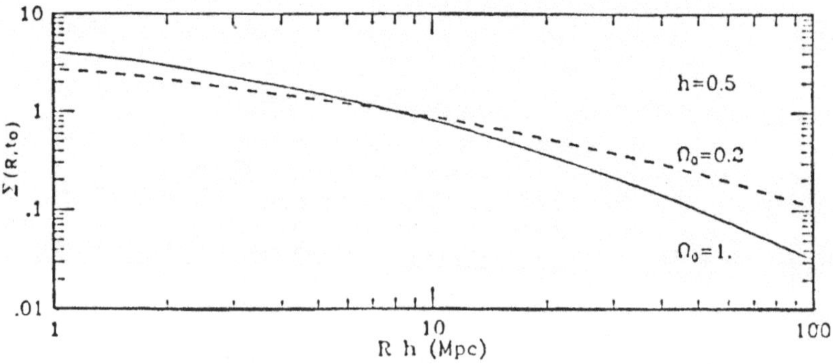

$\Sigma(R,t_0)$ is plotted for CDM dominated universes with $\Omega_0 = 1$ and 0.2 (h=0.5), using this normalization. Note that for a fixed normalization, lowering Ω_0 increases the amplitude of the large scale ($\gg L_{EQ}$) density fluctuations.

It is not unreasonable, however, that galaxies do not form everywhere, but only in the denser regions of the perturbed density field (so-called "biased" galaxy formation: Kaiser, 1986; Bardeen et al., 1986). From one side, we do not know the detailed process of galaxy formation and the influence of the environmental effects; on the other hand, if galaxy formation occurred in the maxima of the density field, it must be rembered that the tipical maximum has a density contrast higher than the rms value (Peacock and Heavens, 1985). In this case, $\delta N/N$ is not a good measure of the expected Σ on a given scale. If galaxies form only where $\Delta \sim \nu\Sigma$, then the normalization must be done by requiring $\delta N/N(R) \sim \nu\Sigma(R,t_0)$. This implies that the rms amplitude of the overall density field is in this case reduced by a factor $\sim \nu^{-1}$ (Kaiser, 1984). We parametrize the results in terms of the galaxy formation threshold ν.

III.The Cosmic Microwave Background

a) Small scale anisotropy (< 1°)

Since photons travelled unimpeded from the last scattering surface up to now, they carry out informations about the status of the universe at redshifts far greater than that of any individually detectable object. The perturbations in the density and velocity fields at the epoch of matter radiation decoupling should necessarely induce fluctuations in the brightness of the CMB radiation at the last scattering surface. For adiabatic fluctuations, since $\rho_b(t) \propto a^{-3}(t) \propto T_\gamma(t)^3$, we expect

$$\frac{\delta T}{T} = \frac{1}{3}\frac{\delta\rho}{\rho}|_{b,*} \tag{10a}$$

where $\delta\rho/\rho|_{b,*}$ is the baryonic density fluctuation at matter radiation decoupling. Observations of the small scale CMB anisotropy constrain the amplitude of fluctuations at the last scattering epoch.

For having an order of magnitude of the expected anisotropy, let us consider a fluctuation on scale l. A comoving length l on the last scattering surface subtends an angle

$$\theta(l) = \frac{\Omega_0 H_0}{2c} l \simeq 30''\Omega_0 h\ l \tag{11}$$

If $l \sim 10h^{-1}Mpc$, this fluctuation is just gone non linear today [cf. Sect. IIc] . The density contrast at decoupling (using Eq.6) was then $\sim 10^{-3}\Omega_0^{-1}$. This is the amplitude of the density fluctuation in the CDM component. The density contrast in the baryonic component is smaller roughly by a factor 10. In fact, CDM density fluctuations can grow from the matter radiation equality ($z_{EQ} \sim 10^4$), while baryonic density fluctuations can growth only after decoupling ($z_{rec} \sim 1000$) . So the actual amplitude of the baryonic component is $\sim 10^{-4}\Omega_0^{-1}$. This implies $\delta T/T \approx 3 \cdot 10^{-5}\Omega_0^{-1}$ on an angular scale $\sim 5'\Omega_0$. This is not so far as we shall see from detailed results. Note that observational upper limits on the CMB anisotropy are expected to impose a lower bound on Ω_0.

The actual calculation of the small scale anisotropy is complicated from the need of quantitatively evaluate the growth of matter density and radiation brightness fluctuations through the epoch of matter–radiation decoupling. Recombination is not an istantaneous process: direct recombination to the ground state is strongly inhibited due to the presence of photons with a mean free time for photoionization much shorter than the expansion time (Peebles,1968). This implies that the last scattering surface has a finite thickness $\Delta z/z \sim 10\%$, corresponding to a comoving length of $l_* \sim 10(\Omega_0 h^2)^{-1/2}$ Mpc. Any imprint on the CMB due to perturbations of wavenumber $\lambda << l_*$ is smeared out (Silk,1968;Weinberg, 1971). One might expect observations at angular scale $\theta(l_0)$ to constrain fluctuations primarly on angular scales l_0: this is incorrect since a wide region of the fluctuation spectrum, almost an order of magnitude in wavenumber , contributes to the anisotropy on a given angular scales.

There is also an isothermal contribution (only matter inhomogeneities are involved) due to the peculiar velocity $v_{b,*}$ of the baryonic matter at the last scattering surface:

$$\frac{\delta T}{T} = \frac{v_{b,*}}{c} \tag{10b}$$

The most stringent limit on fine-scale angular scales is provided by Uson and Wilkinson (1984) on an angular scale of 4'.5 (antenna beam size $\sigma = 1'.5$) $\Delta T/T(4'.5; 1'.5) < 3 \cdot 10^{-5}$. The observational technique in this experiment involved beam switching between three position in the sky. The outer fields of view are 4'.5 apart from the central beam. The quoted upper limit refer to the following quantity: $< |T_0 - (T_1 + T_2)/2|^2 >^{1/2}$ $/ < T >$, where the subscript 0 refer to the central beam. Such an experiment put an upper limit on the expected second derivative of the CMB temperature field.

b) Quadrupole and intermediate angular scale anisotropy (> 1°)

On these scales the main contribution to the CMB anisotropy is due to potential fluctuations, associated with density fluctuations, at the matter radiation decoupling. Photons which climb out of the potential wells on the last scattering surface are red-shifted and we expect: $\frac{\delta T}{T} \sim \frac{\delta\phi}{c^2}$. This effect (Sachs and Wolfe ,1967) determines a radiation fractional brightness at decoupling

$$i_{\gamma SW} = \frac{H_0^2 \Omega_0}{2c^2} \frac{\delta(k, t_0)}{k^2} F(\Omega_0^{-1} - 1) \tag{12}$$

The function $F(y) = 2y/[5 + 15y^{-1} + 15\sqrt{1+y}\, y^{-3/2} ln(\sqrt{1+y} - \sqrt{y})]$, with $y = \Omega^{-1} - 1$, takes into account the reduced growth of fluctuations in an open universe (Peebles 1980). Experiments on these (large) angular scales poses upper limit on the first derivative of the CMB temperature distribution, i.e., $< |T_1 - T_2|^2 >^{1/2} / < T >$, where the two fields of view (1 and 2) are an angle α far away. If the used antenna has a beam width σ, we found the following expression (Vittorio and Silk,1985):

$$\frac{\delta T}{T}|^2_{rms}(\alpha, \sigma) = F^2(\Omega_0^{-1} - 1) \frac{A}{\pi^2} \frac{\Omega_0^2 H_0^4}{8c^4} \sum_{m=1}^{m=\infty} \frac{(-1)^{(m-1)}}{m^2 (m-1)!} \left(\frac{\alpha}{2\sigma}\right)^{2m} . \tag{13}$$

Eq.(13) is valid for $\sigma \sim \alpha < 10^o$ and for an initial Harrison- Zeldovich power spectrum. The scale invariance of the spectrum makes the predicted $\delta T/T$ independent of α and σ, at constant α/σ. Also, for $\Omega_0 < 1$ we should properly take into account the global space time curvature. The curvature radius subtends an angle $\theta_c \sim 0.5\Omega_0/\sqrt{1 - \Omega_0}$ rad. For $\alpha < 10^o$, and for moderately low values of the density parameter, we can neglect this complication, since $\alpha < \theta_c$. We have to take into account only the deficit in the density fluctuation growth and the change in the comoving length- angular scale relation. Computation of temperature anisotropy in open cosmological models over angular scales larger than that subtended by the curvature radius, and in particular the quadrupole anisotropy (Peebles, 1981; Wilson, 1983; Traschen and Eardley, 1986;

Abott and Schaefer, 1986) show that the effects of curvature tend to reduce the expected $\delta T/T$ below the values computed for a flat universe. We will restrict ourselves here to calculate the quadrupole anisotropy in the flat case. For a scale-invariant spectral index, we expect a quadrupole anisotropy (Peebles,1982): $Q = 0.5\sqrt{A\pi/3}(H/c)^2$.

The detection of CMB anisotropies on large angular scales would be the strongest evidence for large scale inhomogeneities of the Universe. Unfortunately only upper limits are available: $Q < 7 \cdot 10^{-5}$ (Lubin at al.,1982; Fixsen at al.,1982), for the quadrupole anisotropy, and $\Delta T/T < 4 \cdot 10^{-5}$ on the angular scale of 6° ($\sigma = 2^\circ.2$) (Melchiorri et al., 1981).

IV. The Peculiar Velocity Field

a) The Virgocentric Infall and the Determination of Ω_0

The Galaxy belongs to the Local Group of galaxies. The Local Group is at a distance of $R \sim 13h^{-1}$ Mpc from the Virgo Cluster and radially falls towards this cluster with a peculiar velocity $|\vec{v}_{inf}| = 250 \pm 50$ km s^{-1} (see, e.g., Yahil , 1985). From the galaxy counts one infers, assuming that galaxies are a good tracer of the overall mass distribution, that the density fluctuation averaged over a sphere of $\sim 13h^{-1}$ Mpc radius centered on the Virgo cluster and with the Local Group on the border is $\Delta \sim 2$. From the linear theory, one obtains (see., e.g., Peebles, 1980): $v_p = (1/3)\Omega_0^{0.6}H_0 R\Delta$, which implies that, for a fixed density contrast, in an open universe smaller peculiar velocities are expected. In fact in an open universe the expansion at the present is essentially undecelerated (cf. Eq.3c), the growth of fluctuations strongly suppressed and, because of this, all the peculiar velocities decay away. Since v_p, H_0R, and Δ are observables, we are left only with one unknown, Ω_0. It is apparent that any overestime of the true density contrast implies an underestime of Ω_0. An overestime of the density contrast by factor ~ 3 (as it would be the case in a biased galaxy formation with $\nu \sim 3$) would result in an underestime of the density parameter Ω_0 by a factor ~ 5. Then, with this technique we would measure $\Omega_0 \sim 0.2$ also if the true value of the density parameter Ω_0 is unity.

b) Large scale drifts

The CMB dipole anisotropy ($\delta T/T \sim 3$ mK) provides the better evidence for our peculiar motion relative to the CMB and implies a Local Group velocity $|\vec{v}_{LG}| = 610 \pm 50$ km s^{-1}, in a direction which is 45° away from the Virgo Cluster (for a recent review see Lubin and Villela, 1986). By subtracting $\vec{v}_{inf}$ from $\vec{v}_{LG}$, one evaluates the velocity of the Virgo Cluster as a whole relative to the CMB: $|\vec{v}_{VC}| = 470 \pm 70$ km s^{-1}. From this simple analysis one concludes that the Local Group velocity relative to the CMB arises as the combined effect of the infall into Virgo and of the motion of the Virgo Cluster as a whole. We will compare $|\vec{v}_{VC}|$ with our model prediction.

Determining how large a volume one must consider so that the matter whithin this volume is at rest respect to the CMB is crucial for comparing different galaxy formation scenarios. An ingenous method of claryfing the locality of the CMB dipole anisotropy involves comparing $\vec{v}_{LG}$ with the measured velocity of the Local Group relative to a given sample of galaxies. Also, assume that the galaxies are selected in a volume big enough not to be strongly affected by local non-linearity. If these galaxies are unperturbed tracers of the Hubble flow, the peculiar velocity of the Local Group relative to the sample should be equal to the velocity of the Local Group relative to the CMB. In other words, the sample and the CMB are at rest and the observed dipole anisotropy is generated by density inhomogeneities inside the sample. If the two velocities are different, a coherent motion relative to the CMB of all the sample is implied. Also, since this bulk motion is determined by inhomogeneities on scales *larger* than the sample itself, we have in principle a direct measure of the amplitude of the large scale density fluctuations.

Different results have appeared in the literature. A decade ago, Rubin et al. (1976) reported evidence of anisotropy of the Hubble flow on surprisingly large scales. Most recently, Collins et al. (1986), have reobserved half of the original Rubin et al. sample of galaxies and confirmed the original effect: a sample of spiral galaxies at a mean redshift of 5100 km s^{-1} has a peculiar velocity relative to the CMB of 1000 ± 300 km s^{-1}. A lower but still large amplitude bulk motion (~ 700 km s^{-1}) for a sample of elliptical galaxies at a similar distance has been found by Burnstein et al. (1986). Aaronson et al. (1986) have done extensive study of 10 clusters of galaxies at comparable distance and conclude that their mean velocity with respect to the CMB is consistent with zero. However, the selected clusters are drawn from the Arecibo observing band, 30> δ >0 (here δ is the declination); as the Rubin and Ford vector is nearly orthogonal to this band this result does not necessarely cotradict the others.

In the linear regime, the local fluctuation in the matter velocity (relative to the Hubble flow) are directly related to the local fluctuations in the mass density (Peebles, 1980): $\mathbf{v_k} = \Omega_0^{0.6}\, H_0\, (\delta_k/k^2)\, \mathbf{k}$. For simplicity we will assume that the galaxy sample is spherically symmetric with number density $n(r) \propto exp(-r^2/R^2)$. Then, the expected rms bulk velocity relative to the CMB of the sample is (Clutton-Brock and Peebles, 1981; Kaiser, 1983):

$$v_{rms}^2(R) = \frac{1}{2\pi^2} \int_0^\infty k^2\, dk |\mathbf{v}_k|^2\, exp(-k^2R^2) \tag{14}$$

As we expected, only perturbations of wavenumber $k < r^{-1}$ contribute significantly to the integral in Eq.12. For $R \to \infty$ the integral reduces to a simple analytical expression: $v_{rms}(R) \to A^{\frac{1}{2}} H_0 \Omega_0^{0.6}/(2\pi R)$.

For inflation produced fluctuations the primordial density perturbations are gaussian distributed , ad so each component of the peculiar velocity is gaussian distributed. The probability of measuring a peculiar velocity with modulus v in the interval $v_1 \to v_2$ is given by

$$P = \sqrt{\frac{54}{\pi}} \int_{v_1}^{v_2} \left(\frac{v}{v_{rms}}\right)^2 e^{-\frac{3}{2}(\frac{v}{v_{rms}})^2} \frac{dv}{v_{rms}} \tag{15}$$

From this follows that there is a probability of 90% of measuring $1/3 < \frac{v}{v_{rms}} < 1.6$.

V. Model Predictions

A CDM dominated universe is very promising in reconciling the observed degree of clustering with the lack of detection of CMB anisotropy. In fact, since in this scenario the growth of fluctuations is most efficient, we can start with the minimum initial density fluctuation amplitude required for forming galaxies.

We can distinguish four effects that contribute to the dependence of the predicted small scale anisotropy on Ω_0 and h: i) as we already mentionned, decreasing Ω_0 and/or h increases L_{EQ}, flattens the fluctuation spectrum on small scales and increases the large scale amplitude for a fixed normalization (cf. Fig.2); ii) reducing h increases the normalization scale, further enhancing the large scale power, roughly by a factor h^{-1}; iii) Because of the comoving length- angular diameter relation (Eq.11), reducing Ω_0 and/or h implies that larger comoving scales contribute to the anisotropy at a fixed angular scale; iv) reducing Ω_0 cuts the growth period since recombination by a factor $\sim \Omega_0^{-1}$ and the amplitude of fluctuations must increase by the same factor. Effects i) and iii) roughly cancel each other. Then, we have for the small scale anisotropy:

$$\frac{\Delta T}{T}(4'.5; 1'.5) \sim \frac{6 \cdot 10^{-6}}{\Omega_0 h \nu} \tag{16a}$$

This is an analytical fit to numerical results (Bond and Efstathiou, 1984; Vittorio and Silk, 1984). Note that in Eq.(13) for the intermediate angular scale anisotropy, $F(\Omega_0^{-1} - 1)\Omega_0 \sim const$ for $\Omega_0 << 1$. Then, the only Ω_0 dependence is provided, in this case, by the overall mormalization constant A [effect (i) discussed above]. We find:

$$\frac{\Delta T}{T}(6^\circ; 2^\circ.2) \approx \frac{3.5 \cdot 10^{-6}}{\Omega_0 h \nu} \tag{16b}$$

Finally for an $\Omega_0 = 1$, CDM dominated universe , the expected quadrupole anisotropy is

$$Q = 3.5 \cdot 10^{-6} h^{-0.93} \nu^{-1} \tag{16c}$$

These values have to be compared with the observational upper limits given in SectIII.

The large scale peculiar velocity field may be evaluated in this scenario using Eq.(14). A good fit to the results of numerical integration gives (Vittorio and Turner, 1986):

$$v_{VC} = 322 \; km \; s^{-1} \Omega_{NR}^{+0.03} h^{-0.57} \nu^{-1} \tag{17a}$$

$$v_{50} = 83 \; km \; s^{-1} \Omega_{NR}^{-0.33} h^{-0.92} \nu^{-1} \tag{17b}$$

where v_{VC} and v_{50} are the expected bulk velocity of the Virgo Cluster and of a sample of galaxies at a mean redshift $\sim$ 5000 km s^{-1}. We note that the dependence on Ω_0 is such that in a low density universe the expected large scale drift has a higher amplitude than in the flat case. This is due to two competing effects. From one side the peculiar velocity decay away as $\Omega^{0.6}$. On the other hand for a CDM scenario the amplitude of density fluctuations on scale $R >> L_{EQ}$ goes as Ω^{-1} [effect (i) above], in such a way to explain the behaviour of Eq.17b. On very small scales, the amplitude of density fluctuations, for a fixed normalization, is independent of L_{EQ}, i.e., of Ω_0, which implies $v \propto \Omega^{0.6}$. At $10h^{-1}$ Mpc there is roughly the transition from one regime to the other which makes the prediction for v_{VC} basically independent of Ω_0. One can hope to increase the predicted peculiar velocity by lowering the value of the density parameter. With the present normalization, using Eq.15, one has $110 < v_{VC}(km\, s^{-1})\, h^{0.57}\nu < 520$ and $30 < v_{50}(km\, s^{-1})\, \Omega_{NR}^{0.33}\, h^{0.92}\nu < 135$, at the 90% confidence level.

It is clear that the predicted value for both CMB anisotropies and peculiar velocities are proportional to the amplitude A of the initial fluctuation spectrum, normalized as discussed in Sect.IIc. We could always use the alternative approach of using some other observables, e.g. the CMB dipole anisotropy or the peculiar velocity field for fixing A. In particular, since the large scale drifts are observed, we have at our disposal a direct measure of the initial amplitude of the density fluctutions. The difficulty in applying such a strategy is that we do not know if the measurements are significative of rms values or not.

Let us consider a flat model ($\Omega_0 = 1$), where galaxies are a fair tracer of the overall density field ($\nu = 1$). There is consistency with the CMB anisotropy upper limits, but there is not a match with the peculiar velocity field. We have infact that there is a probability of less than 5% of measuring a peculiar velocity > 250 km s^{-1}, for h=0.5, on 50 h^{-1} Mpc scale. Moreover, since cold dark matter is able to cluster on all scales, we should always measure $\Omega_0 = 1$: in this case the Ω_0 problem still remains a problem. Also, N-body simulations of the large scale structure in a CDM scenario succeded in matching the observations if the universe is low density (Davis et al., 1985). The same authors find a reasonable match with the observations in flat cosmological model if galaxies did not form everywhere, but only in the denser regions of the density field. As mentionned in Sect.IVa, this is also a way for resolving the Ω problem. In this case, however, since the effective amplitude of the density flucutations is reduced proportionally to the inverse of the galaxy formation threshold ν, the universe should be more uniforme. The observable consequence of this is that the peculiar velocities are drammatically too low. If $\Omega_0 = 1$, h=0.5, and $\nu = 3$, we have at the 90% confidence level $60 < v_{VC}(km\ s^{-1}) < 250$, and $15 < v_{50}(km\ s^{-1}) < 80$. This is a problem for this scenario, since the large scale drift and the Virgo Cluster bulk motion are both predicted to be substantially smaller than the observed values. The prediction of the CMB anisotropies are, obviously, in even better agreement with the observed upper limits.

Since $\Omega_{obs} \sim 0.2$, one can also consider a low density CDM universe, where galaxies form everywhere ($\nu = 1$). From the inflationary scenario point of view, that would imply that the early accelerated expansion phase has not been efficient enough in scratching out the curvature scale in our patch of the universe. From Eq.(16) and (17) it is apparent that lowering Ω_0 increases both the small scale anisotropy and the peculiar velocity field prediction. The present observational upper limit on the CMB small scale anisotropy results in a lower bound on the density parameter: $\Omega_0 h > 0.2$ (Bond and Efstathiou, 1984; Vittorio and Silk, 1984). With this bound, the probability of measuring a peculiar velocity $> 340 kms^{-1}$ on 50 h^{-1} Mpc scale is $< 5\%$, if h=0.5. If we relaxe the CMB small scale anisotropy constraint, we find that in order to have a probability $> 5\%$ of measuring bulk motions $\sim$ 600 km s^{-1} on 50 h^{-1}Mpc scale, we need $\Omega_0 \lesssim 0.1$ (h=0.5).

One other way of resolving the Ω_0 problem is to invoke a relic cosmological constant. In the inflationary scenario, the cosmological constant is not a free parameter, in the sense that for a given Ω_0, $\Lambda = 1 - \Omega_0$ in order to have a flat space-time. It is not clear from the particle physics perspective why such a small but finite value of the cosmological constant should survive. However, since one of the major unsolved problems in particle physics cosmology is to understand why Λ is at least 120 orders of magnitude below its natural scale, we should not to be unduly perturbed at requiring it not to be zero. A non zero cosmological constant does not change the shape of the spectrum, still shaped by Ω_0 and h. In fact, the distorsion of the primordial density flucutation spectrum happened before the matter radiation equivalence, when the cosmological constant was not important at all. Introducing a positive cosmological constant $\Lambda = 1 - \Omega_0$ has two main effects: i) the growth of the density fluctuations is more efficient than in an open universe with the same Ω_0 ; ii) the angular diameter comoving length relation now reads (Vittorio and Silk, 1985):

$$\theta(l) \sim \frac{\Omega_0^{0.4} H_0}{2c} l \approx 30'' \Omega_0^{0.4} h\, l \tag{18}$$

Both these effects conspire in reducing the expected small scale anisotropy and we found (Vittorio and Silk, 1985):

$$\frac{\Delta T}{T}(4'.5; 1'.5) \sim \frac{1.5 \cdot 10^{-6}}{\Omega_0 h \nu} \tag{19a}$$

The peculiar velocity predictions are the same for universes with the same Ω_0, independently of Λ beeing zero or equal to $1 - \Omega_0$. In fact, from one side the amplitude of the density fluctuations is exactly the same in the two cases , because of the normalization used. On the other hand, the peculiar velocities decay away with the same factor ($\propto \Omega_0^{0.6}$), also in the $\Lambda = 1 - \Omega_0$ case (Vittorio and Turner, 1986). So for the peculiar velocity still remains valid the discussion we had for the open universe. However, in this case, $\Omega_0 \sim 0.1$ is compatible with the Uson and Wilkinson upper limit.

For given Ω_0 and h a universe with a cosmological constant $\Lambda = 1 - \Omega_0$ is older than the corresponding one with $\Lambda = 0$. In order to decide if low density models that may be either open or flat because of a relic cosmological constant are consistent with the estimates of the globular cluster ages one has to wait for an independent determination of the Hubble constant.

If the large scale drifts are confirmed, we have to conclude that most of the CDM dominated universes discussed here are not sufficiently inhomogeneous on large scales. Possible ways out are: i) relaxing the assumption that density fluctuations are initially gaussian distributed; ii) assuming an initial density fluctuation spectrum flatter than the Harrison-Zel'dovich one [i.e., $0 < n < 1$; we already know that a white noise initial spectrum (n=0) results in excessive temperature anisotropy on large scale (Vittorio and Silk, 1984)];iii) decoupling the formation of galaxies from the formation of larger structures, by relaxing the assumption of a single power law initial density fluctuation spectrum. These are possibilities which deserve in any case further investigations.

References

Aaronson, M., Bothun, G., Mould, J., Huchra, J., Schommer, R.A., Cornell, M.E. 1986, *Ap.J.*, **302**, 536.

Abott, L.F., and Schaefer, R.K. 1986, *Ap. J.*, in press.

Bardeen, J., Bond, J., Kaiser, N., and Szalay, A. 1986, *Ap.J.*,**304**, 15

Bardeen, J., Steinhardt, P.J., and Turner, M.S. 1983, *Phys.Rev.D* ,**28**, 679.

Bond, J. and Efstathiou, G. 1984, *Ap.J.*, **285**, L44.

Burstein, D., Davies, R.L., Dressler, A., Faber, S.M., Lynden-Bell, D., Terlevich, R., Wegner, G. 1986, in *Galaxy Distances and Deviations from the Universal Expansion*,Eds. Madore, B.F., and Tully, R.B., NATO Brussels, 1986

Clutton-Brock, M., and Peebles, P.J.E. 1981, *Astron.J.*, **86**,

Collins, A., Joseph, R.D., and Robertson 1986, N.A., *Nature*,**320**,506

Davis, M., Efstathiou, G., Frenk, C., and White, S. D. M. 1985, *Ap.J.*, **292**, 371.

Dicus, D., Kolb, E.W., and Teplitz, V. 1977, *Phys.Rev.Lett.*, **39**, 168.

Efstathiou, G. and Bond, J.R. 1986, *Mon.Not.r.Astron.Soc.* **218**, 103.

Faber, S.M., and Gallagher, J., 1979, *Ann. Rev. Astron. Astrophys.*,**17** , 135

Gelmini, G., Schramm, D.N., and Valle, J. 1984, *Phys. Letters B*, **146**, 311.

Guth, A. and Pi, S.Y. 1982, *Phys.Rev.Lett.*, 49, 1110.

Hawking, S.W. 1982, *Phys. Letters B*, **115**, 295.

Kaiser, N. 1983, *Ap.J. (Letters)*, **273**, L17.

Kaiser, N. 1984, *Ap.J. (Letters)*, **284**, L9

Kaiser, N. 1986, in *Inner Space/Outer Space*, eds. E.W. Kolb et al. (Chicago: University of Chicago Press).

Kolb, E.W., Turner, M.S. 1986, *Ann.Rev.Nuc.Sci.*,**33** , 645

Lubin, P., Epstein, G.L., Villela, T.,Smooth , G.F., *Phys Rev.Lett.*, **50**, 616

Lubin, P., and Villela, T., in *Galaxy Distances and Deviations from the Universal Expansion*,Eds. Madore, B.F., and Tully, R.B., NATO Brussels, 1986

Melchiorri, F., Melchiorri, B., Cecarelli, C., and Pietranera, L., 1981, *Ap.J.*, **250**, L1

Mezsaros, P. 1976, *Astron.Astrophys*, **37**, 225

Olive, K.A., Seckel, D., and Vishniac, E.T. 1985, *Ap.J.*, **292**, 1.

Peacock, J.A., and Heavens 1985, *Mon.Not.R.Astr.Soc.*, **217**, 805.

Peebles, P.J.E. 1968, *Ap.J.*, **153**,1

Peebles, P.J.E. 1980, *Large-Scale Structure of the Universe* (Pri nceton: Princeton University Press).

Peebles, P. J. E. 1981, *Ap.J.*, **248**, 885

Peebles, P.J.E. 1982, *Ap.J.*, **263**, L1.

Peebles, P.J.E. 1984, *Ap.J.*, **284**, 439.

Rubin, V., Ford, W.K., Thonnard, N., and Roberts, M.S. 1976, *A.J.* , **81**, 687.

Sachs, R.W., and Wolfe,A.M. 1967, *Ap.J.*, **147**, 73.

Silk, 1968, *Ap.J.*, **151**, 459

Starobinskii, A.A. 1982, *Phys. Letters B*, **117**, 175.

Traschen, J. and Eardley,D. 1986, preprint.

Turner, M.S. 1986b, in *Dark Matter in the Universe*, eds. J. Kormendy and J. Knapp (Dordrecht: Reidel).

Turner, M.S., Steigman, G., and Krauss, L.L. 1984, *Phys.Rev.Lett.*, **52**, 2090.

Uson, J. and Wilkinson, D.T. 1984, *Ap.J.*, **277**, L1.

Vilenkin, A. 1984, *Phys.Rev.Lett.*, **53**, 1016.

Vittorio, N. and Silk, J. 1984, *Ap.J. (Letters)*, **285**, L39.

Vittorio, N. and Silk, J. 1985a, *Ap.J. (Letters)*, **293**, L1.

Vittorio, N. and Silk, J. 1985b, *Ap.J. (Letters)*, **297**, L1.

Vittorio, N. and Turner, M.S. 1986, *Ap.J.*, in press

Yahil, A. 1985, in *The Virgo Cluster of Galaxies*, eds. O.G. Richter and B. Bineggeli (Munich: ESO).

Yang, J., Turner, M.S., Steigman, G., Schramm, D.N., and Olive, K.A. 1984, *Ap.J.*, **281**, 493.

Weinberg, 1971, *Ap.J.*, **168**, 75

Wilson, M. 1983, *Ap.J.*, **272**, 2

Phenomenology And Cosmology
With Superstring Motivated Models

Qaisar Shafi *†
Bartol Research Foundation
University of Delaware
Newark, DE 19716

Abstract

We discuss how realistic low energy physics can arise from the $E_8 \times E_8$ superstring. A key role is played by the discrete symmetries which typically arise after compactification. Cosmological problems are avoided because the discrete symmetry we employ is shown to effectively be embedded in a Peccei-Quinn symmetry. The models we present seem to satisfy the known phenomenological and cosmological constraints. In particular they possess a harmless axion. The existence of topologically stable superconducting vortices (strings), surviving until today, is predicted. Other topics that are discussed include proton decay, the baryon asymmetry in the Universe and n - n̄ oscillations.

* Based on work done in collaboration with G. Lazarides and C. Panagiotakopoulos.

† Supported in part by the Department of Energy under Contract Grant Number DE-AC02-78ER05007.

P. Galeotti and D. N. Schramm (eds.), Gauge Theory and the Early Universe, 119–148.

Superstring Model Building

The discovery of anomaly free superstring theories[1] and the subsequent development of the $E_8 \times E_8$ heterotic theory[2] have fueled expectations that we may have at hand a theory which provides a consistent unification of gravity with the other three forces. Compactification of the $E_8 \times E_8$ theory on Calabi-Yau (C-Y) spaces can lead to chiral fermion families, a must for any realistic theory, and even delivers an unbroken N=1 supersymmetry[3]. However, attempts to obtain a realistic low energy phenomenology and a consistent cosmology must face up to some formidable problems. For instance, there are potential dangers from sources such as rapid proton decay, $\sin^2\theta_W$, flavor changing neutral currents, neutrino masses, domain walls, visible axions etc. Moreover, the model independent axion in superstring theories[4] does not satisfy the standard astrophysical and cosmological constraints.

In these lectures I wish to present a unified approach to these and some other related questions. The aim is to come up with models which, in addition to satisfying all of the above constraints, make some predictions. One of the most striking ones, in the class of models considered, is the presence of topologically stable vortices which are superconducting and possibly detectable in our galaxy through their astrophysical effects.

Before embarking on the actual model building and the ensuing consequences, we need to clarify a number of points. First, let us state some of our main assumptions:

i) The compactification of the ten dimensional $E_8 \times E_8$ superstring theory on a suitable C-Y space gives rise to three generations of chiral massless fermions (contained in the

$\underline{27}$'s of E_6). In addition, there will be a certain number of chiral matter multiplets arising from incomplete $\underline{27}$'s and $\overline{27}$'s of E_6.[5,6]

ii) The C-Y space is non-simply connected thus allowing for non-trivial Wilson loops[3]. The effective gauge group below the compactification scale M_C is a subgroup G of E_6 and, depending on the details of the symmetry breaking, possesses rank five or six.[5] Clearly, G must contain $SU(3)_C \times SU(2)_L \times U(1)_Y$ (3-2-1) as a subgroup.

iii) There exists some mechanism of spontaneous supersymmetry breaking, presumably triggered by the hidden E_8 sector[7], which gives rise to the weak scale M_W. All scalars are assumed to get $(\text{mass})^2$ terms of order $M_W{}^2$ through gravitational interactions.

Next, in order to obtain a phenomenologically acceptable theory, it seems necessary, even in the case of a three family model, that many of the extra fields in the $\underline{27}$ acquire an intermediate mass M_I, which is much larger than M_W. Some reasons for such an intermediate scale are:

a) The requirement of perturbative unification.

b) The possibility of creating some mixing between the Higgs doublets $\overline{H}$ and H which give masses to the up quarks and the down quarks (and the charged leptons). The mixing is important in order to obtain tree level electroweak breaking without the need for large Yukawa couplings that certainly are absent for H.

c) The possibility of having acceptable proton lifetime without imposing exact baryon number conservation, which helps in the creation of baryon asymmetry in the universe.

The most plausible intermediate scale in superstring models appears to be $\sim\sqrt{M_W M_C} \equiv M_I \sim 10^9 - 10^{10}$GeV for $M_C \sim 10^{17}$GeV. Its appearance requires the existence of at least one light 3-2-1 singlet from a 27, accompanied by a corresponding singlet from a $\overline{27}$, plus the existence of D- and F- flat directions in the potential of this singlet.[8,9]

Having convinced ourselves (!) that superstring models ought to possess an intermediate scale M_I at which a part of the gauge group G is broken, we will argue that at least one independent global symmetry also be broken at M_I in order that the models be not phenomenologically and cosmologically unacceptable. The reasons for this include:

1) The absence of large flavor changing neutral currents which are generally present in models with several light Higgs doublets. In superstring models there are at least six such doublets. On the basis of their quantum numbers with respect to the local symmetries there is no reason why some of them are heavy and the others light, or why some of them couple to quarks and leptons and the others do not. If, however, there is a suitable global symmetry which differentiates between them the problem can be overcome.

2) The absence of large masses for the known neutrinos. Unless an extra global symmetry is present neutrino masses generally turn out to be much too large[6].

3) Acceptable proton lifetime. The presence of a global symmetry eliminates some of the undesirable couplings which otherwise would

lead to a much too rapid proton decay[5,6,8,9].

An obvious candidate for the global symmetry in a superstring model is the discrete symmetry that typically arises after compactification of a superstring theory on a C-Y space[5]. Such a symmetry may help in avoiding large neutrino masses and a rapid proton decay. It should be remembered, however, that discrete symmetries create severe domain wall problems if they are spontaneously broken.

The domain wall problem would be neatly resolved if the discrete symmetry could effectively be embedded in some continuous (global) symmetry. Before proceeding to a discussion of what we consider an attractive candidate for the continuous symmetry, let us remark that the continuous symmetry could not reasonably be expected to be a symmetry of the complete four dimensional theory, including all the non-renormalizable interactions. It can, at best, be expected to be the symmetry of all those field operators that could affect the relevant physics between M_C and the QCD scale.

In order to motivate the nature of the continuous global symmetry, let us recall the model independent axion in superstring models[4]. The decay constant for such an axion is close to M_C, much larger than the upper bound of 10^{12}GeV allowed by standard cosmological arguments[10]. The presence of an additional global U(1) Peccei-Quinn (PQ) symmetry [11] (in the sense mentioned above) broken at an intermediate scale resolves the axion problem in a neat way. The true axion is now an appropriate linear combination of the two fields, the model independent axion and the PQ axion, and couples to $F\tilde{F}$ with a decay constant characterized by $M_I \sim 10^{10}$GeV [12]. It is only natural to identify the desired continuous global symmetry with the $U(1)_{PQ}$ symmetry.

To summarize, our task now is to construct an acceptable axion model based on the field content suggested by the $E_8 \times E_8$ superstring. We also should identify the relevant discrete symmetry which is effectively embedded in $U(1)_{PQ}$. The problem of constructing C-Y spaces which lead to such a discrete symmetry and, in addition, have all the other desirable features is, of course, a formidable one and is not addressed in this paper.

Several attempts to construct an acceptable superstring axion model based on rank six subgroups of E_6 were unsuccessful. This is largely due to the small number of couplings allowed by the local symmetry. The PQ symmetry is broken completely only at the weak scale, and one ends up with an unwanted visible axion. It is also difficult to arrange for an acceptable proton lifetime together with some baryon number violation in the model, avoid flavor changing neutral currents, and arrange for the decay of all the heavy particles. We therefore turn to a rank five subgroup of E_6.

In the following we present a model based on the subgroup $G \equiv SU(3)_C \times SU(2)_L \times U(1)_L \times U(1)_R$ [5] which appears to have all the desirable features. The standard hypercharge generator is proportional to the direct sum of the generators of the U(1)'s.

In order for this model to develop an intermediate scale the C-Y space must have other (1,1) harmonic forms besides the Kahler form[8]. We also need two 3-2-1 singlets with different global charges in order to break $U(1)_{PQ}$ at M_I. These singlets come from $\underline{27}$'s and are accompanied by their mirrors from $\underline{\overline{27}}$'s. We denote them S_1 and S_2. Their mirrors are denoted $\tilde{S}_1$, $\tilde{S}_2$. The fields in $\underline{27}$'s with the same local quantum numbers as S_1 and S_2, but which do not have light mirrors are denoted by N.

It turns out that we need additional fields which come from incomplete $\underline{27}$'s (and $\underline{\overline{27}}$'s). For reasons that will become clear later we introduce four doublets, H'', H''', $\bar{H}''$, $\bar{H}'''$ and their mirrors $\tilde{H}''$, $\tilde{H}'''$, $\tilde{\bar{H}}''$, $\tilde{\bar{H}}'''$ as well as two color triplets g' and their mirrors $\tilde{g}'$. In table I are given the quantum numbers of the various fields with respect to G and a global Z_9 symmetry which we impose on the model. Also listed is the multiplicity N_m of each field. F denotes the number of families (three).

The superpotential of the model contains (identifying fields with the same quantum numbers) the following dimension three (d=3) terms: $\bar{H}QU_c$, HLE_c, QD_cH, $H\bar{H}N$, gg_cS_1, $g_cg_cU_c$, gD_cN, $H'H'E_c$, $\bar{H}'LS_2$, $H'\bar{H}'S_1$, $\tilde{H}''\tilde{\bar{H}}''\tilde{S}_1$, $\tilde{H}'''\tilde{\bar{H}}'''\tilde{S}_2$. The d=4 terms are: $\tilde{H}''\tilde{H}''H'H'$, $\tilde{H}''\tilde{H}''LH$, $\tilde{H}''\tilde{\bar{H}}''gg_c$, $\tilde{H}''\tilde{\bar{H}}''H'\bar{H}'$, $\tilde{H}''\tilde{\bar{H}}'''L\bar{H}'$, $\tilde{\bar{H}}''\tilde{\bar{H}}''\bar{H}'\bar{H}'$, $\tilde{H}''\tilde{g}'gH$, $\tilde{\bar{H}}'''\tilde{g}'g\bar{H}'$, $\tilde{H}''\tilde{S}_1H'S_1$, $\tilde{H}''\tilde{S}_1LS_2$, $\tilde{H}''\tilde{S}_2HS_1$, $\tilde{H}''\tilde{S}_2H'S_2$, $\tilde{\bar{H}}''\tilde{S}_1H'E_c$, $\tilde{\bar{H}}''\tilde{S}_1\bar{H}'S_1$, $\tilde{\bar{H}}'''\tilde{S}_2H'E_c$, $\tilde{\bar{H}}'''\tilde{S}_2\bar{H}'S_1$, $\tilde{\bar{H}}''\tilde{S}_2HE_c$, $\tilde{\bar{H}}''\tilde{S}_2\bar{H}'S_2$, $\tilde{\bar{H}}'''\tilde{S}_1QD_c$, $\tilde{\bar{H}}'''\tilde{S}_1LE_c$, $\tilde{g}'\tilde{S}_1gS_2$, $\tilde{S}_1\tilde{S}_1S_1S_1$, $\tilde{S}_1\tilde{S}_2S_1S_2$, $\tilde{S}_2\tilde{S}_2S_2S_2$, $\tilde{S}_1\tilde{S}_2NN$.

Although we have imposed only a discrete Z_9 global symmetry, the important thing is to know the actual symmetries of the theory. It turns out that, if we restrict ourselves to terms in the superpotential with $d \leq 4$, the theory possesses a larger global symmetry than Z_9, namely a $U(1)_{PQ}$ in which the Z_9 is embedded. The maximal symmetry of the model is $G \times U(1)_{PQ}$. The $U(1)_{PQ}$ charges of the various fields are the same as the Z_9 ones given in table I. The Z_9 symmetry is also sufficient to guarantee the absence of d>4 terms (e.g. $S_1 S_1 S_1 \tilde{S}_2 \tilde{S}_2 \tilde{S}_2$) which violate the $U(1)_{PQ}$ symmetry and whose absence is important for the PQ mechanism.

At this stage we can also justify the presence of some 3-2-1 non-singlets like $\tilde{H}''$ and $\tilde{\overline{H}}''$ belonging to incomplete $\overline{\underline{27}}$'s. They are needed to create terms like $\tilde{H}''\tilde{\overline{H}}''\tilde{S}_1$ and thereby break the Z_2 symmetry under which only fields belonging to a $\overline{\underline{27}}$ transform non trivially. The $\tilde{H}''$, $\tilde{\overline{H}}''$, $\tilde{H}'''$ and $\tilde{\overline{H}}'''$ are also important for the development of vacuum expectation values (vev's) by $\tilde{S}_1$, $\tilde{S}_2$.

Through the couplings $gg_c S_1$, $H'\overline{H}'S_1$, $\overline{H}'LS_2$, $\tilde{H}''\tilde{\overline{H}}''\tilde{S}_1$, $\tilde{H}'''\tilde{\overline{H}}'''\tilde{S}_2$, the 3-2-1 singlets S_1, S_2, $\tilde{S}_1$, $\tilde{S}_2$ can develop a negative $(\text{mass})^2$ and acquire vev's $\langle S_1\rangle = \langle\tilde{S}_1\rangle$, $\langle S_2\rangle = \langle\tilde{S}_2\rangle$, all of the order of M_I. These vev's break $U(1)_{L-R} \times U(1)_{PQ}$ down to a Z_{40} generated by $(e^{-\frac{3\pi i}{40}}, e^{\frac{2\pi i}{10}})$. The couplings $NN\tilde{S}_1\tilde{S}_2$ and $H\overline{H}N$ create mixing between H and $\overline{H}$ $(M_I{}^2 M_C{}^{-1} H\overline{H}N^*)$ necessary for tree level electroweak breaking. For suitable values of the parameters of the theory, H, $\overline{H}$ and N acquire vev's of order M_W.[9] These vev's break the Z_{40} down to a Z_{20} generated by $(e^{-\frac{3\pi i}{20}}, e^{\frac{2\pi i}{5}})$. No other field is supposed to develop a vev.

We now justify the introduction of the extra (light) color

triplets g', $\tilde{g}'$. Let us assume for the moment that they are absent. The $U(1)_{PQ}$ is known to be explicitly broken at the QCD scale, due to its color anomaly, down to a Z_N discrete subgroup[13] ($N=\Sigma Q_{PQ}$ where the sum is over all fermion color triplets (and antitriplets) and Q_{PQ} is their $U(1)_{PQ}$ charge). In our model (without g', $\tilde{g}'$), $N=-15$ and the global symmetry is Z_{15}, which is broken by $\langle S_1\rangle$, $\langle S_2\rangle$. We will see later that a Z_5 subgroup of $U(1)_{PQ}$ can be embedded in G. Therefore the breaking of the discrete Z_{15} leads to the formation of Z_3 domain walls (associated with the breaking of Z_{15} to Z_5).

The problem of PQ domain walls is not new and is sometimes solved by inflating away the PQ strings formed at the phase transition at which $U(1)_{PQ}$ breaks spontaneously. This method does not seem to be applicable in our case. Due to the fact that the fields (S_1,S_2) responsible for the intermediate scale have masses $\sim M_W$, there is no phase transition until the temperature falls to $T_C \sim M_W$. The $U(1)_{PQ}$ phase transition is therefore unlikely to be inflationary. For the same reason one is not allowed to break any discrete symmetry in this class of models. Another way of solving the PQ domain wall problem[14] is to add new color triplets and modify the PQ color anomaly such that instanton effects break $U(1)_{PQ}$ down to Z_5 which is then embedded in G. This explains the role of g', $\tilde{g}'$.

To finish with the PQ domain walls we show that the Z_5 subgroup of $U(1)_{PQ}$ is embedded in G. Consider the element $(e^{2\pi i N/3}, e^{2\pi i M/2}, e^{i\alpha}, e^{i\beta}, e^{i\theta})$ of $G \times U(1)_{PQ}$. One can verify that its action on all the fields is equal to the identity if $\theta=\frac{2\pi}{5}k \pmod{2\pi}$, $\alpha=\frac{2\pi}{6}\ell \pmod{2\pi}$, $\beta=\frac{2\pi}{6}\ell-\frac{2\pi}{5}k+\pi r \pmod{2\pi}$, $M=\ell \pmod 2$, $N=\ell \pmod 3$, $(k,\ell,r$

integers).

The model has baryon and lepton number non-conservation due to the presence of the couplings $g_c g_c U_c$ and $H'H'E_c$. For reasonable values of the couplings ($\sim 10^{-1}$-10^{-2}) the proton lifetime is acceptable and possibly experimentally accessible. The dominant modes are of the type $\ell^+ \ell^- M^+ \nu\ (\bar{\nu})$ (M denotes a meson and the rest are leptons) (See fig. 1).

The $U(1)_{PQ}$ also ensures the absence of unacceptably large neutrino masses. Their actual values depend on the unknown parameters of the underlying theory.

The (one loop) renormalization group equations are consistent with $M_C \simeq 10^{17}$GeV, $M_I \simeq 10^{10}$GeV, $\alpha_s(M_W) \simeq 0.11$ and $\sin^2\theta_W(M_W) \simeq 0.22$. For F=3, $\alpha_G \simeq 0.14$ and the calculation therefore is reliable.

The spontaneous breaking of $U(1)_{L-R}$ symmetry produces topologically stable vortices (strings). They have mass per unit length of order $M_I^2 \sim (10^{10}\text{GeV})^2$ and thickness $\sim M_W^{-1}$. They are superconducting by virtue of the fact that there are charged fermions (including the known quarks and leptons and g, g_c quarks) which couple to Higgs fields (H, $\bar{H}$, S_1) whose phases change by 2π around the vortex. The fermions are trapped in zero modes along the vortex which leads to superconductivity.[15]

Because of the unusual nature of the $U(1)_{L-R}$ phase transition (i.e. it does not occur till temperatures of order M_W are reached) the strings do not experience an inflationary phase, if there was one, and so should be present in the universe. It has been suggested that they might be observable as synchrotron sources.[15,16]

Baryogenesis In Superstring Models

In this section we consider two important and often related cosmological problems in the context of superstring models. These are i) the origin of the baryon asymmetry in the universe, and ii) the gravitino problem usually encountered in models in which the gravitino mass is on the order of the electroweak scale. We wish to show that in a class of superstring models, examples of which were presented above, these two problems could be resolved in a novel way.

The merits of superstring models which possess an intermediate scale $M_I \sim 10^9$ GeV have been extolled above. It turns out that not only the existence but also the mechanism whereby M_I arises play an essential role in resolving both the baryogenesis issue and the gravitino problem in superstring models.

In order to see how baryogenesis can arise in superstring models, let us recall that the compactification of the $E_8 \times E_8$ heterotic theory on suitable Calabi-Yau spaces predicts, besides the 'known' quark, lepton and higgs superfields, the existence of additional fields which include color triplet, SU(2) singlet, superfields g_i. For simplicity we restrict our attention to the boson component of g (also denoted by the same symbol with the family index i not explicitly displayed). We would like to show that the out of equilibrium decay of massive g bosons at temperatures close to the electroweak scale can produce the desired baryon asymmetry. This presumes, of course, that the other two requirements for successfully generating the baryon asymmetry, the existence of baryon number violating couplings of g and the presence of CP violation in these couplings, are fulfilled. We will display models in which all three conditions can be met.

We first present a simplified account of the model independent features of the baryogenesis scenario. The g boson has a positive mass squared term $\sim M_s^2 g^* g$ ($M_s \simeq 1$ TeV is of the order of the supersymmetry breaking scale in the known sector) which presumably arises from supersymmetry breaking in the hidden sector. The important point is that it also acquires a mass through its coupling $\sim g^* g\, \phi^* \phi$ to a $SU(3)_C \times SU(2)_L \times U(1)_Y$ singlet scalar field ϕ, whose zero temperature effective potential is of the form(8)

$$V_0(\phi) = -\frac{M_s^2}{2}\phi^*\phi + \frac{\lambda}{6M_c^2}(\phi^*\phi)^3 \qquad (1)$$

Here $M_c \sim 10^{17}$ GeV denotes the compactification scale. The zero temperature expectation value of ϕ is given by $|\langle\phi\rangle| \equiv M_I \sim (\lambda^{-1} M_s M_c)^{1/2}$. Hence, the g supermultiplet, as well as the gauge supermultiplet to which ϕ couples, acquire masses of order M_I.

For non-zero temperatures, we will add to the zero temperature effective potential $V_0(\phi)$ the following one loop contribution(17)

$$V_T(\phi) = \frac{T^4}{2\pi^2}\sum_i (-1)^F \int_c^\infty dx\, x^2 \ln\left[1 - (-1)^F \exp\left\{-(x^2 + M_i^2(\phi)/T^2)^{1/2}\right\}\right] \qquad (2)$$

The sum in (2) is over all helicity states, $(-1)^F$ is ± 1 for bosonic and fermionic states respectively, and $M_i(\phi)$ is the field dependent mass of the ith state.

The salient features of the effective potential $V(\phi) = V_0(\phi) + V_T(\phi)$ are as follows. For $\phi \ll T$, there is a temperature dependent mass term $\sigma T^2 \phi\phi^*$. Hence, $V(\phi)$ has a minimum at $\phi = 0$ for $T > T_c \equiv \sigma^{-1/2} M_s$. For $\phi \gtrsim T$, the temperature dependent mass term is exponentially suppressed, and $V(\phi)$ develops a second minimum at $\phi \simeq M_I$ for $T \lesssim M_I$.

The minimum at $\phi = 0$ is the absolute minimum for $M_I \gtrsim T \gtrsim \mu \equiv (M_s M_I)^{1/2} \sim 10^6$ GeV. This follows because i) in this temperature range, the radiation energy density ($\propto T^4$) dominates over the zero temperature vacuum energy density $\sim \mu^4$ in the $\phi = 0$ phase, and ii) the number of massless degrees of freedom in the $\phi = 0$ phase exceeds that in the $\phi \simeq M_I$ phase.

For $T \lesssim \mu$, the $T = 0$ vacuum energy density dominates and the $\phi \simeq M_I$ phase becomes the absolute minimum of $V(\phi)$. This minimum is separated from the local minimum at $\phi = 0$ by a barrier of height $\sim T^4$ and width $\gtrsim T$. For $T_c < T < \mu$, the phase transition from the false vacuum at $\phi = 0$ to the true one at $\phi \simeq M_I$ could take place, in principle, through barrier penetration. Detailed analytical and numerical studies (18) of such processes with the Coleman-Weinberg (C-W) potential have revealed that a huge amount of supercooling precedes the phase transition. Since the width of the potential $V(\phi)$ exceeds that in the C-W case, we can safely assume that the universe remains in the $\phi = 0$ phase for $T > T_c$. Note that in the temperature range $T_c \lesssim T \lesssim \mu$ the vacuum energy density $\sim \mu^4$ dominates over the radiation energy density and the universe experiences a modest amount of inflation.

When T reaches T_c the minimum at $\phi = 0$ disappears together with the barrier, and ϕ starts to roll towards the minimum at $\phi \simeq M_I$.

The g boson has mass $\sim M_S$ and its number density $n_g = n_{g^*} = n_\gamma \sim T^3$ for $T > T_c$. As ϕ starts the rollover, g acquires an additional mass $\sim |\langle\phi\rangle|$. To compute the mass M_g^* of g when it decays, we compare its lifetime with the time δt needed for $\langle\phi\rangle$ to grow from $\langle\phi\rangle \sim T_c$ to $\langle\phi\rangle \sim M_g^*$. [The reason we are not interested in the time needed for ϕ to grow from zero to T_c is the fact that n_g remains of order n_γ for ϕ in this range].

The classical evolution of the ϕ field is governed by the equation

$$\ddot{\phi} + 3H\dot{\phi} = -\frac{dV}{d\phi} \qquad (3)$$

For $\phi > T$ the temperature dependent mass terms in $V(\phi)$ can be safely ignored, and for $\phi < M_I$, the $\lambda(\phi^*\phi)^3/6M_c^2$ term in eq. (1) can be neglected. Also, the Hubble constant $H \sim \mu^2/M_p \sim 10^{-7}$ GeV $\ll M_s$ ($M_p \simeq 1.2 \times 10^{19}$ GeV is the Planck mass). Eq. (3) then becomes

$$\ddot{\phi} \simeq M_s^2 \phi \qquad (4)$$

which gives, for the time dependent mass of g,

$$M_g \sim \phi(\delta t) \simeq T_c \exp(M_s \delta t) \qquad (5)$$

The probability p for g to decay at time δt is given by

$$p = 1 - \exp\left(-\int_0^{\delta t} dt/\tau_g\right) \qquad (6)$$

Here $\tau_g \sim f^{-2} M_g^{-1}$ is the lifetime of g and f denotes the appropriate coupling constant. It follows that the mass of g at decay is given by

$$M_g^* \simeq T_c \left(1 + f^{-2} M_s/T_c\right) \qquad (7)$$

With $f \sim 1/5$ and $M_s/T_c \simeq 1/3$ for instance, $M_g^* \sim 10\,T_c$.

It follows from the above discussion that the g's are well out of equilibrium when they decay. Thus, at least one of the crucial ingredients for successful baryogenesis[19] is present in superstring models with an intermediate scale.

For consistency we must ensure that the g's do not annihilate before they can decay. The time needed for g-g^* annihilation through the color gauge interactions is $\tau_{an} \sim \alpha_s^{-2} M_g^2 / T^3$, where $\alpha_s \simeq 0.1$ denotes the QCD coupling at T ~ T_c. Clearly, $\tau_{an} \gg \delta t$ for $M_g \gtrsim T_c$ and so there is no g-g^* annihilation.

The field ϕ reaches the bottom of the potential $V(\phi)$ in a time which is estimated from eq. (5) to be roughly of order $M_s^{-1} \ln (M_I / T_c) \sim 10 M_s^{-1}$. It then oscillates about the minimum with a frequency M_s which is much greater than the expansion rate of the universe. The coupling of ϕ to other fields in the theory will result in the conversion of the scalar field energy into radiation. The entropy production dilutes the baryon asymmetry produced in the g decays by a factor Δ which turns out to be at least as large as ~ 10^6. This provides an important constraint on model building. One needs to ensure that the baryon asymmetry initially produced is sufficiently large to sustain the subsequent dilution.

We now discuss the baryon asymmetry produced in models based on the gauge group $G = SU(3)_C \times SU(2)_L \times U(1)_L \times U(1)_R$ which can arise from the compactification of the $E_8 \times E_8$ superstring theory on a suitable C-Y space. It was shown above that models based on G (supplemented by an additional discrete symmetry) possess many desirable properties. For our purposes here, the relevant terms in the superpotential are gD_cN, gg_cS_1 and $g_cg_cU_c$. Here g (g_c) denote $SU(3)_c$ triplet (antitriplet), $SU(2)_L$ singlet superfields, while U_c and D_c carry the quantum numbers of anti-up and anti-down quarks respectively. The N and S_1 fields are singlets with respect to $SU(3)_c \times SU(2)_L \times U(1)_Y$, and transform nontrivially only with respect to $U(1)_Y{}'$, where Y' denotes the generator othogonal to Y.

The scalar component of S_1 has a vev of order $M_I \sim (M_s M_c)^{1/2}$ at zero temperature and plays the role of the scalar field ϕ from our earlier discussion. The masses of S_1 and N particles are of order M_s.

Consider the decay of the scalar g. There are two baryon number violating channels, $g(b) \rightarrow \bar{D}_c(f)\bar{N}(f)$ and $g(b) \rightarrow U_c(b)D_c(b)N(b)$, corresponding to figs. 2a and 2b. Here b and f denote bosonic and fermionic fields respectively. CP violation can be introduced through an interference of these diagrams with those in figs. 3a and 3b. It is important to note that these diagrams give rise to CP violation only if

Although we only discussed the scalar bosons g, it should be clear that their fermionic partners provide a comparable contribution to the baryon asymmetry. This also holds for any other relevant superfields in the theory (in our case the g_c superfields).

A rough estimate shows that n_b/s initially cannot be much larger than about 10^{-5}. With a favorable choice of the parameters the decay width of S_1 can be arranged to be of the order of the Hubble constant. This minimizes the dilution of the baryon asymmetry due to entropy production.[20] The reheat temperature T_r is estimated to be $T_r \sim 3 \cdot 10^5$ GeV. Taking $T_c \simeq 3$ TeV say, the dilution factor $\Delta \simeq 10^6$. The baryon asymmetry consequently is estimated to be less than or of order 10^{-11}. In order that the baryon asymmetry does not undergo further dilution it is important that all subsequent phase transitions do not produce any significant entropy. Of course, it may be possible to construct alternative models which produce a much larger initial baryon asymmetry.

Finally, one also may check that the characteristic times for all baryon number violating scatterings are much greater than $\delta t \sim M_s^{-1} \ln(M_g/T_c)$. Hence the baryon asymmetry generated cannot be erased.

It should now be evident that the gravitino problem is neatly resolved in the present superstring models. For the convenience of the reader we briefly recall what the problem is. In many supergravity models (and also presumably in some superstring models) the gravitino is not the lightest supersymmetric particle and has mass on the order of the electroweak scale. Cosmological arguments rule out this possibility,[21] unless inflation is invoked to dilute away the primordial gravitinos.[22] However, one also must require that the reheat temperature T_R after inflation be below $0\ (10^8$ GeV) to avoid the regeneration of too many gravitinos.[23] Since the particles that produce the baryon asymmetry typically possess masses much larger than T_R, the last requirement often makes it difficult to produce sufficient baryon asymmetry in supersymmetric theories.

The problem is easily evaded in the present models since the baryon asymmetry is generated by the decays of g's which remain in abundance till temperatures of order T_c. The temperature T_R can therefore be as low as is necessary to suppress the gravitino number density to acceptable levels.

To conclude, we have shown that superstring models with an intermediate mass scale possess a novel mechanism for generating the observed baryon asymmetry in the universe. They also neatly evade the gravitino problem. Thus, a gravitino mass on the order of the electroweak scale, which may well turn out to be the case in many superstring models, is perfectly acceptable.

Models With n-$\bar{n}$ Oscillations

In this section we wish to further pursue this phenomenological (experimental) approach to supertsring model building. Following the strategy outlined above, we construct a new class of phenomenologically and cosmologically acceptable models which have some interesting experimental consequences. These include a stable proton, massless or extremely light (<< eV) neutrinos, and experimentally accessible n-$\bar{n}$ oscillations. The models also lead to a baryon asymmetry in accord with observations. A new feature here, absent in earlier models, is the appearance of an (accidental) unbroken global U(1) symmetry which corresponds to lepton number for the "known" particles.

For reasons explained earlier we base our model on the rank five subgroup $G \equiv SU(3)_c \times SU(2)_L \times U(1)_L \times U(1)_R$ of E_6. We also impose a (discrete subgroup of a) global $U(1)_{PQ}$ symmetry. The $U(1)_L \times U(1)_R \times U(1)_{PQ}$ symmetry spontaneously breaks down to $U(1)_Y$ at an intermediate scale M_I ($\sim 10^9$ GeV). The model consists of three <u>27</u>'s of E_6, two $SU(3)_c \times SU(2)_L \times U(1)_Y$ singlet fields S_1, S_2 (plus their mirrors $\tilde{S}_1$, $\tilde{S}_2$) which acquire v.e.v's of order M_I, a set of extra doublets H'', $\bar{H}''$ (plus mirrors $\tilde{H},\tilde{\bar{H}}$), and an extra color triplet g' (plus mirror $\tilde{g}$) with mass on the order of the electroweak scale.

The superpotential of the model contains the following degree three and four terms (For quantum numbers of the fields see Table 2. Note that we indentify H" with H'; $\bar{H}$" with $\bar{H}$' and g' with g): $\bar{H}QU_c$, HQD_c, HLE_c, $\bar{H}HN$, $g_cD_cU_c$, gg_cS_1, $\bar{H}'LS_2$, $H'\bar{H}'S_1$, gD_cN,
$\tilde{H}\tilde{\bar{H}}\tilde{S}_2$, $\tilde{H}\tilde{H}H'H'$, $\tilde{H}\tilde{\bar{H}}L\bar{H}'$, $\tilde{\bar{H}}\tilde{g}g\bar{H}'$, $\tilde{H}\tilde{S}_1H'S_1$, $\tilde{H}\tilde{S}_1LS_2$, $\tilde{H}\tilde{S}_2H'S_2$,
$\tilde{\bar{H}}\tilde{S}_2\bar{H}'S_1$, $\tilde{g}\tilde{S}_1gS_2$, $\tilde{S}_1\tilde{S}_1S_1S_1$, $\tilde{S}_2\tilde{S}_2S_2S_2$,
$\tilde{S}_1\tilde{S}_2S_1S_2$, $\tilde{S}_1\tilde{S}_2NN$.

The above superpotential possesses two global U(1) symmetries (in addition to the local symmetry G) which can be identified with the anomalous $U(1)_{PQ}$ and the non-anomalous $U(1)_{\mathcal{L}}$ where the subscript $\mathcal{L}$ denotes generalized lepton number. The quantum numbers pertaining to thse two symmetries are listed in Table 2. Note the important point that $U(1)_{\mathcal{L}}$ is an automatic symmetry although we only imposed $G \times U(1)_{PQ}$ (more precisely $G \times Z_n$, where Z_n is an appropriately large subgroup of $U(1)_{PQ}$) on the superpotential to begin with.

The axion domain wall problem can be taken care of in essentially the same manner as previously discussed and will not be repeated here.

The one loop renormalization group equations for the low energy couplings are consistent with the compactification scale $M_c = 10^{17}$ GeV, $M_I = 10^9$ GeV, $\sin^2 \Theta_W = 0.22$, $\alpha_s(M_W) = 0.11$ and $\alpha_G = 0.09$.

It was shown in the previous section that the out of equillibrium decays of the heavy colored fields (g, g_c) at temperatures as low as M_s, the SUSY breaking scale, can lead to the observed baryon asymmetry in the Universe. The relevant decay channels for the g bosons are displayed in figs. 4a and 4 b, with their corresponding one loop radiative corrections shown in figs. 5a and 5b.

The out of equilibrium condition requires that the couplings f and j (f stands for the gD_cN coupling and j for the $U_cD_cg_c$ coupling) be smaller than or equal to about 1/5. Taking into account the dilution factor of about 10^6 the final baryon asymmetry turns out to be roughly of order 10^{-10} which is highly satisfactory.

If we assign baryon number (1/3(-1/3) to a color triplet (antitriplet), the sole baryon number violating vertex involving light quarks (mass eigenstates) is $q\, d_c d_c u_c$, where the effective coupling constant $q \sim fj \langle N\rangle/M_g$. Here $\langle N\rangle \sim$ electroweak (SUSY) breaking scale, and $M_g(\sim M_I)$ denotes the mass of the heavy colored fields g and g_c.

An important constraint on q arises from considerations related to the baryon asymmetry in the Universe. We must require that the baryon asymmetry produced at $T \gtrsim M_s$ is not washed out due to the presence of the baryon number violating coupling $q\, d_c d_c u_c$. The rate of $2 \leftrightarrow 2$ baryon number violating scatterings at temperature $T \sim M_s$ is of order $q^2 T$. For these scatterings to be out of thermal equilibrium, their rate must be smaller than the expansion rate of the Universe. In a radiation dominated Universe this gives $q^2\, T \lesssim 30\, T^2/M_p$, where $M_p \simeq 1.2 \times 10^{19}$ GeV is the Planck mass. For $T \sim M_s \sim$ few hundred GeV, $q^2 \lesssim 10^{-15}$. This constraint on q is readily satisfied in our class of models because of the suppression factor $\langle N\rangle/M_I \sim M_s/M_I$ which appears in the definition of q.

Due to the absence of lepton number violating couplings, the proton is effectively stable in these models provided the gauginos and higgsinos are heavier than a GeV[24]. The proton may decay through the exchange of superheavy (~ 10^{17} GeV) particles in which case its lifetime would be ~ 10^{38-40} yrs., well beyond the scope of any forseeable experiment.

A prediction of the model which may be most amenable to experimental searches concerns n-$\bar{n}$ (neutron-antineutron) oscillations[25]. A relevant diagram is shown in fig 6. Using standard estimates, the oscillation time is expected to be given by

$$\tau^{-1}_{n-\bar{n}} \simeq q^2 \alpha^2 M_s^{-5} |\Psi(0)|^4$$

Here Ψ denotes the nuclear wave function. For $|\Psi(0)|^4 \simeq 10^{-3}\text{GeV}^6$, $q^2 \sim 10^{-16}$ and $M_s \sim 250$ GeV,

$$\tau_{n-\bar{n}} \sim 10^8 \text{ sec}$$

This may be within the reach of ongoing experiments searching for n-$\bar{n}$ oscillations. We should emphasize, however, that the value of $|\Psi(0)|^4$ is not accurately known and could even be one or two orders of magnitude smaller.

The unbroken $U(1)_{\mathcal{L}}$ symmetry implies the absence of any neutrino masses. Higher order (degree >4) terms in the superpotential need not conserve lepton number. The neutrinos may acquire a small (<<eV) mass as a consequence.

It is interesting to speculate on potential dark matter candidate(s) in this model. It cannot be the neutrino or the photino which decays into a baryon plus meson. One interesting possibility is mini (~ $10^{-2}M_{\odot}$) black holes which may arise due to large fluctuations in the axion field caused by the extended structures appearing at the QCD Phase transition.

Acknowledgements:

The work described in these lectures was done in collaboration with George Lazarides and Costas Panagiotakopoulos. It has previously been reported in the following publications:

Phys. Rev. Lett. 56, 432 (1986).

Phys. Rev. Lett. 56, 557 (1986).

University of Thessaloniki preprint UT-STPD-3/86 (1986).

Rockefeller University preprint RU86/B/154.

It is a pleasure to thank the organizers for arranging an extremely stimulating Workshop in beautiful surroundings.

Note Added:

We have now succeeded in constructing realistic gauge models based on a rank six subgroup $G = SU(3)_c \times SU(2)_L \times U(1)_L \times U(1)_R \times U(1)_{B-L}$ of E_6. They possess several interesting features. Thus, an accidental unbroken lepton number symmetry leads to an essentially stable proton (lifetime $\gtrsim 10^{38}$ yr.). Neutron-Antineutron oscillations are predicted to occur at a rate which may be experimentally accessible. One of the neutral gauge bosons which couples to B-L may have a mass below a TeV. Further details can be found in the Bartol Research Foundation preprint BA-86-53 (by G. Lazarides, C. Panagiotakopoulos and Q. Shafi).

References

1. M.B. Green and J.H. Schwarz, Phys. Lett. 149B 117 (1984).
2. D.J. Gross, J. Harvey, E. Martinec and R. Rohm, Phys. Rev. Lett. 54 502 (1985).
3. P. Candelas, G. Horowitz, A. Strominger and E. Witten, Nucl. Phys. B258 46 (1985).
4. E. Witten, Phys. Lett. 149B 351 (1984); Phys. Lett. 153B 243 (1985).
5. E. Witten, Nucl. Phys. B258 75 (1985).
6. J.D. Breit, B.A. Ovrut, and G.C. Segrè, Phys. Lett. 158B 33 (1985).
7. M. Dine, R. Rohm, N. Seiberg and E. Witten, Phys. Lett. 156B 55 (1985).
J.P. Derendinger, L.E. Ibanez and H.P. Nilles, Phys. Lett. 155B 65 (1985).
8. M. Dine, V. Kaplunovsky, M. Mangano, C. Nappi and N. Seiberg, Nucl. Phys. B259, 549 (1985).
9. M. Mangano, "Low energy aspects of superstring theories" Princeton preprint (1985).
10. J. Preskill, M. Wise and F. Wilczek, Phys. Lett. 120B 127 (1983), L. Abbott and P. Sikivie, ibid. 133, M. Dine and W.Fischler ibid. 137.
11. R. Peccei and H. Quinn, Phys. Rev. Lett. 38 1440 (1977); S. Weinberg, Phys. Rev. Lett. 40 223 (1978); F. Wilczek, Phys. Rev. Lett. 40 279 (1978).
12. K. Choi and J. Kim, Phys. Lett. 154B 393 (1985).
13. P. Sikivie, Phys. Rev. lett. 48, 1156 (1982).

14. G. Lazarides and Q. Shafi, Phys. Lett. 115B, 21 (1982); S. Barr, D. Reiss and A. Zee, Phys. Lett. 116B, 227 (1982); H. Georgi and M. B. Wise, ibid 116B, 123 (1982).

15. E. Witten, Mucl. Phys. B249, 557 (1985); Phys. Lett. 153B 243 (1985).

16. E. Chudnovsky, G. Field, D. Spergel and A. Vilenkin, Harvard preprint, 1986.

17. L. Dolan and R. Jackiw, Phys. Rev. D9 3320 (1974).
S. Weinberg, Phys. Rev. D9 3357 (1974).

18. E. Witten, Nucl. Phys. B177 477 (1981).
A. Billoire and K. Tamvakis, Nucl. Phys. B200 329 (1982).

19. A. D. Sakharov, Zh. Eksp. Teor. Fiz. Pisma 5:32, JETP Lett. 5 24 (1967. For a recent review and other useful references see E. Kolb andM. Turner, Ann. Rev. Nucl. Science 33 645 (1983).

20. M. Turner, Phys. Rev. Lett. 48 1303 (1982).

21. S. Weinberg, Phys. Rev. Lett. 48 1303 (1982)
H. Pagels and J. Primack, Phys. Rev. Lett 48 223 (1982).

22. J. Ellis, A. Linde and D. Nanopoulos, Phys. Lett. 118B 59 (1982).

23. J. Ellis, D. Nanopoulos and S. Sarkar, CERN preprint TH4057 (1984) and references therein.

24. F. Zwirner, Phys. Lett. 132B, 103 (1983).
R. Barbieri, and A. Masiero, Ecole Normale Superieure Preprint LPTENS 85/20 (1985).

25. V. Kuzmin, Pis'ma Zh. Eksp. Teor, Fiz. 13, 335 (1970) [JETP Lett. 12, 228 (1970)]; S.L. Glashow, in Quarks and Leptons: Cargese 1979, edited by Maurice Levy et. al. (Plenum, New York, 1980), p. 687; R.N. Mohapatra and R.E. Marshak, Phys. Rev. Lett. 44, 1316 (1980), and Phys. Lett. 94B, 183 (1980).

	$SU(3)_C$	$SU(2)_L$	$U(1)_L$	$U(1)_R$	global	N_m
Q	3	2	1	0	0	F
L	1	2	-1	-2	8	F
H	1	2	-1	-2	-12	1
$\bar{H}$	1	2	-1	4	14	1
U_c	$\bar{3}$	1	0	-4	-14	F
D_c	$\bar{3}$	1	0	2	12	F
E_c	1	1	2	4	4	F
g	3	1	-2	0	-10	F
g_c	$\bar{3}$	1	0	2	7	F
H'	1	2	-1	-2	-2	F-1
$\bar{H}'$	1	2	-1	4	-1	F-1
N	1	1	2	-2	-2	2F
S_1	1	1	2	-2	3	1
S_2	1	1	2	-2	-7	1
$\tilde{S}_1$	1	1	-2	2	-3	1
$\tilde{S}_2$	1	1	-2	2	7	1
H''	1	2	-1	-2	-2	1
$\tilde{H}''$	1	2	1	2	2	1
$\bar{H}''$	1	2	-1	4	-1	1
$\tilde{\bar{H}}''$	1	2	1	-4	1	1
H'''	1	2	-1	-2	-2	1
$\tilde{H}'''$	1	2	1	2	2	1
$\bar{H}'''$	1	2	-1	4	-1	1
$\tilde{\bar{H}}'''$	1	2	1	-4	-9	1
g'	3	1	-2	0	-10	2
$\tilde{g}'$	$\bar{3}$	1	2	0	20	2

Table I: Field content, multiplicity and quantum numbers. For Z_9 the (global) charges are clearly defined only mod9.

	$SU(3)_c$	$SU(2)_L$	$U(1)_L$	$U(1)_R$	$U(1)_{PQ}$	$U(1)_{\mathcal{L}}$
Q	3	2	1	0	0	0
L	1	2	-1	-2	2	1
H	1	2	-1	-2	-2	0
$\bar{H}$	1	2	-1	4	-1	0
U_c	$\bar{3}$	1	0	-4	1	0
D_c	$\bar{3}$	1	0	2	2	0
E_c	1	1	2	4	0	-1
g	3	1	-2	0	-5	0
g_c	$\bar{3}$	1	0	2	-3	0
H'	1	2	-1	-2	-8	1
$\bar{H}'$	1	2	-1	4	0	-1
N	1	1	2	-2	3	0
S_1	1	1	2	-2	8	0
S_2	1	1	2	-2	-2	0
$\tilde{S}_1$	1	1	-2	2	-8	0
$\tilde{S}_2$	1	1	-2	2	2	0
H"	1	2	-1	-2	-8	1
$\tilde{H}$	1	2	1	2	8	-1
$\bar{H}''$	1	2	-1	4	0	-1
$\tilde{\bar{H}}$	1	2	1	-4	-10	1
g'	3	1	-2	0	-5	0
$\tilde{g}$	$\bar{3}$	1	2	0	15	0

Table 2: Field Content and Quantum Numbers

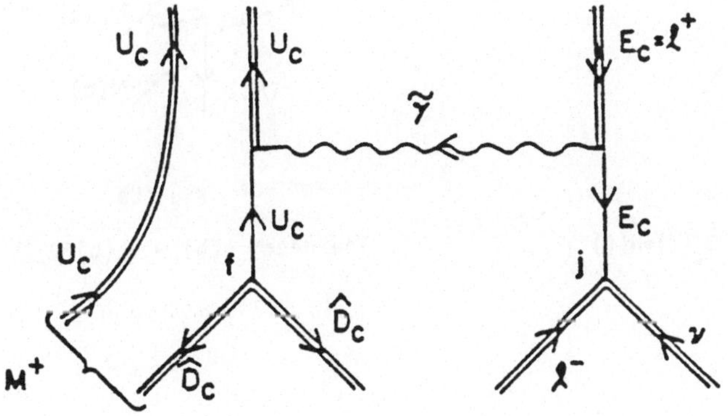

Fig. 1. One of the dominant diagrams for proton decay.

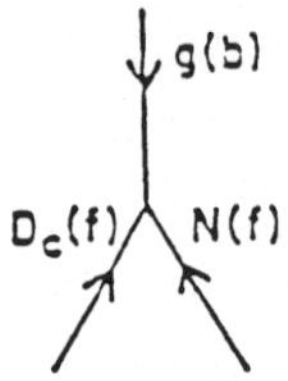

Fig. 2a

The decay $g(b) \to \bar{D}_c(f)+\bar{N}(f)$

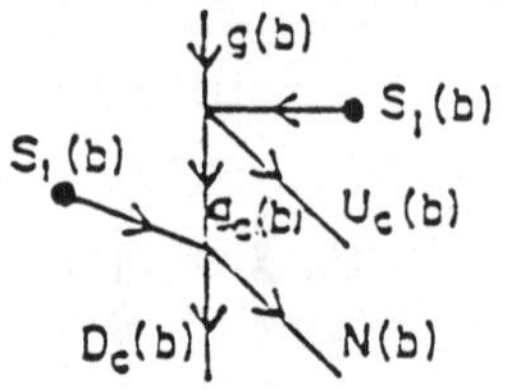

Fig. 2b

The decay $g(b) \to U_c(b)+D_c(b)+N(b)$

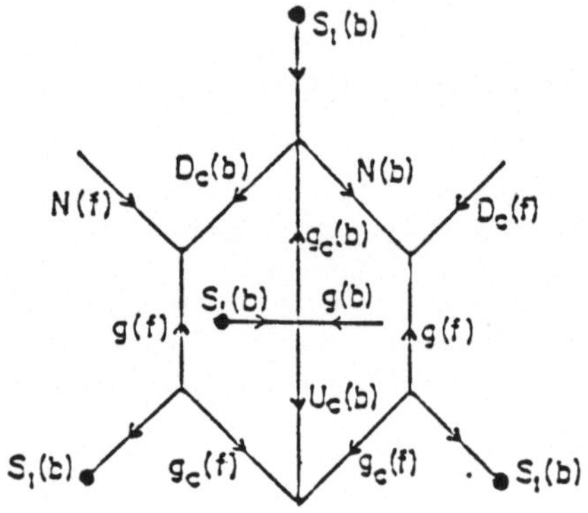

Fig. 3a

Radiative correction to Fig. 2a

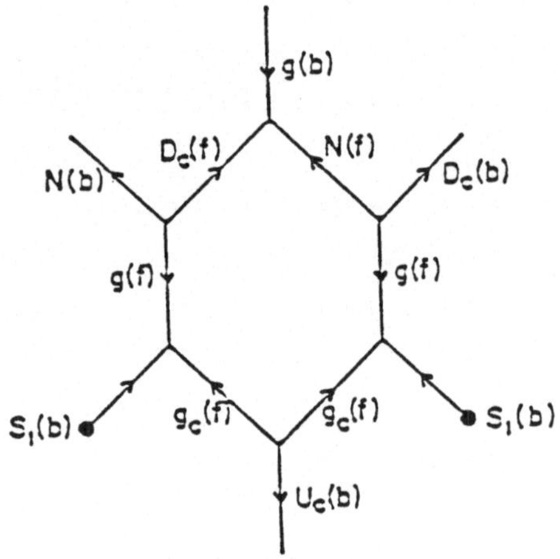

Fig. 3b

Radiative correction to Fig. 2b

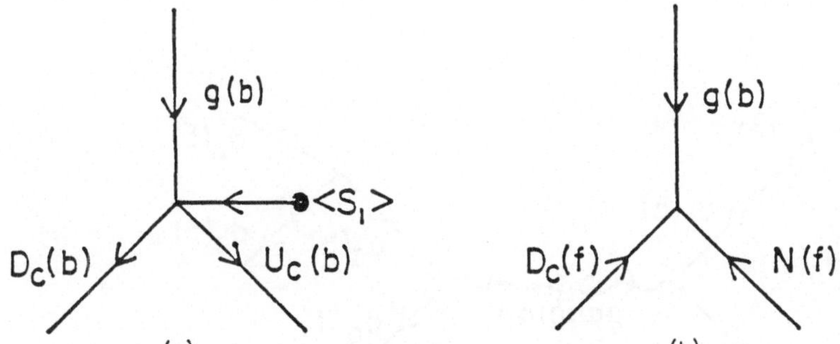

FIGURE 4

(a) The decay $g(b) \longrightarrow D_c(b) + U_c(b)$; (b) the decay $g(b) \longrightarrow \overline{D_c}(f) + \overline{N}(f)$.

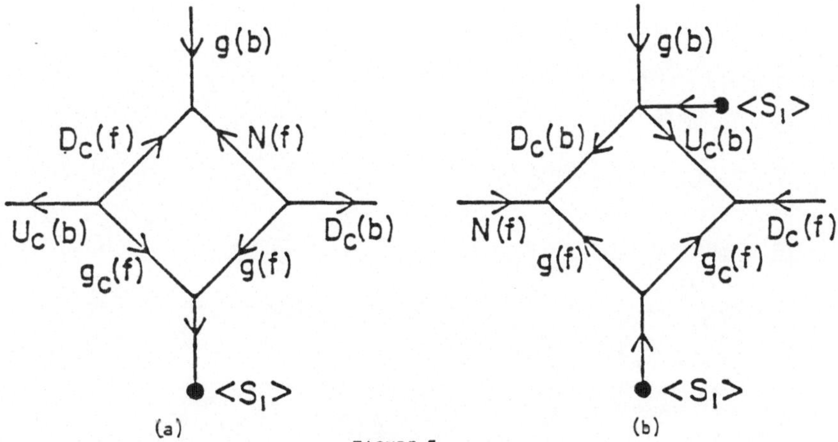

FIGURE 5

Radiative corrections to (a) Fig. 4(a) and (b) Fig. 4(b).

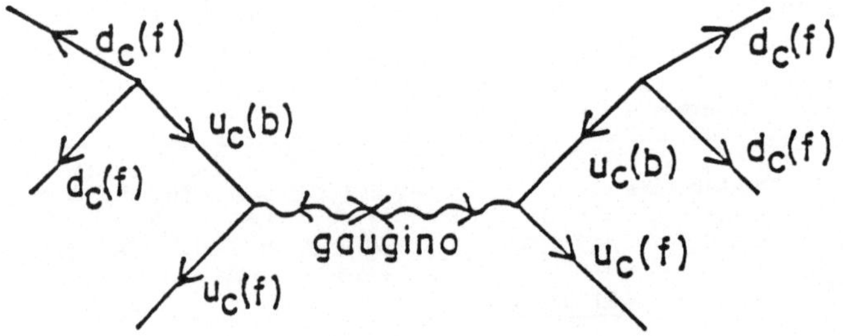

FIGURE 6

Diagram for n-$\bar{n}$ oscillations.

MONOPOLES AND AXIONS IN AN INFLATIONARY UNIVERSE

Q. Shafi*†
Bartol Research Foundation
of The Franklin Institute
University of Delaware
Newark, DE U.S.A.

Abstract

In an inflationary cosmology the phase transitions associated with intermediate mass scales may experience a large though not huge number of e-foldings before being finally completed. Extended structures (monopoles, strings, etc.) arising from such transitions are not necessarily inflated away. In an SO(10) model, for example, magnetic monopoles with mass $\sim 10^{13} - 10^{14}$ GeV can occur at a level which is experimentally detectable. The topologically unstable string - wall system of axion models provides an important source of density fluctuations as a consequence of inflation.

* Supported in part by Department of Energy Grant DE-FG02-84ER40176.

† Work done in collaboration with G. Lazarides and published in Phys. Lett. 148B, (1984) 35

P. Galeotti and D. N. Schramm (eds.), Gauge Theory and the Early Universe, 149–158.

A successful implementation of the inflationary scenario [1] in the context of the standard non-super-symmetric grand unified theories (GUTS) has been presented [2] which employs a gauge-singlet scalar field ϕ. The slow "rollover" of ϕ to its ground state vacuum expectation value (VEV) is responsible for the inflationary phase. Due to the coupling of ϕ to the usual Higgs fields of the theory, other phase transitions can, and in some cases will occur during the inflationary phase. Thus, the phase transition from the GUT symmetry G (e.g. SU(5)) to some subgroup H_0 (SU(3) x SU(2) x U(1), also referred to as 3-2-1) occurs early in the inflationary phase. However, the relevant Higgs field (24-plet) acquires its full VEV only at the end of the inflationary era. The GUT phase transition is forced to be inflationary [2], and any extended structures produced when G breaks to H_0, especially the monopoles, are inflated away [3].

The presence of the inflationary phase is felt not only by the GUT symmetry breaking but also by the symmetry breakings with scales lying between the GUT scale and the Hawking temperature. For definiteness, we will work with a GUT scale of order 10^{15} GeV and a vacuum energy density of order $(10^{15}\ \mathrm{GeV})^4$. The scales under discussion then lie between 10^{15} GeV and 10^{10} GeV. As we will see, the number of e-foldings experienced by the phase transitions occurring at scales between 10^{12} and 10^{13} GeV can be large without being completely inflationary. This has important implications for the extended structures produced at such scales.

For example, the breaking of SO(10)to $H_0 = SU(4) \times SU(2)_L \times SU(2)_R$ at a scale $\sim 10^{15}$ GeV produces monopoles [4] that are inflated away. The subsequent breaking of H_0 to 3-2-1 at a scale between 10^{12} and 10^{13} GeV produces new, "lighter" monopoles [4] that carry two units of the Dirac charge. If the number of e-foldings experienced during this phase transition is $\gtrsim 25$, and we will see that particle physics and cosmological considerations do allow this possibility, then monopoles with mass between 10^{13} and 10^{14} GeV are expected to occur at or below the Parker limit in our galaxy.

Another interesting application of partially inflationary phase transitions is provided by the axion models [5]. The global U(1) Peccei-Quinn symmetry breaking at a scale $\sim 10^{12}$ GeV may experience a large number of e-foldings before the phase transition is finally completed. Although the scale of the string network inflates during this period, for a reasonable choice of parameters it is not completely inflated away. At the QCD transition the strings get connected to domain walls [6]. If the number of e-foldings experienced by the U(1) transition is close to 50, the scale of the string-wall system is the largest allowed by the constraint that one does not end up with a wall dominated universe. These structures will then provide an important source of density fluctuations [7] in an axion dominated universe.

Consider an SO(10) model which breaks to SU(3) x SU(2) x U(1) via H_0 = SU(4) x $SU(2)_L$ x $SU(2)_R$ (this is the Pati-Salam subgroup [8] of SO(10)). The first step of the summetry breaking can be achieved either with a 54-plet or a 210 of Higgs fields. If the 54 is employed then, in addition to the Z_2 monopoles [4], topologically stable Z_2 strings are produced [9], which in the subsequent transition to the 3-2-1 phase get connected to domain walls [9]. The Z_2 strings do not arise if the 210 is employed [10]. This difference will be important, as we shall soon see.

As pointed out earlier, the inflationary phase starts when a gauge singlet scalar field ϕ begins a slow rollover towards its ground vacuum expectation value $M \sim 10^{18}$ GeV [2]. When the classical evolution takes over the time dependence of ϕ is given by (see [2] and references therein)

$$\phi^2(t) \simeq (3H/2\lambda_0)(\tau - t)^{-1}, \quad \tau - t \gg H^{-1}. \tag{1}$$

Here $H \sim 10^{11}$ GeV is the Hubble constant for a GUT scale $M_X \sim 10^{15}$ GeV, $\tau \sim \lambda_0^{-1/2} H^{-1}$ is the rollover time, and $\lambda_0 \lesssim 4 \times 10^{-12}$ from the requirement that $(\delta\rho/\rho)_{hor} \lesssim 10^{-4}$.

The SO(10) symmetry breaks to H_0 when the relevant negative mass squared term $-1/2 \langle\phi\rangle^2 \lambda_{eff} \Phi^2$ in the effective potential ($\lambda_{eff} \sim 10^{-6}$ and Φ is either 54 or 210) becomes of order $T_H^2 \Phi^2$, $T_H = H/2\pi \simeq 10^{10}$ GeV, being the Hawking temperature. It is readily checked [2] that this occurs in the early stages of the inflationary phase, although Φ acquires its full VEV only at the end of the inflationary era. Consequently the Z_2 monopoles and, if the 54 is employed, also the Z_2 strings are inflated away.

Let $\chi(126)$ denote the field whose VEV breaks H_0 down to 3-2-1 at the scale M_c. The quartic coupling $-1/2\, c\phi^2 \chi^+\chi$ has strength $c \sim (M_c/M)^2$, and χ begins to develop a non-zero VEV only when ϕ has evolved sufficiently which makes $c\langle\phi\rangle^2 \gtrsim T_H^2$. The breaking of H_0 to 3-2-1 produces new monopoles through the Kibble mechanism [11]. These topological structures are "frozen" in when the Higgs mass squared $m_X^2 = 2c\,[\phi^2 - (2\sigma_X/c)T_H^2] \sim H^2$. (This comes about by comparing the energy in a typical fluctuation region of radius H^{-1} with T_H.) Here $\sigma_X \simeq 1$ is the coefficient of $T_H^2 X^+ X$ which appears as one of the terms in the effective potential. The necessary condition for monopoles to form is thus $2c\phi^2 \simeq H^2$. Since $\phi^2 \sim 3H/2\lambda_0(\tau - t)$, the time t_X at which monopole formation occurs is given by

$$H(\tau - t_X) \equiv \eta \sim 3c/\lambda_0. \tag{2}$$

At production the monopole number density is $\sim H^3$, which is diluted to $n_M \sim H^3 \exp(-3\eta)$ by the time inflation is over. The universe reheats to a temperature T_R ($\sim 10^9 - 10^{10}$ GeV for this class of models) [2], and the relative monopole number density r is given by

$$r \equiv n_M/T_R^3 \simeq (H/T_R)^3 \exp(-3\eta). \tag{3}$$

The requirement that r should not exceed 10^{-30} (the Parker limit)[12] gives a lower limit on the number of e-foldings experienced in the transition from H_0 to 3-2-1,

$$\eta \simeq 3c/\lambda_0 \gtrsim 23 + \ln(H/T_R) \simeq 25\text{-}27, \tag{4}$$

which, since λ_0 is essentially fixed from considerations of $\delta\rho/\rho$, is a constraint on c. For $\lambda_0 \sim (1\text{-}4) \times 10^{-12}$, the equality sign in (4) holds for

$$M_c \simeq (3\text{-}6) \times 10^{12}\ \text{GeV}. \tag{5}$$

Remarkably enough, an intermediate scale of this order of magnitude is allowed by the one-loop renormalization group equations if $M_X \sim 10^{15}$ GeV, $\sin^2\theta_W$ is between 0.23 and 0.24, and $\alpha_s \sim 0.1$. Thus, we have shown that inflation and an observable number density of monopoles can coexist. Note that even if these monopoles catalyze nucleon decay with strong cross section[13], there is no new limit on their number density from neutron star considerations since their mass $\lesssim 10^{14}$ GeV[14].

The breaking of SO(10) to 3-2-1 with a 54 and 126 leads to the production of domain walls [9] which must be inflated away. However, this would also inflate away the lighter monopoles. This state of affairs is easily remedied by employing the 210 of Higgs instead of the 54 [10]. The C-symmetry (which interchanges left and right and conjugates the representations) is broken by a non-zero vacuum expectation of the 210. Thus, there are no Z_2 strings and no subsequent domain wall production.

As another interesting application of these ideas let us consider a grand unified axion model in which the Peccei-Quinn U(1) symmetry is broken at some scale f_a. Astrophysical [15] and cosmological [16] considerations imply 10^8 GeV $\lesssim f_a \lesssim 10^{12}$GeV. With $f_a \sim 10^{12}$ GeV, the axions contribute significantly to the dark matter in the universe.

It is well known that axion models lead to extended structures called walls bounded by strings (WBS)[6]. With f_a on the order of 10^{12} GeV, we may expect the U(1) phase transition to participate in the inflationary phase. Provided the strings are not completely inflated away, they would reenter the particle horizon at some later time. They get connected to domain walls when the QCD phase transition takes place. Large WBS provide an important source of density fluctuations in the early universe [7].

Let us now make this discussion a bit more quantitative. The initial scale of the string network is of order H^{-1}. It inflates and becomes $\sim H^{-1} e^{\zeta} = 1$ at time τ, where ζ denotes the number of e-foldings experienced by the U(1) phase transition, and the time τ signals the end of the inflationary era. Thereafter, l grows like $t^{1/2}$, where t denotes the cosmic time in the radiation dominated era. At $t = \tau$,

$$(1/d_H)_\tau = H^{-1} e^{\zeta} /\tau, \tag{6}$$

where d_H denotes the particle horizon, For $t > \tau$,

$$(1/d_H)_{t>\tau} = (H^{-1} e^{\zeta}/\tau)(\tau/t)^{1/2}. \tag{7}$$

The sting network enters the particle horizon at time t_s when the ratio in (7) becomes unity. This gives

$$(\tau t_s) = H^{-2} e^{2\zeta}. \tag{8}$$

Let us next impose the requirement that the time t_s be no later than $(G\sigma)^{-1} \sim (\Lambda^2 f_a/M_p^2)^{-1}$, where $\Lambda \sim$ 100 MeV is the QCD scale and $M_p \simeq 1.2 \times 10^{19}$

GeV is the Planck mass. This corresponds to the time at which the wall energy density equals the radiation energy density ($\sigma \sim \Lambda^2 f_a$ is the mass per unit area of the wall). The time required by the WBS to oscillate and radiate away its energy in gravitational waves is also $(G\sigma)^{-1}$. The above requirement gives

$$2\zeta \lesssim \ln\,[3H\lambda_0^{-1/2}(G\sigma)^{-1}]. \tag{9}$$

Also, from previous considerations,

$$\zeta \simeq (3/\lambda_0)(f_a/M)^2. \tag{10}$$

With $\lambda \sim 10^{-13}$, the equality sign in (9) holds for $f_a \simeq 1.3 \times 10^{12}$ GeV which corresponds to $\zeta \simeq 52$. We therefore obtain the following scenario. On scales up to $\sim 10^6 M_\odot$ (the horizon mass in axions at $t = (G\sigma)^{-1}$), density fluctuations of order unity are generated by the WBS [7]. On scales larger than this $(\delta\rho/\rho)_{hor} \sim 3 \times 10^{-5}$. These two sources of density fluctuations presumably lead to a satisfactory scenario of galaxy formation, with axions providing a significant amount of the dark matter. Note that the presence of an inflated string-wall system of the type discussed here can also alleviate the difficulty in the growth of axion perturbations encountered in ref. [17] due to the presence of a cosmologically significant neutrino component.

To conclude, extended structures produced during phase transitions at intermediate mass scales in an inflationary universe will not necessarily be inflated away. Indeed, for a reasonable choice of the parameters, monopoles could be present in our galaxy at a level which is experimentally accessible. Extended structures that inevitably occur in axion models can provide an important source of density fluctuations.

References

[1] A. Guth, Phys, Rev. D23 (1981) 347;
K. Sato, Phys. Lett. 99B (1981) 66;
A. Linde, Phys. Lett. 108B (1982) 389;
A. Albrecht and P. Steinhardt, Phys. Rev. Lett. 48 (1982) 1220.

[2] Q. Shafi and A. Vilenkin, Phys. Rev. Lett. 52 (1984) 691.

[3] G. Lazarides, Q. Shafi and P. Trower, Phys. Rev. Lett. 49 (1982) 1756;
M. Turner, Phys. Lett. 115B (1982) 95.

[4] G. Lazarides, Q. Shafi and M. Magg, Phys. Lett. 97B (1980) 87;
S. Dawson and A. Schellekens, Phys. Rev. D27 (1983) 2119;
E. Weinberg, D. London and J. Rosner, University of Chicago preprint (1983).

[5] R. Peccei and H. Quinn, Phys. Rev. Lett. 38 (1977) 1440;
S. Weinberg, Phys. Rev. Lett. 40 (1978) 223;
F. Wilczek, Phys. Rev. Lett. 40 (1978) 279.

[6] A. Vilenkin and A. Everett, Phys. Rev. Lett. 48 (1982) 1867;
G. Lazarides and Q. Shafi, Phys. Lett. 115B (1982) 21.

[7] F. Stecker and Q. Shafi, Phys. Rev. Lett. 50 (1983) 928.

[8] J. C. Pati and A. Salam, Phys. Rev. D10 (1974) 275.

[9] T. W. Kibble, G. Lazarides and Q. Shafi, Phys. Lett. 113B (1982) 237.

[10] D. Chang, R. Mohapatra and K. Parida, University of Maryland preprint (1984).

[11] T. W. Kibble, J. Phys. A9 (1976) 1387.

[12] E. Parker, Cosmic magnetic fields (Clarendon, Oxford, 1979).
G. Lazarides, Q. Shafi and T. Walsh, Phys. Lett. 100B (1980)21;
E. Parker, M. Turner and T. Bogdan, Phys. Rev. D26 (1982) 1296.

[13] V. Rubakov, JETP Lett. 33 (1981) 644;
C. Callan, Phys. Rev. D26 (1982) 2058;
See also A. Sen, Fermilab preprint PUB-84/42-T (1984).

[14] V. Kuzmin, Talk Intern. High Energy Physics Conf. (Leipzig, July 1984), and references therein.

[15] D. Dicus, E. Kolb, V. Teplitz and R. Wagoner, Phys. Rev. D18 (1978) 1829.

[16] J. Preskill, M. Wise and F. Wilczek, Phys. Lett. 120B (1983) 127;
L. F. Abbott and P. Sikivie, Phys. Lett. 120B (1983) 133;
M. Dine and W. Fischler, Phys. Lett. 120B (1983) 137.

[17] Q. Shafi and F. Stecker Phys. Rev. Lett. 53 (1984) 1292.

COSMIC INFORMATION AND THE "ANTHROPIC " PRINCIPLE.

Hubert Reeves
C.E N.S Saclay .
France

Erice School , May 1986

ABSTRACT. The emergence of structures and complexity in the expanding universe is related to the existence of sources of entropy and information (stocked entropy). The physical processes responsible for these phenomena are reviewed and situated in their chronological order.

A "PRINCIPLE OF COMPLEXITY".

The subject-matter usually covered under the title 'the anthropic principle" does not ,in my opinion, justify this denomination. The "initial conditions" required for the advent of our species in the universe are also required for the advent of the sea-urchins or the black-capped chicadees. In fact, the mere existence of giant molecules (proteins) sets almost the same constraints on the numerical parameters describing both the global properties of the hot primordial soup and the constants of the laws of physics governing this matter.

I prefer the expression " A principle of complexity". If they could speak, the numberless species of plants and animals sharing our terrestrial habitat would also prefer this less chauvinistic point of view.

THE MOST ANCIENT IMAGE OF THE WORLD.

The" fossil radiation" detected by Penzias and Wilson in 1965 gives us

P. Galeotti and D. N. Schramm (eds.), Gauge Theory and the Early Universe, 159–191.

an image of the ancient universe . Over the recent years , thanks to the work of many observers , (summarized in the Proceedings of this Erice 1986 School on Cosmology) the quality of this picture has gradually improved . With a fine degree of resolution , it describes the cosmos of fifteen billions years past as a hot "fluid", extraordinarily homogeneous in texture. No "granularity", no trace of the numberless structures so familiar in our contemporaneous universe (galaxies, molecules , atoms, etc.).

The question is our mind: how could such a completely amorphous soup of elementary particles give rise to the highly sophisticated variety of complex systems, including the human brain wondering over these mysteries?

The elements of our discussion will involve the concepts of *entropy* and *information*. The second law of thermodynamics requires the entropy of the universe to increase at each event (or events) leading to the birth of a new structure (nuclei, atoms, organisms , stars etc.). I shal call them: "organizing events" Clearly they require "irreversible processes" which , in turn require states of "disequilibrium". (In states of equilibrium, all processes are reversible, and any organizing event would be necessarily be followed by a reverse disorganizing event obliterating the new structure). Irreversible processes always lead to an increase of entropy.

The problem is that the early fluid, detected through the fossil radiation, appears, at first sight , to be in a state of maximum entropy. How can we, globally, further increase the entropy of this totally disorganized matter, in order to decrease the entropy of a given region for structuring purposes?

A GOLDEN RULE.

To simplify our discussion, we shall use a very useful "golden rule": *the maximum entropy associated wih a given substance is proportional to the number of particles in this substance.* This rule is exact for an isothermal gas of photons (black-body radiation). The entropy is the number of photons,(proportional to the third power of the temperature) multiplied by 7.1.(For convenience, I shall put this factor equal to one.) The rule is also exact for all relativistic gases of free particles ($kT > mc^2$). It is approximately valid in many other situations covering the needs of our present discussion.

The most entropic matter is thermal light. A given amount of energy can be split in a maximum number of particles, since no rest mass has to be provided. In the course of cosmic evolution, emission of thermal (or thermalized) photons will turn out to be the favourite way of nature to produce the extra entropy accompanying the organizing events.

This brings us back to our previous question: how can we increase the number of photons over those contained in the fossil radiation?

New photons can be made in two different ways. First, by the *destruction* of a given amount of *massive* material and transformation of this mass in radiation. Second, by the creation of *temperature gradients* and the transportation of thermal energy throughout these gradients. The stars are presently the main agents of these two types of processes.

Although the emission of new thermal photons appears to be a *necessary* condition for the growth of complexity, it is by no means a *sufficient* condition. One of my aims in this paper is to explore the list of the required conditions.

To illustrate this point, let us consider the process of matter annihilation occuring each time the cosmic temperature (kT) becomes lower than the rest mass (mc^2) of a given species of particles. Around 0.5 MeV, at a few seconds in the standard chronology of the Big-Bang, electrons and positrons annihilate, creating new photons, rapidly thermalized in the background radiation. These events however creates no bound systems leading to organization.

The forces of nature play a fundamental role in the growth of complexity. Three of the forces: the nuclear (N), the electromagnetic (EM) and the gravitational (G) are able to bind particles and thereby to generate stable systems. A fraction of the mass of the initial constituents (the binding energy) is emitted away in the form of photons carrying both energy and entropy. The weak force (W) is unable to generate bound systems. The reason is the very large mass of its vector particles: the intermediate bosons W and Z. However, as we shall discuss later, the numerical value of this mass is an essential gradient in the birth of out-of-equilibrium nuclear processes in the universe.

Even if complexity is always borne by many-body bound systems, binding new systems is *not* a sufficient way to generate complexity. Black holes do incorporate many particles bound together by gravity, but they can hardly be said to represent complexity.

As a matter of fact, how can we define "complexity"? We all have an intuitive image but can we be more accurate? "Complexity" can be associated with" variety" , "diversity" and "performance". A system reveals its internal degree of complexity through the quality and specificity of its operations . In a scale of increasing complexity, we find, in succession, heavy nuclei interacting with radiation through a complicated network of excited levels, bacteria moving bodily toward a source of light , human people formulating the laws of quantum electrodynamics.

NUCLEAR INFORMATION.

The role of the nuclear force will be treated in details ; it will be of general pedagogical value for the rest of our story . It will allow us to introduce the important notion of *information*, extended on a cosmic scale.

The nucleus of iron-56 is the most strongly bound structure of the nuclear force. The binding energy is 8.6 MeV per nucleon or about one percent of the rest mass. In the hot initial soup the mass excess liberated by this association is transformed in thermal photons . In the following pages, the words *nuclear entropy* will mean : *the entropy added to the cosmic background by nuclear processes.* A better expression would be " entropy of nuclear origin" .

If the association of 56 nucleons in one iron nucleus increases the radiation entropy, (each photon of the soup has a mean energy of three KT). it decreases the nucleon entropy since, for each iron generated, it reduces their number by a factor of fifty-six. At high temperature the net entropy balance is in favor of the nucleon gas which , then , represents the state of maximum entropy. As kT decreases, the number of photons obtained from the association increases and the state of iron-plus-radiation becomes the state of maximum entropy.

Following our golden rule, we shall consider a group of N free nucleons in the primordial soup. At high temperature, their entropy is proportional to N , while at low temperature it becomes proportional to: $\{N/56 + N \times (8.6\ \mathrm{MeV})/3kT)\}$. As the universe reached the temperature of 2.7 MeV, the iron gas became suddenly the most entropic state. (figure 1). Nuclear entropy became *available.* At that historical moment, the nuclear force *could* have transmuted everything in iron thereby increasing the cosmic entropy to its new maximum value

It *could* but it *did* not. Or , more exactly , it went only part of the way toward this "goal". Only twenty -five percent of the nucleons were involved in the process called "primordial nucleosynthesis". And the end-product was not iron, but helium, corresponding to only 6.8 MeV (and not the full 8.6 MeV) per nucleon.

As a result , hydrogen is still present in the universe, despite the fact that , entropywise ,the iron state is favored. This fact will serve to define quantitatively the notion of ***nuclear information*** . Following the theory of Brillouin, the amount of information in a system will be defined as ***the difference between its state of maximum possible entropy and its present actual entropy state*** .

According to this definition, the nuclear information is zero above the temperature of 2.7 MeV since both the nuclear maximum entropy and the nuclear real entropy are equal to zero (figure 2). At 2.7 MeV, the maximum entropy rises slowly (as 8.6 MeV/kT). The real nuclear entropy however remains zero as no new nuclei are actually formed.

This situation lasts until the temperature of 0.1 MeV, when the first deuterons manage to resist the photodisintegrating effect of the photon gas. Rapidly these deuterons are transformed in helium nuclei but proceed no further up in the mass scale . As mentionned before, three fourth of the nucleons managed to escape these nuclear processes. In the following paragraphs we shall investigate in details the reasons for this incompleted job of the nuclear force . Here I want to stress the fact that this "incompletion " is the essential factor giving rise to nuclear information in the universe . The figure 2 shows its numerical value as a function of time.

Today the background-radiation is at 3K, (one milli-electron volt per photon). The amount of nuclear information is equivalent to six billion photons for each nucleon. Stars are busily spending this information, thereby increasing the nuclear entropy in the universe .The flow of entropy emerging from these stars can be "tapped" to organize matter in their vicinity. Numerical examples will be worked out later.

At this point I want to come back to the reasons for the incompletion of the nuclear binding task at the time of nucleosynthesis. One key-element involves the confrontation of the ***times-scales.***

On the one side, there is the basic fact of the expansion of the universe. The rate of expansion is governed by the force of gravity acting on the bulk of the universe (global scale). The expansion time can be defined as the time required for the temperature to fall by a factor of two. Einstein's General

Relativity, applied to the cosmological model, shows that this expansion-time, inversely proportional to the square root of the total density, increases gradually . In the early moments of the universe, it grows inversely with the *square* of the cosmic temperature (figure 3).

Then there is the time-scales for particle interactions. Reactions take time. The reaction rates are proportional a) to the strength of the interactions , b) to the temperature at which they take place and c) to the density of partners with which a given particle can undergo an interaction. During most of the expansion, the particle-density decreases with the *cube* of the temperature. The reaction cross-sections are proportional to some power of the energy which is almost always positive , typically *two*. The interaction rate , (the product of the density times the cross-section) is then proportional to the *fifth* (3+2) power of the temperature. As a result, the interaction-timescale increases faster with decreasing temperature than the expansion-timescale.(fig 3).

In the very early moments, the interaction-timescales are *shorter* than the expansion-timescales . All reactions of the type (A+B → C+D) are then in equilibrium with their inverses (C+D → A+B). The microphysics is reversible.

Sooner or later, however , the two curves intersect. At lower temperatures, the cosmic temperature falls too rapidly for the reactions to remain in equilibrium. States of disequilibrium set in successively for each force, each type of reactions.

Let us come back to the building of nuclear structures. There are several reasons why iron nuclei do not emerge from the primordial nucleosynthesis . To build an iron nucleus, one must first associate one proton with one neutron and generate a deuteron. Then new nucleons are progressively added, all the way to 56. But deuterons are very weakly bound and do not survive the cosmic heat until the temperature has fallen to about 0.1MeV.

We meet here the "nucleation" problem of very general occurence in nature. In order for a state to occur, one must first undergo a sequence of intermediate physical processes. These processes may turn out be impossible, or very slow, at the crucial moments when this state become suddenly entropy-favored . It is well known that the water of mountain lakes often remains liquid well below the zero degree "freezing point", if it is very pure and if the air temperature falls rapidly

As mentionned before , although the weak force is unable to bind any material structure, it plays indirectly a crucial role in the build-up of nuclear information . At temperature larger than one hundred GeV, the electromagnetic and weak force have approximately the same strength. Then comes a crucial event, analogous to a phase transition, by which the W force becomes progressively weaker, due to the very large mass (≅ 100 GeV) of the particle responsible for carrying this force.

One major effect of the weak force is the transformation of protons into neutrons, and vice versa. On the other hand , neutrons turn out to play a major role in primordial nucleosynthesis. Essentially all the neutrons present in the universe at temperatures at which the deuterons survive the heat (below 0.1 MeV) will manage to capture a proton and will be transformed, after a few more reactions, into helium nuclei. Thus the building of nuclear information is directly related to the fact that the population of neutrons is *smaller* than the population of protons at the moment of primordial nucleosynthesis , (the later transmutation of helium into iron yields less than two MeV per nucleon). Nuclear information will then be borne by the protons remaining in excess after the pairing.

ELECTROWEAK SYMMETRY BREAKING AND NUCLEAR INFORMATION.

Here I want to study the effect of the mass of the W bosons (≈84 GeV) on the advent of nuclear information. To do this I shall arbitrarily change the value of this mass and compute the corresponding alteration on the cosmological model . Such procedures are generally done in the context of the study of the "anthropic "principle. However a word of caution is appropriate here.

Our understanding of physics , in particular in the nuclear sector, is not good enough to guarantee that we can adequately estimated all the effects of such an alteration. The numerical values of the nuclear masses and excited energy levels would certainly be changed if the parameters of the EM and weak forces were altered but by what fraction?

The strength of the weak force is related to the mass of the W boson. The empirical Fermi constant is inversely proportional to the square of this mass: ($G_F = \gamma_W / M_W^2$) . The lifetime of the neutron (G_F^{-2}) is proportional to its fourth power : ($t_n = M_W^4$) and the strength of the weak force is inversely proportional to the fourth power of this mass.

At temperatures larger than the equivalent of this mass ($T = 10^{15}$K) the WF is comparable in strength to the EMF. Later, the WF becomes gradually weaker, as the large masses of its vector particles (W and Z) impede the transport of the force . What would be the effect of increasing the mass of the W ? The corresponding decrease in the W F strength ($G_F^{\circ} < G_F$) would result in a decoupling of the weak interaction at a higher temperature T° (decoupling) (see figure 3).

Before decoupling, the relative population of neutrons and protons is given by the Boltzman equilibrium formula : (where ΔM is the mass difference between the neutron and the proton).

$$n/p = \exp(- \Delta M / kT)$$

(With our present limited knowledge of fundamental physics, it is not possible to evaluate the alteration of this mass difference which would be induced by the change of the M_W mass. This is an example of the difficulties plaguing this whole discussion. However the following conclusions would probably remain valid.)

After decoupling , the neutrons will decay in the usual fashion . However the increase in the neutron lifetime will guarantee that essentially all the existing neutrons will survive until the period of primordial nucleosynthesis , to be incorporated in helium . The amount of free protons after this period will thereby be sharply reduced as compared to the real case.

As a simple numerical example an increase of a factor of six in the mass of the W would increase the lifetime of the neutron by one thousand, (to three months) and increase the decoupling temperature by ten (to ten MeV or so). At this temperature, the neutron -proton ratio is 0.95, and all the neutrons can survive until the n-p capture period, leaving very few free protons to later form hydrogen (figure 4).

Because of the reduced weak interaction rate ,the p-p stellar hydrogen burning (of whatever H is left) would be very slow. In fact, He burning would probably occur before H burning, generating some ^{12}C to catalyse the H burning through the ordinary Bethe cycle (incidentally implying a cooling of the stellar core to adjust to the fast rate of this cycle.) The net result would be a the drastic shortening of stellar lifetimes, probably incompatible with the developpement of life in planetary habitats.

In a recent article in Physics Today, Steven Weinberg says: "the great mystery is not why the weak interactions are so weak, but why are they so strong. That is, why is the electroweak scale of 300 GeV (responsible for the mass of the W) so small with the really fundamental Planck or GUT or Kaluza-Klein or string scale of around 10^{17} GeV? *A posteriori*, this fact appears of importance for the growth of complexity.

Lowering the mass of the M_W (i.e retarding the spontaneous symetry breaking between the WF and the EM forces) would keep these two forces more similar in their properties. As a result, the mutual tranformation of proton in neutron (and inversely) would be easier and through n-p captures, the nuclear information would be dissipated at a faster rate. It would only be protected by the n-p mass difference .

This time the stellar p-p burning rate would be much faster than the real one . It would take place in stars of much lower temperatures(infra red objects) leading to a very slow release of nuclear entropy. The world would indeed be very different . It is not clear whether life could evolve in these conditions.

THE SCALE OF COSMIC ENTROPY.

Before discussing the birth and evolution of electromagnetic and gravitational information , in order to provide quantitative estimates of their cosmic contributions , we shall discuss here the meaning and status of the *cosmic entropy.*

The natural unit in which to measure the cosmic entropy is the "comoving" volume of space (covolume): a large volume containing many galaxies and expanding with the universe. The volume of this covolume increases with the cube of the temperature ($V=R^3$). Except at particular moments of entropy producing events, the cosmic temperature decreases proportionally to the increase of R ($T=R^{-1}$) The thermal photon population per unit fixed volume is proportional to the cube of the temperature (N(photon)= $T^3=R^{-3}$). Hence the entropy of the covolume (proportional to the population of thermal photons) is conserved during the expansion .

This statement merely reflects the fact that the expansion is a reversible process which by itself generates no entropy. *However its presence turns out to be a major factor for information and entropy generation* . We have already seen how the nuclear information emerges from the fact that the expansion rate becomes, after a few minutes, too fast for the WF to maintain equilibrium between particles.

Today the cosmic entropy is represented by the 400 photons per cc, of the fossil radiation (3 degree background) to which we have to add some 450 neutrinos and perhaps other unknown particles . The total entropy enclosed in the observable universe (15 billion light-years) in of the order of 10^{87}.

Traditionally cosmic entropy is evaluated in another way , based on the fact that the population of nucleons remains essentially constant in a covolume ; while the creation of photons is nature's preferred way of generating new entropy. The ratio of the number of photons to nucleons can then be use as a good measure of the cosmic entropy.

This ratio is approximately 10^9 in our present universe. Compared to typical stellar entropy(neglecting ,for the moment ,black holes) this valueis very large . The question has often been raised of the entropy creating events which could have generated such a large value in the early universe.

This question does not appear to be so relevant any more . Indeed if we go backward in time until we reach temperatures of a few hundred MeV, the ratio of quarks to photons is about one , due to the rapid phenomena of pair creation. But the photon number divided by the difference between the number of quarks and quarks does not change, as quarks and antiquarks are always created in pairs.

According to the ideas presented by Sakharov to explain the baryon asymetry of the universe (the fact that the baryonic number is not identically zero), reactions violating the conservation of this number,(and violating also parity P conservation and parity times charge conjugation CP) did take place much earlier , close to the period where the three forces NF, WF, and EM were unified. While the exact nature of these events are far from being understood. it is postulated that they gave rise to the present value of the baryonic number.

The consequence is that the source of the present large value of the entropy per nucleon is more likely to be explained in terms of these subtle effects of particle physics than through the occurence of entropy non-conserving reactions in the early universe. More correctly, both factors are likely to have played roles whose relative contributions we can not presently evaluate. As mentioned later, compactification of extra geometrical dimensions may also have played a role in this respect

Be it what it may, the present photon to baryon ratio is the appropriate scale on which to evaluate the importance of the various additions to the cosmic entropy.

Consider again the case of nuclear entropy. The best estimate of the helium abundance immediately after the primordial nucleosynthesis is about 24 % in mass. The present cosmic abundance is about 28 % in mass.

The implication is that all the stellar activity in the universe have increased the helium abundance by a mere 4 %. The gamma rays emitted during the corresponding nuclear reactions have been degraded, by the opacity of stellar interiors, to starlight photons of about one eV. This correspond to about 3×10^5 photons for each nucleon or an added contribution of less than one per mil to the cosmic entropy.

The contribution could be much larger if there were ways of "splitting" these stellar photons all the way to the thermal energy of the cosmic background (one milli-electron volt , corresponding to 3kT at 3 degrees) .

Planetary surfaces go part of the way in this direction . Our Earth is at a temperature of about 300 degrees K, some twenty times cooler than the solar surface, at 6000 degrees or so. Yellow (one eV) photons coming from the Sun are absorbed by the ground surface to be later emitted in the form of infrared (one twentieth of an eV) photon. As a result, the Earth is a photon multiplicator by a factor of twenty. The flux of entropy thereby created is "tapped " by all organizing events on the surface of our planets: winds, ocean currents as well as all biological manifestations. If the solar and terrestrial surfaces were at the same temperature, complexity would probably not have developped on earth, whatever would be the solar luminous flux.

Winds are observed in distant planets , as Jupiter and Saturn ,where the surface temperature is colder than on the Earth, and the entropy generation consequently larger. Comets in the Oorts cloud would go even further in this respect . However their extremely low temperatures makes it highly unlikely that they could effectively use their formidable flux of entropy to reach a high degree of organisation .

Quite generally , although the "break-up" of stellar photons all the way to the thermal fossil radiation (for instance by their being absorbed by interstellar dusts) could generate immense amount of entropy we do not know how these photons could be used to generate organisation in the extreme cold of space. As far as we know, terrestrial planets are still the best sites in this respect.

ELECTROMAGNETIC INFORMATION.

When the universe reached about three thousand degrees, some five hundred thousand years after the Big-Bang in the conventional chronology, protons and electrons recombined to form hydrogen atoms. The binding energy (13.6 eV) was emitted under the form of U.V. photons rapidly thermalized to the background temperature . Since the process took place in conditions of kinematic equilibrium, all the protons were bound in atoms and the electromagnetic entropy created at this moment(13.6 eV per nucleon) was immediately released, resulting in the generation of no *electromagnetic information* .

Later on, stars started to generate heavy nuclei with much higher electromagnetic (atomic) binding energy per nucleon . The full dressing with electrons of an iron atom , for instance, releases in space about 500 eV of photons.

When stellar nucleosynthesis takes place, the universe is already at least a few hundred million years old and the fossil radiation is far too cold and diluted to interact reversebly with the bound atomic electrons. Binding energies can be emitted in out-of-equilibrium conditions, thereby providing appropriate conditions for irreversible processes such as photosynthesis.

Thus it would be correct to say that electromagnetic information appears when heavy nuclei are ejected in cold interstellar space at the moment of stellar disruptions in the form of red giants, planetary nebulae, novae or supernovae. On earth, EM information is stocked in the ground, under the form of oil or coal, the product of ancient photosynthesis , ready to release their energy and entropy in appropriate conditions as provided by furnaces or carburators. Figure 5 shows the growth of EM information as a function of time . As in the case of nuclear information, only a small fraction of this electromagnetic information can be usefully released by natural processes.

GRAVITATIONAL ENTROPY (CLASSICAL).

Let us consider the entropy variation of a gas of $N = 10^{57}$ free particles of mass m (one solar mass of nucleons) enclosed in a volume V with temperature T, as this volume shrinks progressively. For the moment we neglect the presence of the gravitational field except as an agent of volume reduction.

The volume entropy is given by the formula of Sackur-Tetrod:

$$S = kN \left(5/2 + \ln N/V \left(2\pi mkT/h^2\right)^{3/2}\right)$$

Let us define $d = (V/N)^{1/3}$ as the average distance between particles in the volume.

Define also: $l_{th} = h/ m\, v_{th}$, with $3kT = m(v_{th}{}^2)/2$. Here l_{th} is the De Broglie thermal wave length of the particles at temperature T . Then:

$$S = kN \{ 5/2 + 3 \ln (d/ l_{th}) \}$$

The minimum entropy per particle is reached when $d = l_{th}$ which is then $l_{(Fermi)}$ since we have then reached degeneracy. Then:

$$S = 5/2\ kN.$$

In table 1 a few typical astrophysical examples are considered: a) an interstellar cloud, b) the solar center, c) a white dwarf, d) a neutron star, .(S^e and S^p are associated with the electrons and the protons respectively.)

TABLE NO 1

	kT (eV)	d	$l_{th}(e)$	$l_{th}(p)$	S^e/kN	S^p/kN
		(fermi=10^{-13}cm)				
Cloud	10^{-2}	10^{13}	$4X10^6$	$6X10^4$	47	60
Sun	10^3	10^5	1200	200	16	21
W.D	10^6	10^3	400	6	6	18
N.Star	10^8	10	400	0.6	2.5	10

TABLE NO 2

	$S/kN/(S/kN)^{min}$	S(total)	Gravitational entropy emitted per N.
Cloud	42	10^{59}	=1
Sun	15		10^3
W.D.	10		10^6
N.Star	5	10^{58}	10^8

The entropy decrease, going down the columns of the table, is related to the fact that we have an increasingly good localisation of the stellar object under study. The entropy per nucleon goes down by about a factor of ten, going from a typical interstellar cloud to a neutron star configuration.

This computation is of course highly incomplete since it takes no account of the gravitational field itself, nor of the stellar temperature gradients. Its real merit is that we can obtain numbers which illustrate qualitatively a part of the story.

GRAVOTHERMAL CATASTROPHY.

Consider N particles in an adiabatic box. If we "turn off" gravity, they tend toward an isothermal distribution. This equilibrium is stable. It corresponds to a state of maximum entropy. This is also the case in all situations where the gravitational energy content is smaller than other energy contents.

In cases where gravitational energy is important, the isothermal state is no more the state of maximum entropy. The system is unstable to density fluctuations. Contraction occurs which, by the virial theorem, leads to the creation of temperature gradients and to a flux of photons from the warmer to the cooler regions. These photons are carrying what we shall "classical gravitational entropy" originating from the gradually increasing binding energy of the central condensation.

The physical reason behind the strange behaviour of the gravitational force, in relation with entropy generation, is the long range nature of its span. While the EM ,W and N forces are only effective between close neighbours, the G force extends throughout the entire system. Each particle interacts with all the other particles. As a result, while the electromagnetic and nuclear energy contents of a given mass are proportional to this mass, the gravitational energy content is proportional to the *square* of this mass.

Fluctuations in density have a tendancy to grow, thereby increasing their gravity field and further accelerating the contraction. The local infall of matter generates heat and induces a temperature gradient toward the center. Radiation emission removes part of this heat , thereby reducing the thermal pressure which tries to resist further collapse. No stableequilibrium can be reached and the system moves farther and farther away from isothermy. Hence the words " gravothermal catastrophy". *By this process, the particular behaviour of the gravitational force is responsible for the birth of thermal gradients in the universe.*

Consider the Sun when it reached the Main Sequence. It's gravitational binding energy was (and still is) about one keV per nucleon. This binding energy had been emitted throughout the Pre-Main-Sequence phase under the form of infra-red (one eV) photons. The gravitational entropy generated by this starlight (one thousand units per nucleon) far exceeds the "localisation entropy decrease "computed previously . In table 2, the gravitational entropy (per nucleon) emitted by various stellar structures is evaluated in the last column. Clouds are almost transparent to their own infrared emission , thus very little entropy is produced . For white dwarves and neutron stars , I have, for simplicity, assumed that the binding energy of one MeV (for w.d.) and one hundred MeV (n.s.) is emitted in the form of one eV photons.

GRAVITATIONAL INFORMATION.

In the super-hot fluid of the early universe, no condensation could occur. In terms of our previous discussion, the thermal energy content was larger than the gravitational energy content, so that the isothermal state was a stable state of equilibrium.

More accurately we should ask: how big a chunk of matter should we have considered at any given time (and any given temperature and density) to insure that , in this object, the gravitational energy is larger than thermal energy (remember that gravitational energy increases as the square of the mass, while thermal energy increases linearly with mass) This of course is merely the Jean's problem, the search for the Jean's mass.

In a standard Friedman model of the early expanding universe, the Jean's mass is very closely equal to the mass within the cosmological horizon. The situation changes abruptly at the moment of recombination when the universe is approximately one million years old. Two almost simultaneous events are responsible for the alteration . First, the fact that, at that moment, the matter density becomes larger than the radiation density. Second, the capture of electrons by protons to form bound hydrogen atoms leading to a strong reduction in the matter-radiation interaction.

After this important moment, galaxy and (still later) star formation becomes possible, just as iron production became possible around ten billion degrees. Pursuing the parallel, the same difficulties are met with timescales and nucleation problems. Small condensations, as natural seeds of big condensations, may not be possible even when large condensations are possible.

Stellar formation take times. The numberless many-colored stars that shine in dark nights are slowly dissipating in space the gravitational (and nuclear) entropy made available at the moment of recombination, when gravitational collapse became possible.

The figure 6 illustrates the behaviour of gravitational information (classical) in the expanding universe. We consider an object of stellar size at various stages of its evolution. The slanted curves shows the maximum entropy released (per nucleon) by , for instance , a sunlike object, if all the photons emitted were thermalized to background radiation. In the region A the number S^{max}_G / N is less than one. This is analogous to the region A of the nuclear information diagram (fig 1). In region B this factor is larger than one but the Jean's mass is much larger than the size of our stellar object, so that the thermal pressure is able to prevent its condensation. As the universe reaches the recombination period (3kT - one eV) gravitational collapse becomes possible at least for globular cluster size objects. Further condensation and fragmentation gives rise to stellar structures, with the release of the corresponding gravitational (and nuclear) entropy.

Gravitational information is zero before recombination time. It appears suddenly when the collapse becomes possible at this very moment.

GRAVITATIONAL ENTROPY (QUANTUM EFFECTS)

In classical gravitational theory, a contracting body can emit a limited amount of entropy. The entropy flux is quenched when the gravitational binding energy becomes comparable to the rest mass, that is when the structure becomes a black hole. After that stage, nothing comes out of the condensing object.

The study of quantum effects in the curved space surrounding a black hole have brought about a different conclusion. One can associate to a black hole a quantum gravitational entropy, which can eventually be released in the universe, throught a mechanism called" black hole Hawking evaporation".

An order of magnitude estimate can be obtained by an heuristic argument. First, a temperature is associated to the black hole, in analogy with a standard thermodynamical "black body", stating that photons emitted by such a body have a mean energy:

$$kT_{b.h.} = hc/ R_{b.h.}$$

Next, remember that the radius $R_{b.h.}$ of the black hole is the smallest distance on which we may have any information about the black hole. In consequence, the wave lenght emitted by this object must be (at least) of the same order.

The radius of the B.H. is $R = 2GM/c^2$.

Hence : $kT = hc^3/2GM$, is inversely proportionnal to the mass. Numerically we obtain:

$$kT = M_{pl}^2/ M_{b.h.} = 10^{-7}(M_o/M_{b.h.})\,K$$

(where M_{pl} =(hc/G) $^{1/2}$, is the Planck mass, equal to 10^{19}GeV or 22 microgram and M_o is one solar mass),

The entropy emitted can now be obtained through our" golden rule". How many photons do we need to exhaust the mass M? (neglecting in this approximation the fact that as the mass decreases, the temperature increases). The answer is :

$$S = M/kT = k(M_{b.h.}/ M_{pl})^2$$

$$S/k = 10^{77}(M_{b.h.}/M_o)^2$$

Note that the entropy of a b.h. of onesolar mass is 10^{77}, while the entropy of the initial ensemble of elementary particles in the dispersed state was approximately 10^{57}. The large entropy increased is due to the very many ways (microscopically) by which the b.h. can be prepared .

Notice that the b.h. entropy is proportional to the square of the mass, ,just as the gravitational energy , and for the same reason .

The lifetime for evaporation of b.h. is given by:

$$t_{b.h.} = t_{pl} (M_{b.h.}/M_{pl})^3,$$

(where $t_{pl} = (Gh /c^5)^{1/2} = 5,4 \times 10^{-44}$sec $\cong 10^{-43}$sec.)

With these data, we can evatuate the growth of available gravitational entropy in the expanding universe. A ***causal volume*** is defined as a volume of spacc in which physical effects have had enough time to take place. In most cosmological models , the distance to the horizon grows linearly with time ($d \propto ct$), hence the causal volume increases with t^3. On the other hand, the energy density (proportional to T^4 in the radiation dominated era) decreases with t^{-2}, so that the total energy in a causal volume increases with t. The largest amount of entropy could be obtained if all this energy would be collapsed in one b.h. Numerically we find:

$$S(max) = k (t/ t_{pl})^2$$

Let us assume that this expression can be extrapolated all the way to the Planck time (most likely a very poor assumption) . Then we find that the maximum entropy is approximately equal to one (in unit of k) at the Planck time , and grows as the *square* ot the elapsed time.

We understand this result if we remenber that at the Planck time, the horizon contains an amount of energy equal to the Planck mass, which evaporates in one Planck unit of time (for this reason this object is called an "instanton").

What is the entropy of a causal region which contains only radiation (no b.h.)? As discussed before , it is given by the number of photons (proportional to T^3 hence to $t^{-3/2}$) in the causal volume ($\propto t^3$). Numerically we have:

$$kT = m_{pl} (t_{pl} / t)^{1/2} \quad \text{and}$$

$$S = k (t / t_{pl})^{3/2}.$$

These equations give us a qualitative understanding of the physical situation. . As we move to the Planck time , we have to consider the gravitational effects of the radiation which are "embedded " in the fact that one photon of Planck energy is indistinguishable from an instanton. In the simple picture presented here , the radiation entropy is comparable to the maximum entropy at the beginning, but grows more slowly than this maximum value as time proceeds.

The gravitational information picture must be reconsidered in the framework of quantum (black hole) gravitational entropy. Again we define the gravitational information as the difference between the maximal entropy and the the real radiation entropy of the universe. Fig. 7 gives us a representation of the growth of gravitational information since the Planck time.

We do not have much information on the amount of b.h. in the universe. A few candidates of stellar mass are known in our galaxy. The center of active galaxies is believed to harbor giant b.h. of hundred of millions of solar masses. All this represents less than one part in one thousand of the mass in the universe. More b.h. may be present in the "dark matter " component needed to explain the flat rotation curve of the galaxies or the binding of clusters of galaxies. But we have no *direct evidence* that b.h. contribute significantly to the mean density of the cosmos.

There is no *a priori* reason why the universe did not incorporate in its early days a population of b.h. of all masses. To discuss this point, we shall consider the fate of a b.h. embedded in an adiabatic volume V filled with thermal radiation at temperature T_r.(Sugimoto, Eriguchi, and Hachisu, 1981)

The total energy E is given by :

$$E = M_{b.h.}c^2 + aV\,T_r^{\,4}$$

and the entropy:

$$S = k\,(M_{b.h.}/M_{pl})^2 + (4aT_r^{\,3}/3)V$$

The b.h. absorbs a flux of photons coming from the thermal radiation, and simultaneously evaporates away his own entropy. The net effect (increase or decrease in mass) can be evaluated by looking at the entropy balance :

$$dS = 2\,M_{bh}\,dM_{bh}\,/(M_{pl})^2 - c^2\,dM_{bh}\,T_r$$

The b.h. will grow if dS is positive, that is: if the mass of the b.h. is larger than a critical mass given by :

$$M_{bh} > M_{bh}^{\ cr} = M_{pl}\,(T_{pl}\,/\,8\pi\,T_r)$$

(The higher the radiation temperature, the lower the number of photons needed to generate one unit of mass.)

We meet again here the problem of the nucleation centers. At any radiation temperature, small b.h. will evaporate away while large b.h. will grow (" one lends only to the wealthy"). This is quite reminiscent to the formation of bubbles in water near the boiling point. Below a certain size, vapor bubbles are resorbed back in the liquid, leading to" superheating "of the fluid.

Today b.h. with masses over 10^{26}gr (the mass of the Moon) would grow while smaller bodies would evaporate away. The expression for the temperature of a b.h. can also be used in this respect: 3 K, the present cosmic temperature, corresponds to a body of lunar size.

Bodies of less than 10^{15} grams would have completed their evaporation today. The larger ones should still be around . The absence of observational evidence for these objects suggests that the initial universe did not contain such b.h.

Thus the initial universe appears in a state of low gravitational entropy compared to what would have been the case if b.h. had been very abundant. At the Planck time this entropy is, nevertheless, very close to its maximum value, (zero gravitational information) . But, as time goes on , the gravitational information grows (fig 7) .

Gravitational collapse of objets is prevented, by thermal pressure, until recombination time. Even then, the thermal pressure (within stellar bodies) is still active in retarding or even preventing the collapse all the way to the b.h.

For stars of approximately one solar mass , the contraction will be stopped by the degenerate pressure of the electrons (white dwarves) while for masses up to three solar masses, the degenerate pressure of the nucleons will play the same role (neutron stars) . Only larger stars could collapse all the way to the b.h. state.

The release of gravitational entropy has two distinct regimes, one associated with the classical grav. entropy and one with the quantum grav.entropy.

The first one has been discussed previously (fig 6). It begun as soon as the first stellar generation was formed. It is still going on in the sky today. It will end up in about one trillion years , when all the diffuse matter will have collapsed in stellar corpses (white dwarves, neutron stars, b.h.) For b.h. the flux of entropy is quenched by gravitational effects when the binding energy reaches a value of about one hundred MeV per nucleon.

The second regime is associated with the evaporation of the b.h. For stellar objects it does not become significant before some 10^{64} years.

CHRONOLOGY OF CRUCIAL EVENTS FOR THE ORIGIN OF ENTROPY AND INFORMATION.

Complexity in the universe is borne by many-body systems appearing during phases of non-equilibrium (thermal and kinetic) reactions. The corresponding processes lead to emission of entropy, usually in the form of photons. This entropy is drawn from a sort of bank of entropy which we call information , stocked for a long time and ready to be used in appropriate conditions. Our aim here has been to identify the main factors responsible for the existence of extra sources of entropy (over and above the entropy of the fossil radiation) and for the fact that these sources were not entirely tapped as soon as they became available.

We have identified in the existence of the natural forces one major factor in origin of entropy , inasmuch as they are able to create bound systems and liberate some photons from the destruction of the corresponding mass difference. Here I plan to go down the chronology of the universe discussing, one by one, the most important epochs and their contributions to the origin of entropy and information.

Planck epoch

Very little is understood about the physics around the Planck time (10^{-43} sec , E = 10^{19} GeV) It has become traditional, in view of our ignorance, to identify as "initial data " the various properties of the universe (bulk properties, laws of physics), at this remote time. (Contemporary cosmology deals with the idea of cosmic superstrings, leading to space with many extra dimensions , the most popular version includes six new dimensions. The compactification of these extra dimensions woould in certain theories, be responsible for the existence of the natural forces and would contribute an amount of photons perhaps responsible for the cosmic entropy.)

One important characteristics of early universe is the absence of a large amount of b.h. of various sizes, or in other words *the very low value of the gravitational entropy.* This fact is of major importance for the birth of gravitational information.

The importance of vacuum energy terms in the early universe is the subject of much recent developments in cosmology. These terms , acting as cosmological constants in the Einstein equation, give rise to chapters of *inflationnary* expansion which in principle could "explain "the very low population of b.h. in the universe, together with many other surprising properties of the cosmos. This subject has been treated by other lecturers during this school.

Grand Unification epoch.

It was usually accepted , up to a few years ago , that at a temperature of about 10^{15} GeV (10^{28} K) the N , EM and W forces were "unified" (one coupling constant) . Recent developments , in particular the development of *supersymmetric* theories tend to indicate that the unification energy may be somewhat larger, up to 10^{17} GeV, but most likely below the Planck energy (10^{19}GeV).

The processes bringing about the differentiation of the forces (spontaneous symmetry breaking : SSB) are responsible for *giving masses* to the quarks and the electrons. This event is if course of major importance in our chronology since the transformation of a part of this mass in binding energy (photons), by gravitational and nuclear processes in stars, is the main source of entropy in our present universe.

Baryogenesis , that is the occurence of phenomena leading to a small excess of quarks over antiquarks, is usually believed to occur at this moment , perhaps somewhat delayed until the end of the accompanying inflation chapter (reheating) . Without this event, essentially all the quarks would have annihilated later on, giving rise to a lot of entropy but no baryonic matter to stock information.

As time went on ,the coupling constants of the three forces evolved differently . The nuclear coupling , rose progressively from a value of about 0.02 to about one. This increase is responsible for the fact that about one percent of the nuclear mass can be released in the form of entropy by stars.

Electroweak epoch.

The temperature at which the W and EM forces differentiate (100 GeV) is related to the value of the masses of the intermediate bosons W and Z . As discussed in the text, an increase of one order of magnitude in these masses would have resulted in an increase in the decoupling temperature of the weak interactions leading to an exhaustion of the free protons during primordial nucleosynthesis , with the simultaneous emission the all nuclear entropy and hence no build-up of nuclear information. To the question: why is the Weinberg -Salam phase transition, leading to the splitting of the W and EM, taking place so long after the GUT transition? or in other words : why are the masses of the intermediate bosons so much smaller than the Planck mass or the GUT boson masses? One could answer that if it had been higher ,the consequent absence of Main -Sequence stars would have severely impaired the advent of complexity.

Epoch of primordial nucleosynthesis.

Many events occur around the first minutes , at temperature around one MeV, first: the decoupling of the neutrino (weak) interaction letting the population of neutrons to freely decay; second: the wholesale annihilation of positrons and electrons, and then the capture of the free neutrons by protons to form deuterons and helium nuclei.

The major factor, with respect of the growth of nuclear information, is the fact that most of the free protons escape nuclear capture at this time and remain available for nuclear burning in stars .

Epoch of recombination.

Two major events take place when the temperature reaches one eV or so. The universe becomes matter-dominated and the electrons combine with the protons , leading to a decoupling between matter and radiation.

The crucial consequence for the birth of information is the onset of thermal gradients. Freed from the homogeneizing effects of the radiation pressure, local condensations of matter can yield to the gravitational pull, thereby emitting gravitational entropy (classical) and reaching the high temperatures needed for the onset of nuclear reactions and the emission of nuclear entropy. The lowering of the cosmic temperature, brought about by the expansion, insures that these entropy emitting processes will take place in conditions far from thermal and kinetic equilibrium.

This moment is the birth-date of gravitational (classical) entropy, inasmuch as star formation becomes possible but is delayed by the interaction times needed for the actual growth of these massive bodies. Quantum gravitational entropy, stocked since the Planck time will only be released after the evaporation of the b.h. a process which for a star of solar mass will last approximately 10^{64} years...

GOLDEN MOMENTS FOR THE ORIGIN OF ENTROPY AND INFORMATION.

Epoch	Entropy	Information
Planck 10^{-43} sec	Cosmic entropy from compactified spaces?	Absence of primordial b.h. Birth of quantum grav. inf. Vacuum energy terms.
	Origin of forces.	
GUT 10^{-35} sec	Masses given to q and e	B, C, CP violating reactions
EW 10^{-10}sec		Mass given to W and Z
BBN 100 sec.		75% of H survives. Origin of nuclear information
Recombination 10^5 years		First T gradients. Birth of class. grav. inform. Birth of EM information.

Bibliography

Brillouin, L. Science and Information Theory. Academic Press, New York, (1962).

Keith, A., Entropy., Am. J. Phys. 52 6 (June 1984).

Sugimoto, D., Eriguchi, Y., Hachisu, I., Prog. Theor. Phys. Supp. 70, (1981).

Zurek, W.H., Phys. Rev. Let. 1689, 49 (1982).

Frautschi, S., Science 593 217, (13 Aug 1982)

Weinberg, S., Physics Today, 35, Aug 86

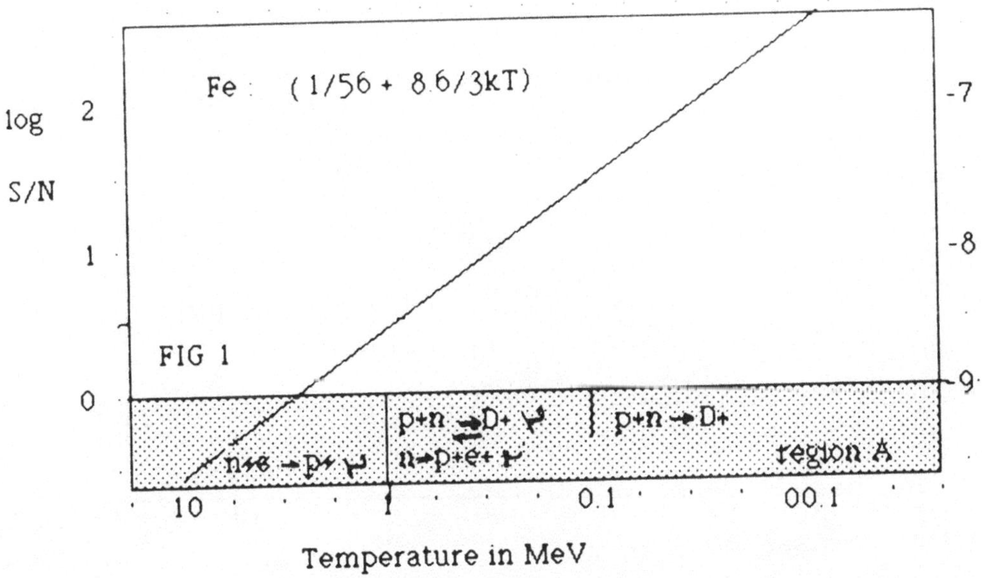

Figure no 1. The birth of nuclear entropy at the moment of primordial nucleosynthesis . In abcissa, the decreasing cosmic temperature, In ordinate the maximum amount of nuclear entropy per nucleon available by the (hypothetical) formation of iron -56 nuclei from the sea of nucleons, assuming that all the released energy would be broken away in thermal photons.

At temperature higher than 2.7 MeV this number is less than one and the state of maximum entropy is associated with free nucleons (region A). At lower temperature the state of maximum entropy becomes the state with all nucleons bound in iron nuclei. However the transmutation can not start before deuterons can survive the thermal bath. .

On the left side, the ordinate is multiplied by 10^9 to give the ratio of nuclear entropy per nucleon over the entropy per nucleon in the cosmic radiation. (For convenience I have put the entropy equal to the number of thermal photons , instead of 7.1 times this number.)

The major kinetic events of this period are also indicated in the diagram.

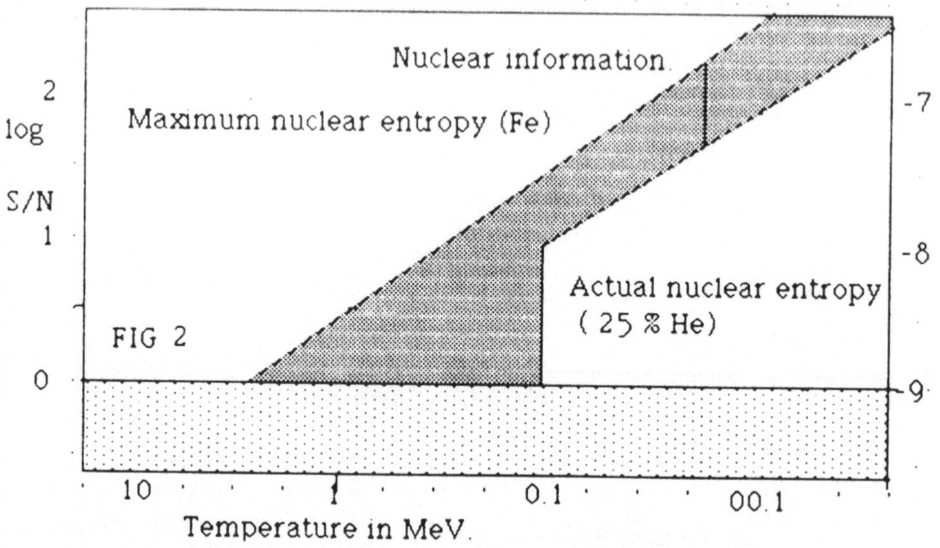

Figure no 2. The birth of nuclear information. Same coordinate as in the figure 1 . Here is shown again the curve of maximum possible nuclear entropy per nucleon and the amount of nuclear entropy per nucleon generated by the actual transformation of some 25 % of the nucleons into helium nuclei. The difference between the two curves is ,by definition, the nuclear information in the universe . This number is zero above a few MeV since both the maximum nuclear entropy and the actual nuclear entropy are zero.

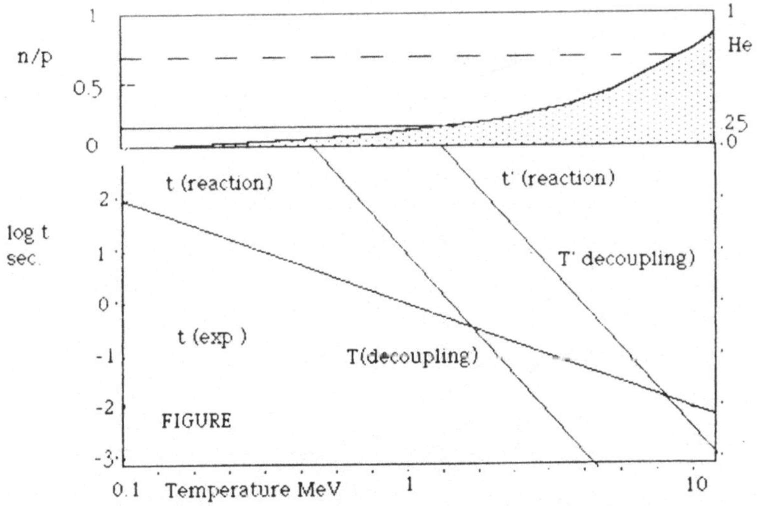

Figure no 3 Decoupling of the neutrino interactions. The abcissa gives the cosmic temperature and the ordinate, the age of the universe in seconds. The curve t (exp) gives the relation between cosmic age and temperature ($t\ (exp) \propto 1/[(g^*G_N)^{1/2}\ T^2)]$ in the standard Big-Bang. The t (reaction) curve is the mean reaction time for weak interactions involving neutrino capture and emission, ($t\ (reaction) \propto 1/(G_F^2\ T^5)$). At temperatures below the crossing of the curves (at the decoupling temperature T(decoupling) $\cong$ 1 MeV), the neutrino interactions are too slow to keep pace with the expansion and the neutron-proton equilibrium abundance is no more insured.

An increase in the mass of the W particle would decrease the value of the Fermi constant ($G_F \propto M_W^{-2}$) and shift the reaction-time curve t'(reaction) to the right in the diagram, leading to a higher decoupling temperature T'(decoupling).

On the upper part of the diagram, the n/p ratio is shown as a function of temperature. On the right side of the figure, the abundances are given by the Boltzman equilibrium equation; on the left side, the neutrons are freely decaying. On the scale at the right is given the resulting helium abundance. The position of the decoupling T can be altered by changing g^*, G_F or G_N as seen from the expressions for the timescales. The BBN turns out to be a very sensitive test of the "constancy" of the coupling constants.

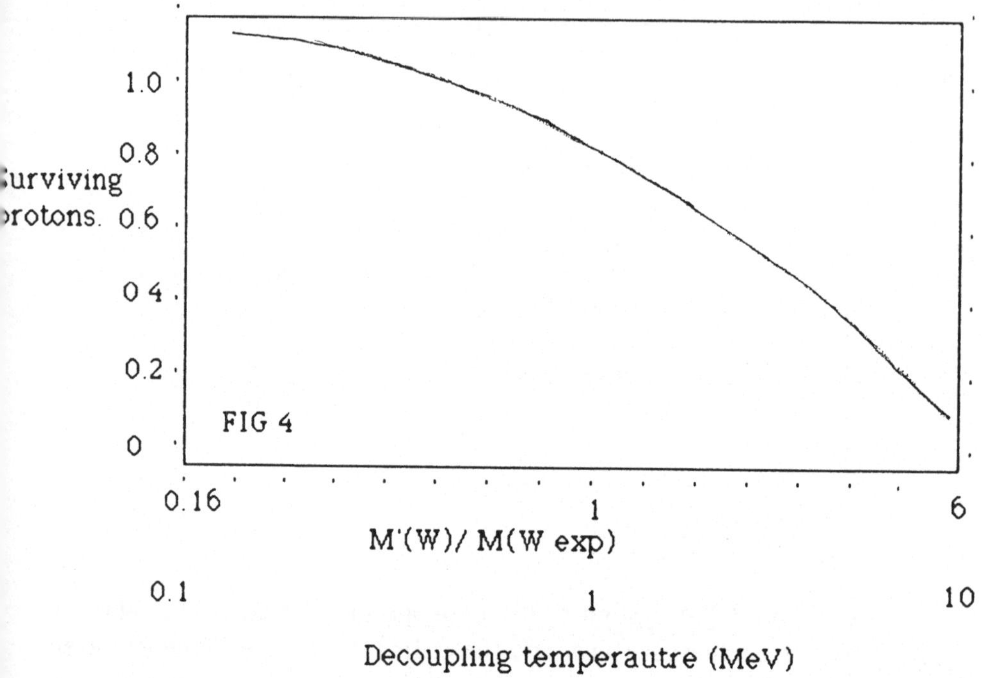

Figure no 4. Fractional surviving proton abundance as a function of the assumed mass of the W particle. In the abcissa the scale is in units of the measured mass. The scale gives the corresponding decoupling temperature (neglecting the effect of the alteration of M_W on the neutron-proton mass difference).

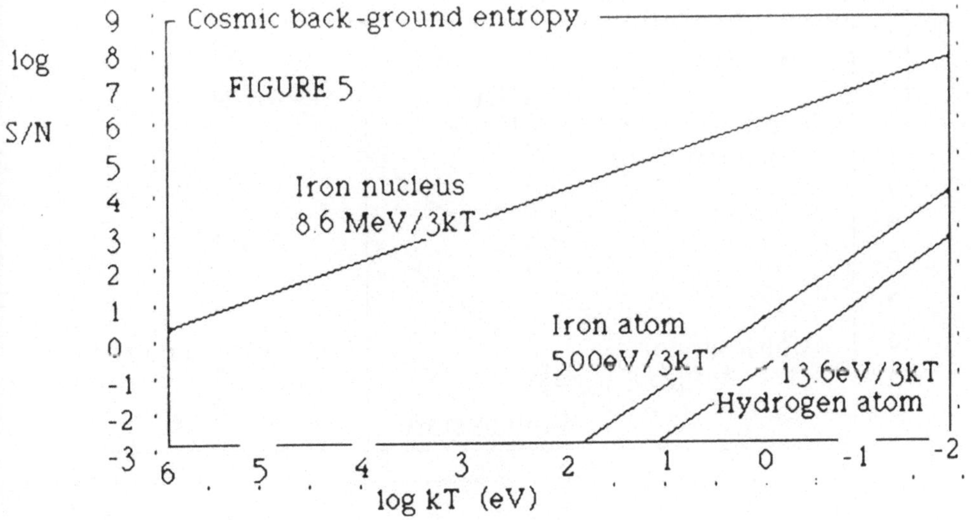

Figure no 5 . The growth of the maximum nuclear and electromagnetic entropy . The units are the same as in figure 1 and 2. The upper curve is the continuation of fig 1 (nuclear entropy for iron), the lower curves shows the maximum electromagnetic entropy per nucleon emitted by the formation of one atom of hydrogen and one neutral atom of iron (taken as a typical example). Here, as before, the factor of 7.1 between entropy and number of thermal photons has been put equal to one.

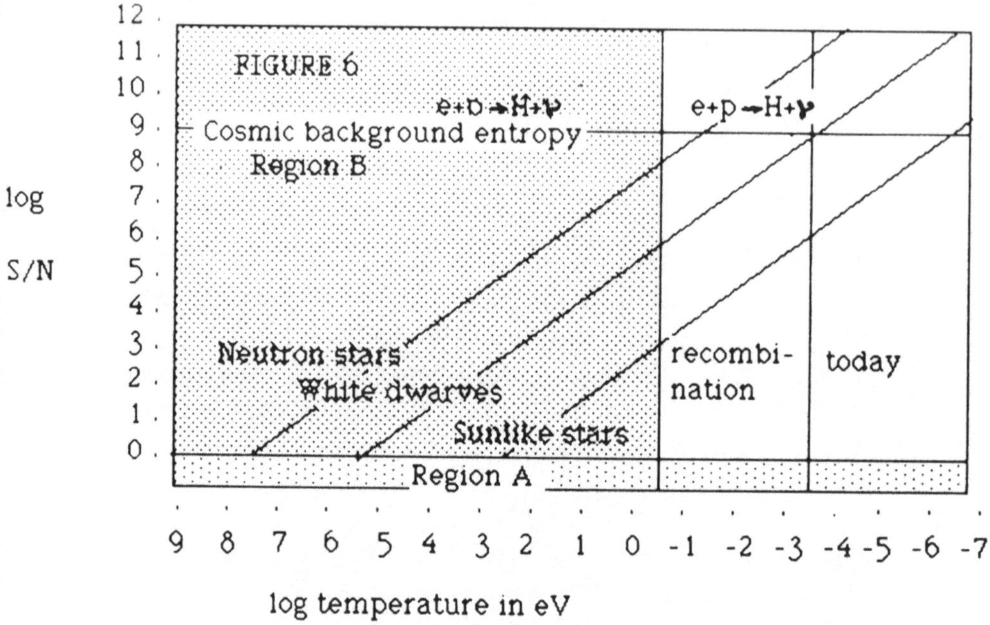

Fig no 6. <u>The growth of the maximum classical gravitational entropy.</u> (The units are the same as in figure 1.) The various curves show the amount of gravitational entropy per nucleon emitted by a star to reach various stellar stages (Main Sequence, white dwarf , neutron star,) assuming this stage *could* be reached at the the corresponding period and that all the emitted photons *could* be thermalized to the temperature of the fossil (back-ground) radiation . Below the recombination temperature (region B) , the collapse is prevented by the thermal radiation pressure.

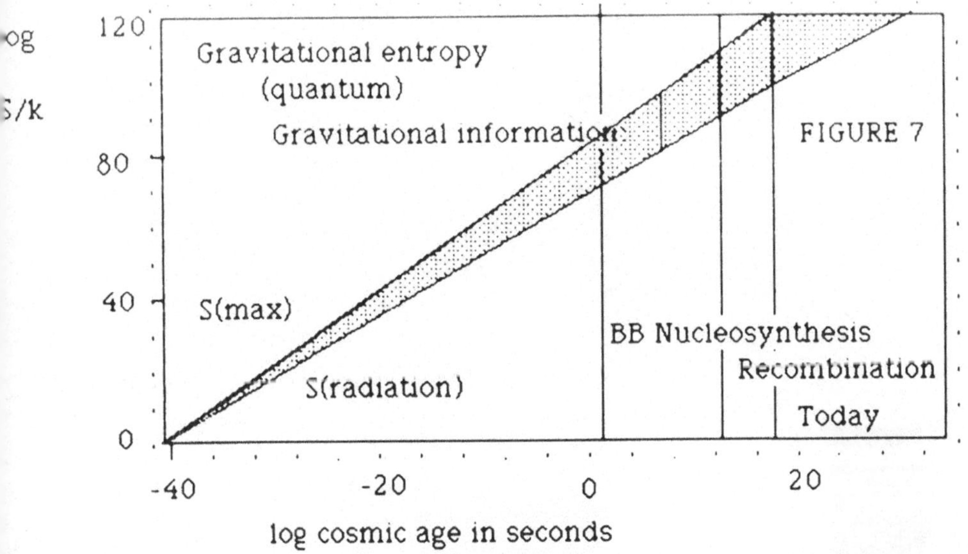

Figure no 7. <u>Growth of quantum gravitational information in the expanding universe.</u> The upper curve gives the maximum amount of entropy within the horizon ($S_{max} \propto t^2$), (corresponding to the collapse of all the energy density into one b.h. and ignoring all form of vacuum energy), in units of k, as a function of cosmic age . The lower curve is the entropy in the cosmic radiation within the horizon ($S_r \propto t^{3/2}$). The difference is the amount of quantum gravitational information in the universe .

TRENDS IN THE ASTROPHYSICS OF LIGHT ELEMENTS: COSMOLOGY AND STELLAR PHYSICS.

Hubert Reeves
Section d'Astrophysique, C.E.N.S. Saclay.
Institut d'Astrophysique de Paris.

Abstract: The properties (nuclear, atomic, molecular) of the light elements H, He, and Li are known well enough to warrant the study of many astrophysical processes through their abundance determinations. In this paper I plan to review three domains in which this situation applies:

A) THE SPECIES MULTIPLICITY OF ELEMENTARY PARTICLES

B) THE STABILITY OF THE EXTRA DIMENSIONS OF THE COSMOS

C) THE KINETICS OF STELLAR SURFACES AND STELLAR WINDS

P. Galeotti and D. N. Schramm (eds.), Gauge Theory and the Early Universe, 193–210.

Primordial nucleosynthesis

The years 1965 -1975 have witnessed the establishment of primordial nucleosynthesis , (also called BBN for Big-Bang nucleosynthesis). The problem of the origin of the light elements from ^{2}H to ^{11}B was gradually solved , and it became clear that the four isotopes ^{2}H, ^{3}He, ^{4}He, and ^{7}Li , were bona- fide relics from these ancient hot times. (1 to 8) . It is worthwhile, at this point, to mention the fact that the solar wind experiment of Johannes Geiss and his Berne collaborators, in giving reliable non-terrestrial values for the helium isotopic ratio, played an important role in the set-up of this achievement(9).

Following the discovery of the fossil radiation(10, 11) (also called cosmic background radiation CRB), which was rapidly interpreted as evidence that the cosmic temperature had been up to, at least, a few thousand degrees, theoretical calculations of primordial nucleosynthesis showed that reasonable account of the abundances of the light isotopes could be obtained if it was futher assumed that the scale of past temperatures had reached the 10^{10} degrees (one MeV) mark.

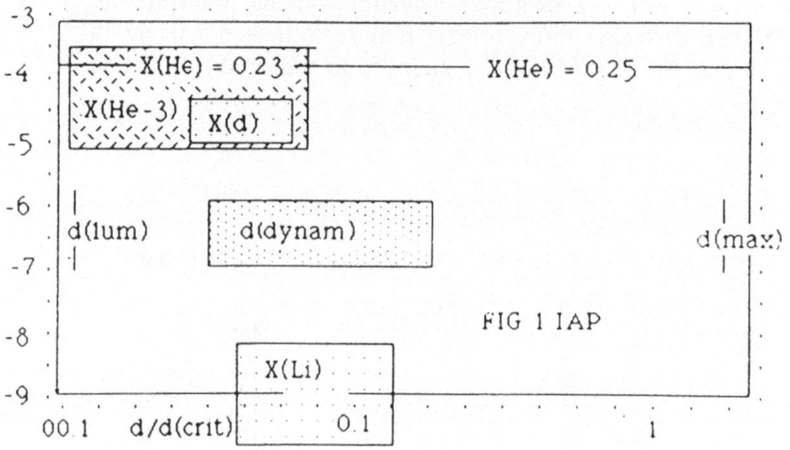

Fig I Estimated cosmic densities . The abcissa is units of the closure density. On the left, d(lum) is the density of luminous matter . The range of d estimated from dynamical effect is enclosed in a box; Also shown is the upper limit from the search of galaxy deceleration. The light isotopes observational uncertainties and the corresponding uncertainties in the calculated BBN densities are shown in boxes.(fractional mass abundances) (scale on the left).

This calculation also provides an estimate of the baryonic(nucleonic) density of the universe. The estimates (fig 1) are between three and twelve percent of the closure density . (The main uncertainty is due to the poorly known distances of remote galaxies). The best estimates of the total (bayonic and non-baryonic) cosmic density, from dynamic effects on galactic motions, yields values around five to twenty per cent of the closure density . Thus there is (in July 1986) no disagreement between the baryonic density given by BBN and the dynamic evaluations. There is thus no sound proof of the existence of non-baryonic matter contributing in a major way to the total density of the universe.

(To avoid confusion one should also make the distinction between luminous and non-luminous matter. The density of shining matter, stars, galaxies, amounts to only one per cent of the closure density. Thus at least ninety per cent of the matter is invisible - the so-called missing mass - but may well be baryonic).

In the last decade, the success of primordial nucleosynthesis has become a major card in cosmological studies and explorations. It has been used as a *ground basis* for testing new hypotheses or even conceptual frameworks. It has served as an "anchor" for theories -and theorists- to keep contact with observations , (a fundamental but sometimes overlooked condition for successful scientific progresses). There is already a large cemetery of cosmological models , brillant or not , which have died because they could not reproduce, in a convincing matter, the success of the simple BBN . In this review I want to discuss two areas of research in which primordial nucleosynthesis has played a proeminent role.

Proliferation of particle families.

In the present standard theory of physics , the elementary particles are grouped in three main families called *electronic*, *muonic* and *tauonic*. Each family contains four members : two leptons and two quarks . To each member of a given family correspond members in the other two families which , as far as we know, differ only in masses. For instance , to the electron (0.5 MeV) corresponds the muon (107 MeV), and the tauon (1.5 GeV).

The families are shown in the table with the best estimates of the masses . We have, so far, no evidence of any neutrino masses, only upper limits are quoted. The existence of the t-quark has not yet been firmly established.

FAMILY BOOK.

electronic	muonic	tauonic	?onic	?onic
electron-neutrino <10 eV	muon-neutrino <.25 MeV	tauon-neutrino <100MeV	?	?
electron 0.5 MeV	muon 0.1 GeV	tauon 1.5 GeV	?	?
quark-u 0.3 GeV	quark-c 1.5 GeV	quark-t 40 GeV (?)	?	?
quark-d 0.3 GeV	quark-s 0.5 GeV	quark-b 5.0 GeV	?	?

The question in every one's mind is *how many more* families are there in our big blue world. As a few years ago, high energy physics had hardly anything to say about this question.

Big-bang nucleosynthesis , on the other hand, was making very definite predictions. The number of extra families could not be very large. At best, one or two . More probalv , in fact, we already have come to the end of the list with our three families.

The experiments leading to the discovery of the W and Z particles (responsible for carrying the weak interaction) have confirmed this prediction of primordial nucleosynthesis in limiting the number of new (extra) families to three at most. In the following paragraphs, I will present the physical arguments behind these statements. Here I want to take the occasion to put some emphasis on the importance of the event I am talking about , because it seems to have passed largely unnoticed in the astrophysics community.

We should keep in mind that Big-Bang is , in the usual scientific context , quite an extravagant theory. Contrary to the scientific paradigm held unequivocally from the time of the ancient Greeks, throughout the Renaissance until quite recently, it places the universe in an *historical framework*. Instead of being the observer of an eternal unchanging realities, the astrophysicist becomes an *historian* exploring the past, in search of events which have given the world the properties it has today.

A similar transition had already taken place in the life sciences ,one century ago , when Darwin denied the "fixity" of the animal and plant species ,to introduce the notion of biological *evolution* With the Big-Bang , this notion is enlarged to the whole physical universe.

The more extravagant, (or out-of-the-beaten- path), a theory is, the stronger should be the proofs in his favour, before it is accepted. Confirmed predictions are of prime values here, as it is always easier to find explanations to known facts than to predict correctly the result of a future observation or experimentation . After correctly predicting the existence of the fossil 3 K radiation, the theory has also passed successfully the test of the family proliferation . This is worth a double mention.

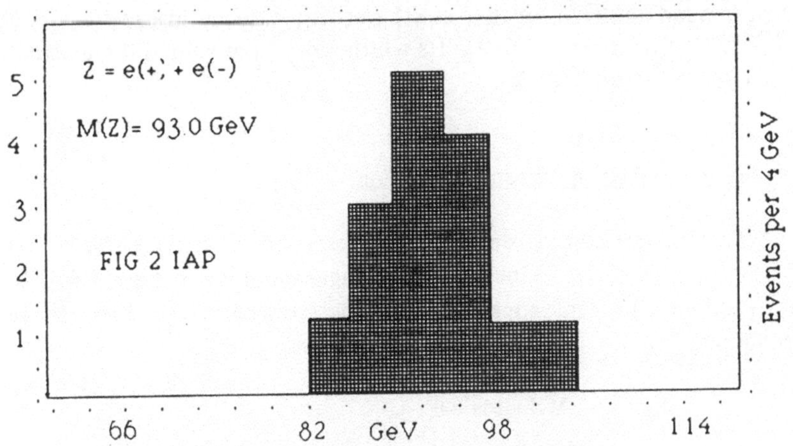

FIG 2 Electron-positron resonance manifesting the existence of the Z particle at CERN. The width of the resonance is a function of the number of decay channel hence of the number of unknown families of elementary particles. The observed width can not accomodate more than three undetected families over and above the three known families.

<u>Evidence from the width of the Z</u>

Figure 2 gives the experimental basis for the detection of the Z particle (carrier of the neutral weak interaction) at CERN in 1984 (12 and 13). The curve shows the large resonance in the electron-positron cross-section around 93 GeV (the mass of the Z). The energy width of this resonance is according to the Heisenberg principle , related to the lifetime of the particle. And the lifetime is related to the number of channels (partial widths) in which the Z can decay.

Contrary to us, the Z "knows" how many families there are : it decays in all possible channels open to him . Hence the observed width of the resonance gives us a measure of the total number of families.

Numerically, the energy width of the Z, computed taking into account the three known families, turns out to be 2.63 GeV. Each unknown neutrino would contribute an additional 180 MeV to the width. Comparison with the increasingly accurate observational data shows that the upper limit on unknown channels corresponds to the equivalent of less than three new neutrino types, with a 90% percent degree confidence.

There are some restrictions . The Z particle cannot decay in righ-handed neutrinos (if they exist). Nor can it decay in particles of more than half of its mass (46 GeV). Its width would not reveal the existence of such hypothetical particles.

Evidence from mass renormalisation

Another test comes from a combination of the masses of the W and the Z , together with the Weinberg mixing angle θ between the EM and the W interactions . In first approximation, these quantities are related by the equation:

$$M_W^2 = M_Z^2 \cos^2\theta \qquad (1)$$

When radiative corrections are taken into account however, this expression is no more exact because of the renormalisation imposed on the masses of these particles. The important corrections come from the inclusions of closed loops of fermions, especially those which are widely split in masses (as the b and the t quark whose mass difference is several tens of GeV).

The experimental ratio of the L.H.S over the R.H.S of equation (1) differs from one by less than five percent. This is enough to exclude new families with quark mass splitting of the same order or larger than the b and the t. Inspection of the data on known particles in the table shows that the quark mass splitting increases as we move from left to right. We can exclude the presence of new families in which this trend would continue.

Evidence from primordial nucleosynthesis.

Most of the story is illustrated in figure 3. Here is plotted first the run of cosmic temperature(in ordinate) as a function of time (in abcissa) as obtained from the Einstein equation describing the early radiation-dominated universe:

$$(\dot{R} / R)^2 = (8\pi/3)\, G_N \rho = (1 / t_{exp})^2 \qquad (2)$$

where R is the distance scale , $\dot{R}$ is its time derivative and ρ is the total energy density.

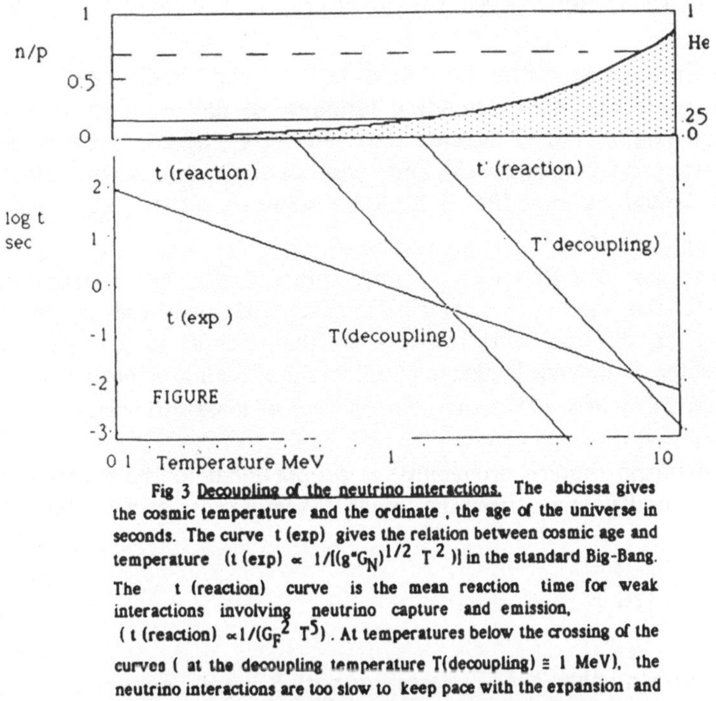

Fig 3 Decoupling of the neutrino interactions. The abcissa gives the cosmic temperature and the ordinate , the age of the universe in seconds. The curve t (exp) gives the relation between cosmic age and temperature ($t\,(exp) \propto 1/[(g^* G_N)^{1/2}\, T^2]$) in the standard Big-Bang. The t (reaction) curve is the mean reaction time for weak interactions involving neutrino capture and emission, ($t\,(reaction) \propto 1/(G_F^2\, T^5)$) . At temperatures below the crossing of the curves (at the decoupling temperature T(decoupling) $\equiv$ 1 MeV), the neutrino interactions are too slow to keep pace with the expansion and the neutron-proton equilibrium abundance is no more insured.

An increase in the mass of the W particle would decrease the value of the Fermi constant ($G_F \propto M_W^{-2}$) and shift the reaction-time curve t'(reaction) to the right in the diagram, leading to a higher decoupling temperature T'(decoupling).

On the upper part of the diagram, the n/p ratio is shown as a function of temperature. On the right side of the figure , the abundances are given by the Boltzman equilibrium equation; on the left side, the neutrons are freely decaying. On the scale at the right is given the resulting helium abundance. The position of the decoupling T can be altered by changing g^*, G_F or G_N as seen from the expressions for the timescales. The BBN turns out to be a very sensitive test of the "constancy " of the coupling constants.

At the high temperatures of the early universe, the particles are relativstic and they contribute in a democratic way to the density; each species being represented by its multiplicity number:

$$\rho = g^* (\pi^2 /30)\ T^4 \tag{3}$$

$$g^* = [\,7/8(\Sigma_f\, g_f) + (\Sigma_b\, g_b)] \tag{4}$$

In these expressions, $h = c = k = 1$; f stands for fermions and b for bosons. Hence we may write:

$$t_{exp} \propto (1 /(g^* G_N)^{1/2})\ T^2. \tag{5}$$

The corresponding curve is plotted in fig 3, with the g^* value corresponding to the three standard families of physics. ($g^* = 9.75$). Adding new families would increase the value of g^*, hence increase the total density through equation (3), and consequently increase the rate of expansion through equation (2) . In fig 3 the resulting curve t'_{exp} (for g^{**} larger than g^*) is shifted below the standard one.

Consider next the timescale of weak interactions. For instance, the timescale for the capture of a neutrino to react with a neutron and give a proton and an electron. This reaction is fundamental in keeping the neutron-proton abundance Boltzman equilibrium at high temperature ($n/p \equiv \exp(-\Delta M/kT)$ where ΔM is the neutron-proton mass difference (1.293 MeV)

The neutrino capture probability is proportional to the number of neutrons per unit volume, times the mean thermal cross-section for this event.

$$P_v = N(n)\ \langle \sigma v \rangle \tag{6}$$

The neutron number-density decreases with R^{-3} and hence with T^3 ($T \propto R^{-1}$ in the expansion). The mean thermal cross-section is proportional to E^2 (hence to T^2) and also to the Fermi coupling constant square (G_F^2).

$$t_{reac} = P_v^{-1} \propto 1/ G_F^2\, T^5 \tag{7}$$

The two curves (t_{exp} and t_{reac}) meet around 1 MeV ,called the decoupling temperature(T_d). Below this temperature, the reaction rate is too slow to follow the expansion rate and the equilibrium is lost. After this time, the neutron essentially freely decay (lifetime of about one thousand sec). Around T = 0.1 MeV, (one hundred seconds later) the deuterons manage to resist photodisintegration. Essentially all the surviving neutrons are then captured by protons and transformed gradually in mostly helium-4 (BBN).

In a nutshell, the junction of the timescales curves fixes the n/p ratio at decoupling (upper part of figure 3). Very few neutrons decay before BBN, and the rest results in helium. Thus, assuming the existence of new families results in an increase in g^* which increases T_d and hence the abundance of He.

The best estimate [14] of the helium cosmic abundance after BB (obtained from observations of galaxies with very low metal abundance) is 0.24 ±0.01. This is best reproduced in the calculations if we assume three families with neutrinos of masses less than 0.5 MeV(to insure that they are relativistic at the time of BBN and "weigh " a full T^4 term in the density balance of equation 3). With the uncertainty on the data , it is possible to include one more family ,perhaps even two , not more . It is on the basis of these arguments that BBN did make its successful prediction on the limitation of the number of families of elementary particles.

Let us discuss again the factor g^*. The expression given in equation (4) should be corrected for a factor to be presently discussed . Below the decoupling temperature, the neutrinos do not interact anymore with the other particles . They live a separate life and are only affected by the expansion, which increases all wavelengths and hence shifts their temperature downward (while respecting their Bose-Einstein energy distribution).

Around 0.5 MeV, the positron-electron annihilation takes place, transforming their masses in radiations, essentially all in thermal photons since the neutrinos cannot their share of the bounty. The corresponding temperature effect in the photon gas can be computed through photon entropy conservation (initial and final) during the annihilation.

$$S_i = g_i^*(T^3_i) = S_f = g_f(T^3_f) \tag{8}$$

$$g^*_i = [(2 \underset{\text{photons}}{} + 7/8 (2+2)] \tag{9}$$

photons positrons and electrons

$$g^*_f = [2] \quad \text{(photons only)} \tag{10}$$

thus:

$$T_f/T_i = (11/4)^{1/3} = 1.31 \qquad (11)$$

This numerical ratio is also the present ratio of the photon to neutrino cosmic background radiation since the neutrino gas did not undergo this temperature effect . With the observed value of 2.7 K for the CMB, the big-bang theory is therefore predicting the existence of a cosmic neutrino background of 2.1 K. Such a detection is presently out of the reach of existing technology.

This discussion introduces the possibility that some species of particle may not weigh as much as others in the cosmic density balance, if they decouple before the onset of massive annihilation chapters.

The expression for g* in equation (4) should thus be corrected in the following way:

$$g = (\Sigma\, g_b [T^4(b)/T^4(\text{photons})] + [7/8\, \Sigma\, (g_f [T^4(f)/T^4(\text{photons})]\,) \qquad (12)$$

At the moment of decoupling, the photons and the neutrinos had the same temperature so that eqn,(4) and (12) are equal. But let assume that there exist very-weakly interacting particles i characterized by a $G_{Fi}^* < G_F$ (an hypothetical right-handed neutrino for example). The corresponding curve in figure 3 would be shifted to the right , leading to a larger decoupling temperature for this particle .

Assume ,for instance, that this decoupling occurs before the muon anti-muon annihilation phase around one hundred MeV. The released energy would be shared amongst the electrons and the left-handed neutrinos (all these particles seeing their temperature go up by a factor of 1.3) , but not with these i particles. In consequence, they should be weighted with a factor of $T^4(i)/T^4(\text{photons})=1/2.95$ in the expression of g* (eqn (12) . A similar computation could be made for particles decoupling before the quark anti-quark annihilation. ; their contribution to the cosmic density would be correspondingly smaller.

Big-bang nucleosynthesis specifies that the value of g* is somewhere between 9 and 13. Although this range limits the number of particles interacting with the standard Fermi interaction, it clearly does not preclude the existence of a large number of other species, provided that their interaction strength is weak enough not to contribute to g* in an ex agerate way.

A multidimensional universe.

Cosmologies with extra dimensions have been the center of much interest in recent years. The most popular version nowaday, in the context of superstrings, involves ten dimensions, thus adding six new compact dimensions, over and above the familiar three-space and one-time dimensions.

The radius of curvature of these extra-dimensions would be of the order of the Planck length (10^{-33}cm) , far smaller than the smallest dimensions within reach of presently operating accelerators (the TeV accelerator of Fermilab can probe to a few 10^{-18} cm) . Energies of the order of the Planck mass (10^{19}GeV) would be required to excite the corresonding modes. This is the reason why we can spend our life without being aware of the existence of these compact dimensions .

However they express themselves in properties that we are just reckognizing as being related to the existence of these extra- dimensions. Here I will discuss one of these manifestations which will bring us back to big-bang nucleosynthesis. In these cosmological models *the values of the coupling constants of the various forces depend upon the radius of curvature of these compact dimensions*. If , as is the case in our familiar expanding 3-D world, these radii are changing with time, the coupling constants would also vary with time.

We have much evidence today that these coupling "constants" are faithful to their name. Again we can take advantage of the success of BBN to study the question . To what extent can we alter the value of these constants without "messing up" the good agreement with the simple theory [15]? The answer is the following : since the universe cooled off to less than one MeV, some ten billion years ago (after the first minutes in the standard chronology), *the coupling constants have not changed by more than one percent.*

We are not used to question the constancy of the "laws of nature". We usually take this fact for granted , or at least as "natural. The situation is changed when the extra-dimensions are added, and the observed constancy become a property of the model. This property severely constrains the choices of acceptable theories, just as renormalisabilities of gauge theories have been most useful guides in elementary physics research.

I will try to illustrate the situation by a simple case : the historical Kaluza-Klein model published in 1921. The aim of this model was to formulate an unified theory of gravitation and electromagnetism. One extra space dimension is added to the standard 3-D geometry. This space, with radius D too small to be detectable, is responsible for the electromagnetic force through which it manifests itself.

Assume a Fourier decomposition of a field $\Phi(x,y)$ in 5-D, where x represents the 4 familiar dimensions and y is the fifth one.

$$\Phi(x,y) = \int \Phi^k(x)\exp(iky/D) \qquad (13)$$

The 5-D (del$_5$) Laplacian applied to this field will give the equivalent of a Klein -Gordon equation for massive particles:

$$(\mathrm{del}_5)\,\Phi^2 = [(\mathrm{del}_4) - m_k{}^2]\,\Phi^2 \qquad (14)$$

where $m_k = k/D$ with $k=0, \pm1, \pm2 ....$

If D is chosen as the Planck length, the different modes corresponds to integers of the Planck mass, needless to say unobservable today.

Our everyday physics involves only the $k = 0$ mode. It is to be derived from the 5-D action integral:

$$S_5=(1/16\,\pi G_5)\int d^5(-g^5)^{1/2}\,R(N)+R(EM) \qquad (15)$$

where G_5 is the fundamental Newton constant of the (real) 5-D world The number 5 on the other letters of this equation are a remainder of the presence of the fifth dimension. R is the curvature tensor which includes both a gravitational (N) and an electromagnetic (EM) term related to Maxwell's equations.

By confining ourselves to the $k=0$ mode of the field , it is clear from equation (13) that the fields are no more functions of y. This dimension gets out of the dynamics. In consequence, we may simply integrate equation (15) over y, to obtain:

$$S_5 = (2\pi D/16\pi\, G_5)\int d_4(-g_4)^{1/2}(\,R_N+R_{EM}) \qquad (16)$$

or

$$S_5 = (1/16\pi G_4)\,[\,S_4\,(N) + S_4(EM)] \qquad (17)$$

Thus we recover the standard 4-D physics if we identify the observed Newton's constant G_4 with G_5/D, where G_5 is the really fundamental constant of the theory. Hence the need to keep D constant to explain the observed constancy of G_4.

Since the years of the Kaluza-Klein theory, the gauge theories of weak and nuclear interactions have been developped successfully. The forces of nature are now seen as resulting from group-symmetry operations in internal (isospin) spaces. The EM force corresponds to an U(1) group, the full electro-weak force to an SU(2) X U(1) and the nuclear force to an SU(3).

In analogy with the K-K model , the present view of multidimensional cosmologies involves the identification of the internal spaces of the standard gauge theories with geometric spatial dimensions on which the corresponding group operators would act. The motivation for this identification is related to the many difficulties of the standard models (divergencies , anomalies etc,) together with the hope of formulating a realistic theory of quantum gravity. There appears to be no other ways, known at the present time, to reach these goals.

The problem of the variability of the coupling constants with the radii of the compact spaces is met in higher-D spaces just as in the 5-D space of K-K. In the Einstein cosmological versions of the expanding universe, all the radii are coupled together so that variations of any one dimension always result in variations of the others. We know that our familar 3-dimensions have expanded to some 10^{60} times the Planck length in the last ten billion years. It is usually assumed that the other six dimensions have contracted upon themselves after a few Planck times (10^{-43}sec). The situation is described in fig 4. The question in everyone's mind is : how did these dimensions managed to remain so amazingly stable (at least after the BBN) while the others underwent such a large modification? It is fair to say that no satisfactory answer have yet (in 1986) be given to that question.

One popular way of stabilizing the compactification of these spaces is to introduce an appropriate cosmological constant in the Einstein equation. This brings reminiscence of the situation met by Einstein in 1915 when he first investigated the cosmological problem.

Realizing that the model implied a global motion of cosmic matter , and having strong dislike for this effect, he introduced the cosmological constant precisely to stop this motion. However it was soon shown that this solution would be of no avail since it is unstable to small perturbations. The problem disappeared some years later, after Hubble observed the recession of galaxies.

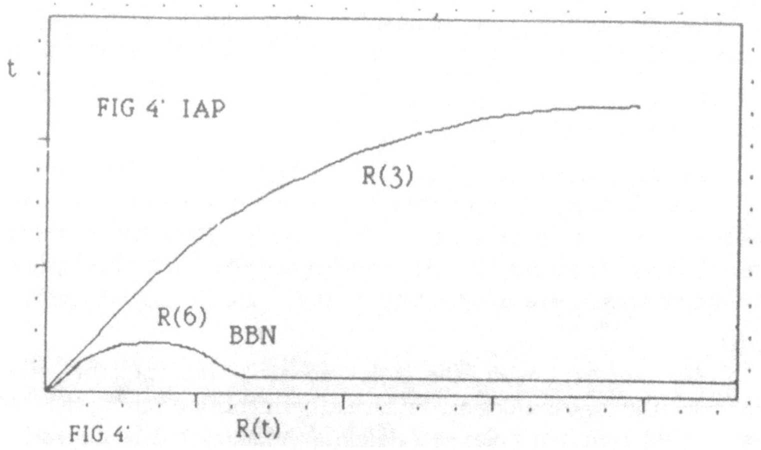

FIG 4'
Cosmological evolution . of space in a 10-D universe. Our familiar 3-D space is expanding with time, while the 6-D compact space rapidly reaches very small radii. The success of Big-Bang nucleosynthesis implies that these radii have not varied by more than a few percent since the first minutes.This constancy imposes severe constraints on superstring theories, and in general on all multidimensional cosmologies.

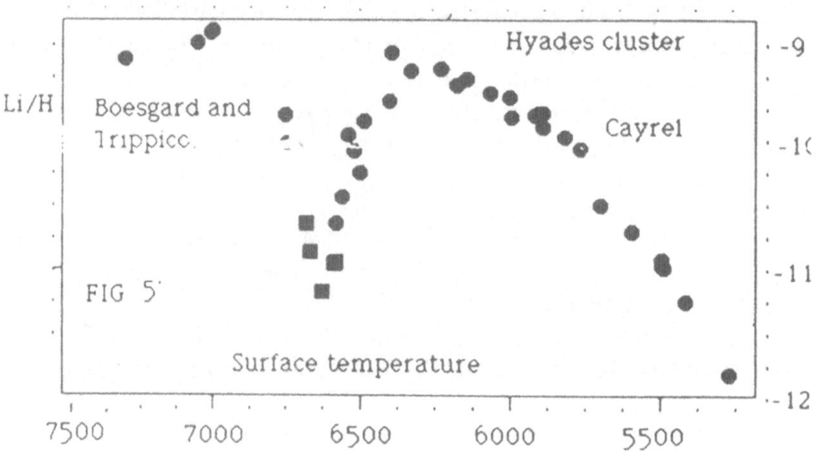

FIG 5
Lithium observations in the Hyades cluster. The data is form Boesgard and Trippico (1986) and also of the Cayrel (1986) Square dots are upper limits.

In ordinate, the abundance Li/H .In abcissa, the stellar surface temperature.

Present efforts to stabilize the extra dimensions by introducing a cosmological constant are facing the same stability requirements , which again should be considered as a constraint that a successful model should meet.

Summary.

I have described two instances in which the good success of standard Big Bang nucleosynthesis in reproducing the abundance observations of the light elements has been used for the developpment of high energy physics and of early cosmological models. The first case is in relation with the families of elementary particles ,the other case is in the formulation of multidimensional cosmologies (as superstring theories) taking into account the observed constancies of the 4-D versions of the coupling constants.

Light elements as probe of stellar physics.

To end up this review, I want to talk shortly on a new development of so great pedagogical interest that it should soon finds its way into physics textbooks.

Boesgard and Tripicco[16)] have recently reported a rather astonishing feature of the abundance curve of Li in the Hyades cluster (age 8×10^8 years) as a function of surface temperature. When completed with the data of the Cayrel[17)], the curve shows a deep dip around 6600 degrees, followed by the well known decreasing slope at lower temperatures (fig 5)

This low-temperature slope is understood as being due to the nuclear destruction of lithium, by proton-induced reactions, at the bottom of the stellar surface convective zone (b.s.c.z.). As we go to the right of the diagram, the b.s.c.z. becomes deeper and its mean temperature increases with decreasing surface T, resulting, after eight hundred million years, in a gradual destruction of Li , more pronounced to the right.

According to Michaud[18)] the dip could be understood in terms of atomic (not nuclear) phenomena, in relation with photon-atom collisions *below* the convective zone. The differential light presure on the atoms in their various ionization states have been computed with realistic stellar models, and compared with the force of gravity.

For surface stellar T above 6800K, the net force upward on the lithium ions situated just below the b.s.c.z. is stronger than the gravity force pushing downward. The ions are pushed up into the convective zone. Below 6800 K, the opposite situation holds and the convective zone is gradually depleted in lithium ions.

Depletion takes time. The deeper the layer (in other words, the lower the T (surf)), the longer is the timescale for appreciable depletion. Fig 6 shows the computed depletion timescale as a function of T(surface). For T < 6600 K or so, no depletion due to the sinking of the ions is expected ,since the timescale is longer than the age of the cluster.

The nuclear timescale of Li destruction at the b.s.c.z. is shown in the same figure (adapted to the data with some degree of overshooting). The remarkable feature is the fact that the two depletion processes (nuclear and atomic) are manifesting themselves side by side with no overlap(fig 7).

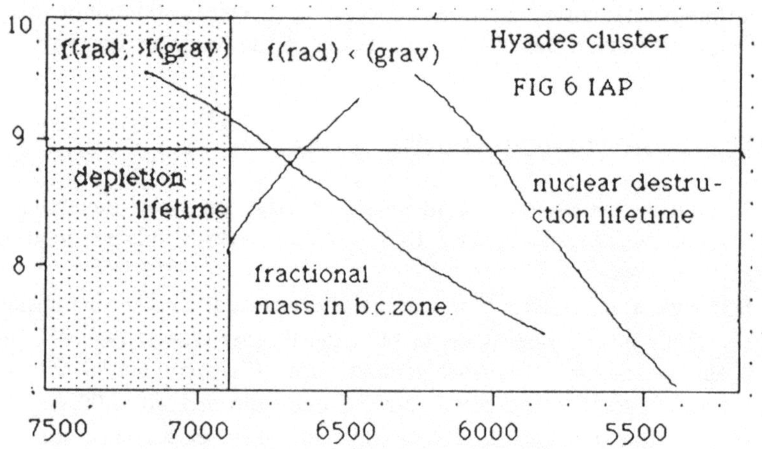

FIG 6 Physical parameters of Hyades stars. On the left (shaded area) thelight pressure on lithium atoms is stronger than the gravitational pull downwars. The depletion lifetime and also the nuclear destrution lifetime are plotted (scale on the left) log of the time in years. On the right, the fractional mass of the star in the convective zone. The age of the cluster is indicated by the horizontal line.

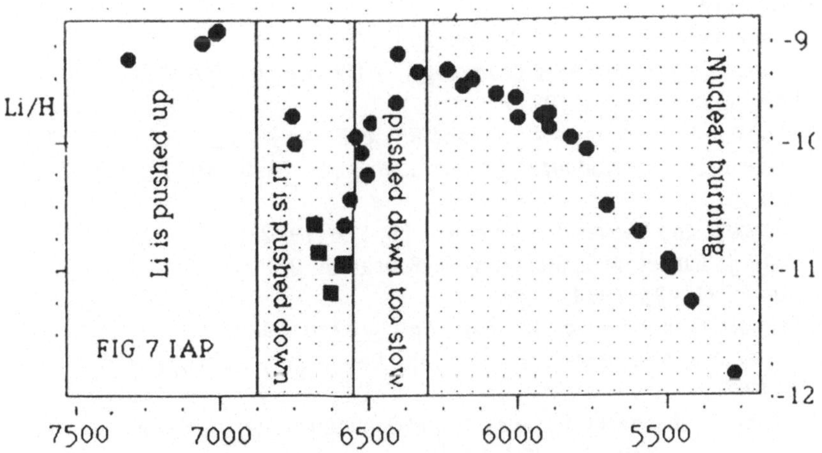

Fig 7 The observational data with the physical phenomena presumed to account for the behaviour of the abundances.

Depletion takes time. The deeper the layer (in other words, the lower the T (surf)), the longer is the timescale for appreciable depletion. Fig 6 shows the computed depletion timescale as a function of T(surface) . For $T < 6600$ K or so, no depletion due to the sinking of the ions is expected ,since the timescale is longer than the age of the cluster.

The nuclear timescale of Li destruction at the b.s.c.z. is shown in the same figure (adapted to the data with some degree of overshooting). The remarkable feature is the fact that the two depletion processes (nuclear and atomic) are manifesting themselves side by side with no overlap(fig 7).

We are left with a question about the left side of the diagram : why are there no *overabundances* of lithium at $T > 6800$ K, where the ions are pushed up? Michaud and his collaborators[19)] provide an answer in terms of mass loss through stellar winds . According to their computations, stellar winds of 10^{-14} to 10^{-15} solar masses per year could account for the observations.

Further studies have shown (op.cit.) that other elements would react differently to stellar winds. This opens up a new field of stellar dynamics through stellar abundance determinations.

Similar analyses of clusters of different ages would be welcome to test the validity of these nice ideas.

Bibliography :

1) Hoyle, F., and Tayler, R.J., Nature 203, 1108, 1964.
2) Peebles, P.J. E., Physical Cosmology (Princeton University Press, Princeton, N.J. 1971).
3) Wagoner, R. V., Fowler, W. A., and Hoyle, F., Ap.J. 148, 3, 1967.
4) Meneguzzi, M., Audouze, J., and Reeves, H., Astr. Ap. 15, 337, 1971.
5) Geiss, J., and Reeves, H., Astr. Ap., 18, 126, 1972.
6) Yang, J., Turner, M.S., Steigman, G., Schramm, D.N. and Olive, K., Ap. J. 281, 493, 1984.
7) Beaudet, G., and Reeves. H., Astr. Ap.134, 240, 1984.
8) Boesgaard. A., and Steigman. G., Ann. Rev. Astron. Astrophys. 1985
9) Buhler, F., Eberhardt, P., Geiss, J., and Schwarzmuller, J., Earth Planet. Sci. Kett. 10, 297, 1971.
10) Penzias, A.A. and Wilson, R.W. Ap.J. 142, 419, 1965.
11) Wilkinson, D., in Inner Space/ Outer Space, eds . Kolb et al. (Univ. of Chicago Press Chicago 1985)
12) Altarelli, G., Ecole d'été de physique des particules.1985. Gif/Yvette 91191.
13) Di Lella , L., Proc. Int. Symp. on Lepton and Photon Interactions at High Energy , Kyoto 1986.
14) Kunth, D., and Sargent, W.L.W. Ap.J. 273, 81, 1983.
15) Kolb. W.E., Proceedings of the VIII Johns Hopkins Workshop on Current Problems in Particle Physics .June 1984 .
16) Boesgaard, A., and Trippico, M.J., Ap.J., 302, L49, 1986.
17) Cayrel, R., Cayrel de Strobel, G., Campbell, B., and Dappen, W. Ap.J., 283 , 205, 1984.
18) Michaud, G., Ap. J. 302, 650, 1986
19) Michaud , G., *preprint* .

THE CONSISTENCY PROBLEMS OF LARGE SCALE STRUCTURE, DARK MATTER, AND GALAXY FORMATION

David N. Schramm
Departments of Astronomy and Astrophysics, and Physics
The University of Chicago
Chicago, Illinois 60637

ABSTRACT

The combined problems of large scale structure, the need for non-baryonic dark matter if $\Omega = 1$, and the need to make galaxies early in the history of the universe seem to be placing severe constraints on cosmological models. In addition, it is shown that the bulk of the baryonic matter is also dark and must be accounted for as well. The nucleosynthesis arguments are now strongly supported by high energy collider experiments as well as astronomical abundance data. The arguments for dark matter are reviewed and it is shown that observational dynamical arguments and nucleosynthesis are all still consistent at $\Omega \sim 0.1$. However, the inflation paradigm requires $\Omega = 1$, thus, the need for non-baryonic dark matter. A non-zero cosmological constant is argued to be an inappropriate solution. Dark matter candidates fall into two categories, hot (neutrino-like) and cold (axion or massive photino-like). New observations of large scale structure in the universe (voids, foam, and large scale velocity fields) seem to be most easily understood if the dominant matter of the universe is in the form of low mass ($9eV \leq m_{\nu} \leq 35eV$) neutrinos. Cold dark matter, even with biasing, seems unable to duplicate the combination of these observations (of particular significance here are the large velocity fields, if real). However, galaxy formation is difficult with hot matter. The potentially fatal problems of galaxy formation with neutrinos may be remedied by combining them with either cosmic strings or explosive galaxy formation. The former naturally gives the scale-free correlation function for galaxies, clusters, and superclusters. The latter requires fine tuning and percolation to get the large scales and the scale-free correlation function. However, combining hot matter and strings reduces the ability of the hot matter to give some of the large scale features and still yield $\Omega = 1$. Questions to be examined are raised.

1. INTRODUCTION

The major confrontation of early universe studies with the "real" universe now focuses on the problems of galaxy formation, dark matter, and the generation of large scale structure. The observable aspects of these problems came into being shortly after recombination; however, the condition of the universe as it approaches recombination are determined by events taking place much earlier, when nuclear and particle physics effects dominated. Since the recombination epoch is the limiting epoch for direct observations, it

P. Galeotti and D. N. Schramm (eds.), Gauge Theory and the Early Universe, 211–224.

is only natural that this epoch serve as the interface between early universe cosmologists and astronomers.

The problems are to produce initial conditions and types of matter which will yield the observable universe, the large scale structure. In particular, the observable universe now appears to have large scale structure on scales of $\sim 40 Mpc$ that looks like foam or at least intersecting sheets and filaments with large voids[1,2,3]. In addition, there appear to be large, coherent motions of 40 Mpc clumps with velocities of $\sim 600 km/sec$[4]. To this very large scale structure must be added the apparent fact that clusters of galaxies cluster with each other more strongly than galaxies cluster[5], or to use the analysis of Szalay and Schramm[6], the clusters and galaxies appear to cluster in a scale-free manner as if laid out in some fractal pattern.

2. THE DYNAMICAL ARGUMENTS

To these large scale observations must be added the dynamical measurements of mass and the so-called dark matter problem. In particular, the dynamics of the visible parts of galaxies imply an Ω of ≤ 0.01 (where $\Omega \equiv \frac{\rho}{\rho_{crit}}$ is the critical density of the universe). However, when galaxies interact with other galaxies in binary pairs or in small groups, they interact with ~ 10 times as much mass, implying an $\Omega \sim 0.1$. When galaxies interact with one another in large clusters they interact with possibly even more mass, implying $\Omega \sim 0.1$ to 0.3. (*No well studied system gives anything near* $\Omega = 1$.)

3. BIG BANG NUCLEOSYNTHESIS

To the dynamical arguments we can add the arguments from Big Bang nucleosynthesis (Yang et al.) which show that observed abundances are consistent only if $\Omega_b \sim 0.1$ (where $\Omega_b \equiv \frac{\rho_b}{\rho_{crit}}$ and ρ_b is the density of baryons).

Thus as Gott et al.[7] pointed out over ten years ago, direct astronomical evidence points towards $\Omega \sim 0.1$ with the dark halos being baryonic and no need for exotic stuff. In particular, it should be noted that the lower bound on Ω_b is $\Omega_b \geq 0.03$[8]. Since this is > 0.01, it implies that the bulk of the baryons are dark. (Note that because of this point, dark halos for dwarf spheroidal galaxies are no problem since they can be baryonic.) Also, it is important to remember that nucleosynthesis contrains $\Omega_b < 0.15$. (This is lower than the 0.19 from Yang et al.[9] due to better current upper limits on the microwave background temperature.) Thus, if $\Omega \sim 1$, the bulk of the universe would be non-baryonic *and* could not cluster with the light emitting galaxies and clusters.

The nucleosynthesis arguments are gaining even greater credence now that their predition[9,10,] that the total number of neutrino types (generations) is small (three or at most four) is being verified by collider experiments[11] with current experimental limits at < 5. From particle physics theory alone any number of generations might be possible. The preliminary verification of the cosmological prediction is the first time that cosmology has made a prediction which has been verified by a high energy accelerator experiment.

4. BARYONIC HALOS?

Can halos of galaxies and dwarf spheriodals really be baryonic? While the coincidence of $\Omega_b \sim 0.1$ and $\Omega_{dynamic} \sim 0.1$ is suggestive, it is certainly not compulsory. Different forms of dark matter can mix with baryons in different ways depending on the mechanism of galaxy formation.

With cold dark matter the halos must be a mixture of $\sim$ 90% cold matter and 10% baryons whereas in hot matter models the halo mixture depends on the galaxy formation scenario.

If the halos do contain significant baryonic materials, what form can it be? Hegyi and Olive and Schramm have argued that most baryonic things do not work. However, they leave two very important loopholes:

1. Black holes left from an early generation of massive stars with the bulk of the stellar material falling into the hole and not producing excess heavy elements. Such black holes are contrained by Big Bang nucleosynthesis baryon limits since they were baryons then (so they count as baryonic material).
2. Low mass objects too dim to be seen in telescope searches. Jupiter-like clumps or even 0.01 $M_\odot$ stars would work. In order for the abundance of such objects to be sufficient, the abundance spectrum for these objects would probably be above the low mass extrapolation of the Salpeter initial mass function. However, that function is strictly empirical and there could certainly be a low mass excess if the initial stellar generation with pure H and He, but more objects low than currently occurs with heavy elements present. (Option 1., of course, requires exactly the opposite behavior for the early stellar mass function.)

5. THE FLATNESS ARGUMENTS

If everything agrees so well with $\Omega \sim 0.1$, why do people continue to think $\Omega = 1$? The only astrophysical evidence for large Ω is clearly weak at the present time. It consists of the following:

1. With Gaussian adiabatic initial density fluctuations of the type described by Zel'dovich and expected from simple inflation models, it is impossible to make galaxies rapidly enough when constrained by limits on microwave background anisotropies unless $\Omega >$ 0.2[12,13].
2. The velocity field of IRAS galaxies on scales of $\sim$ 200Mpc implies a virial mass on these large scales of $\Omega \sim 1$[14].
3. The density of galaxy counts versus redshift is optimally consistent with $\Omega = 1$ geometry[15].

The first of these is clearly removable if galaxies form by something other than Gaussian adiabatic fluctuations with a Zel'dovich spectrum. In particular, string models which are also derivable from grand unified gauge models do not yield such a stringent requirement on Ω, nor do, for that matter, models where galaxy formation is stimulated by early explosions[16].

The second argument has the problem that a reliable way to determine distances to IRAS galaxies has not been established and a complete redshift survey of IRAS galaxies remains to be done. In addition, IRAS counts may have a significant north–south bias due to induced instrumental variations in sensitivity of the satellite in the northern and southern hemispheres.

The third argument, while potentially the strongest, still requires a more detailed analysis of galactic evolution effects and normalization of distant galaxy counts to nearby where different techniques are used.

Thus, while suggestive, these arguments do not yet establish $\Omega = 1$. However, there is a Copernican-like argument which is sufficiently powerful that most theoretical physicist believe $\Omega = 1$. The argument was best articulated by Dicke and Peebles and later provided Guth with a strong motivation for inflation which gave a physical mechanism for yielding the desired Ω. The argument, simply stated, is that Ω is a time changing quantity going to $\Omega < 1$ and to ∞ if $\Omega > 1$, and only remaining constant if $\Omega = 1$. The timescale of change is the expansion rate of the universe. Thus, the only long-lived values are 0, 1, and ∞. Since we are here, Ω is neither 0 nor ∞. The only other long-lived value is 1. To have any finite value below unity today would require that we live at a very special time, the early epoch in cosmic time when Ω was not 1 or 0. Such a value would require the extraordinary fine tuning at the Planck time of $\sim$ 60 decimal places, or at least 17 decimal places at the time of Big Bang nucleosynthesis. Thus, unless we live at a special time and some unknown mechanism tunes Ω to exactly the right amount to fantastic accuracy, Ω is probably unity.

Since any early deSitter phase for the universe produces a flat universe ($\Omega = 1$ if the cosmological constant $\Lambda = 0$) and since inflation means an early deSitter phase, and since most scalar fields yield inflation, it is reasonable to believe $\Omega = 1$. While many have recently focused on the problems many models of inflation have been producing, the right sized initial fluctuations[17] any inflation model which solves the horizon problem, getting a nearly constant background temperature, will also solve the flatness problem.

6. THE COSMOLOGICAL CONSTANT

Some astrophysicists (who shall remain nameless) have focused on the formal mathematical loophole that flatness can also be obtained with a non-zero Λ and $\Omega < 1$. However, such a solution is missing the philosophical motivation (like killing for pacifism). If today we have $\Omega \sim 0.1$ and non-zero Λ yields flatness, that is an epoch- dependent solution since the contribution of Ω and Λ vary differently with epoch. Such a solution would imply that we live at the only epoch where Λ and Ω contributions to curvature are comparable, again requiring amazing fine tuning (tuning Λ to ≥ 120 decimal places). Unfortunately we don't as yet have a nice physically motivated mechanism like inflation to set $\Lambda = 0$, but if we buy the philosophy, I believe we should also assume Λ is negligible. Of course both arguments are philosophical (or theological) rather than based on physical observation, but the Copernical principal of us not being special has held up well for several hundred years.

7. DARK MATTER AND GALAXY AND STRUCTURE FORMATION

As mentioned before, if Ω is 1, then we need non-baryonic dark matter. Such matter has been classified as either hot (neutrino-like with high velocities just prior to the epoch of matter-radiation equality) or cold (low velocities prior to matter-radiation equality).

Initially, hot, low mass, neutrinos were quite popular as candidates for solving the cosmological dark matter problem, since they were the least exotic of the non-baryonic options, and they naturally clustered only on large scales where the dark matter was needed, rather than on the small scales where the contribution of dark matter was known to be minimal[18]. They received a major boost with the preliminary reports of measured mass[19] for ν_e (although probably only the most massive ν is cosmologically important, and that might well be ν_τ (or a nucleosynthesis-allowed 4th generation) which could still have a $\sim 10eV$ mass, even if $m_{\nu_e} \ll 1eV$). Also, they gained strength when it was shown[3] that the neutrino Jean's mass was

$$M_J \sim \frac{3 \times 10^{18} M_\odot}{m_\nu^2(eV)} \text{ or } \lambda_J \sim \frac{1300 Mpc}{m_\nu(eV)}$$

which for $m_\nu \sim 30eV$ yielded $M \sim 3 \times 10^{15} M_\odot$, and $\lambda \sim 40 Mpc$, the mass and scale of large clusters.

Unfortunately, massive neutrinos fell into disrepute as dark matter when it was emphasized[20] that in the standard adiabatic model of galaxy formation with a random phase, Zel'dovich fluctuation spectrum of the type expected by inflation, and with $\delta T/T$ constrained by microwave observations, galaxies did not form until redshift $z \lesssim 1$. This occurred because the initially formed pancakes with mass M_J took a while to fragment down to galaxy size. This contradicted the observations which showed that quasars existed back to $z \sim 3.5$. In addition, if baryons stay in gas form in the potential wells of the large ν pancakes, they light up in the x-rays beyond what is observed[21].

While some[22] have appealed to statistical tails, etc., to escape these conclusions, most cosmologists began abandoning neutrinos and adopting cold dark matter[23], which could enable rapid galaxy formation[24,25].

Cold matter also had its problems[26]. In the standard model, it would all cluster on small scales, and thus be measured by the dynamics of clusters, such as the Virgo infall. Since such measurements implied that $\Omega \sim 0.2 \pm 0.1$ on cluster scales, this meant that $\Omega_{cold} \lesssim 0.3$, and not unity. Remember that $\Omega \sim 0.1$, so observationally, non-baryonic dark matter is not required unless one wants an Ω of unity, so cold matter wasn't naturally solving one problem for which it was postulated. This constraint on cold matter could be escaped if it were *also* assumed that galaxy formation was biased[25,27] and did not occur everywhere. Thus, there could be many clumps of cold matter and baryons that did not shine for some ad hoc reason. Biasing ran into problems when it could not explain the observation[5] of a very large cluster–cluster correlation function, ξ_{cc}, relative to the galaxy–galaxy correlation function[26,27], ξ_{gg}. With biasing $\xi_{cc} \propto \xi_{gg}$ but in all models $\xi_{gg} < 0$ for a few 10's of Mpc, whereas ξ_{cc} was observed to be positive out to scales $\gtrsim 50Mpc$. Hardcore cold matter lovers had to argue that the ξ_{cc} data might be wrong, although no one has been able to disprove it.

A way out of the ξ_{cc} problem was proposed by Szalay and Schramm. There we noted that the correlation functions appear to be scale free, thus implying that large-scale structure is dominated by something other than random noise and gravity, say either percolated explosions or strings. In fact, the scale-free structure is characterized by a fractal of dimension $D \sim 1.2$, not too different from the $D \sim 1$ that naive string theory might yield. String calculations[28] of galaxy formation indeed found support for such a fractal process with the appropriate dimension being valid from galaxy to supercluster scales.

Thus, there were already strong hints that something was wrong with the previous, in vogue, picture of biasing and cold matter with random noise initial fluctuations. To this we now add the new observations of many large voids[1,2] of diameter $50h_{1/2}Mpc$ ($h_{1/2} \equiv H_0/50km/sec/Mpc$), with most galaxies distributed on the walls of the voids, and the observation[4] that our local 40 Mpc region of space is moving with a coherent velocity field of $\sim 600km/sec$ toward Hydra-Centaurus. While at least one large void (in Böotes) had been observed before[3], using a pencil beam approach, until the Harvard redshift[1] survey work, it was not known how ubiquitous voids were. In fact, the Harvard data shows that almost all galaxies are distributed along the "walls" of voids; galaxies and clusters are not randomly distributed, but fit onto a well-ordered pattern.

While the Harvard work only goes out to $\sim 100Mpc$, there is substantial evidence that this sort of pattern persists to redshifts $z \sim 1$ from the Koo and Kron survey[2]. A simple explanation for the peaks and valleys in the distribution of galaxies and quasars with redshift is that one is looking through filaments or shells with voids in between, once again demonstrating that galaxies and clusters are not laid out randomly on the sky, but follow a pattern.

While statistical fluctuations with cold matter might yield a few large voids as well as many small voids[21,25], it is difficult to get all of space filled with large voids and have galaxies appear only at the boundaries unless some special form of "biasing" is used. However, the real killing blow for the cold matter plus biasing scheme comes from the velocity field work. Even if the biasing could be selected so as to give ubiquitous large voids, the velocities of a $40Mpc$ region of galaxies would be relatively small and random, rather than large and coherent[29]. In fact, the more extreme the biasing used to get large voids, the *lower* the large scale velocities. Thus, it appears that the large-scale structure is telling us that we need something that gives us $\sim 40Mpc$ coherent patterns, and cold matter doesn't appear the way to go. (Unless, of course, the large scale velocity field work is in error. In other words, cold matter with gaussian Zel'dovich fluctuation requires *both* ξ_{cc} and the velocity to be completely wrong.

Since neutrinos naturally gave us patterns on this scale, maybe they should be reexamined. In addition, since the voids look rather spherical, and since explosions tend to produce spherical holes after a few expansion times even if the initial explosion is asymmetric, perhaps an explosive mechanism should be considered also. Since the Ostriker–Cowie[16] explosion mechanism by itself cannot yield such large voids, the only way it could work is via a high density network of explosions which percolated[25,30]. However, to get $\Omega = 1$ with an exploding scenario would still require non-baryonic matter that did not cluster with the light emitting stuff. In principal, this could be either neutrinos or cold matter but at least with neutrinos an $\sim 40Mpc$ scale might still be naturally imposed.

8. NEUTRINOS PLUS STRINGS OR EXPLOSIONS

Of course, in order for neutrinos to work as the dominant matter, some mechanism to rapidly form galaxies must be imposed both to enable galaxies to exist at $z \sim 5$, and to condense out the gas before it falls into the forming deep potential wells, and emits x-rays. Two ways that might achieve this rapid formation are either via the aforementioned explosion scheme within the collapsing ν-pancakes, or via cosmic strings[31] which would act as nucleation sites for galaxy formation. Since strings are not free-streamed away by the relativistic neutrinos[32], the galaxy scale fluctuations remain within the ν-pancakes. Notice that since neutrinos are not used by themselves simple arguments based on relating their primordial fluctuation spectrum to observed galaxy velocity and distribution features are not necessarily valid and must be reexamined in the more complete scenario.

It should be noted that even with strings as seeds so that cold matter can cluster in a scale-free way fitting ξ_{cc}, the large scale velocity fields for cold matter are small, and it is difficult to get $\Omega = 1$ while observing $\Omega_{cluster} \sim 0.2$. However, we have the additional problem that the strings might mess up the nice large scale neutrino features and background of ν's will still slow galaxy growth around the strings over how cold matter would form on the strings.

It is interesting that two surviving galaxy formation options, strings and explosions, involve the same two options that the scale-free cluster–cluster correlation function arguments point towards. Let us look at each of these scenarios in a little more detail and see if there might be ways of resolving whether either of them might actually be correct. Also, let us see what each requires for the physics of the early Universe.

Both of these scenarios seem to need hot matter if we want to solve the velocity field, $\Omega = 1$, and large scale problems.If $\Omega = 1$, as is necessary to avoid our living at a special epoch, and as agrees with the recent large-scale galaxy count arguments of Loh and Spillar[15] (but disagrees with the direct dynamical arguments on scales of clusters and smaller, and with the baryonic measurements from nucleosynthesis), then $m_\nu \lesssim 35eV$. Since with $\Omega = 1$ the age of the Universe $t_0 = \frac{2}{3H_0}$, and since globular clusters and nucleochronology require $t_0 \gtrsim 11 \times 10^9 yr$ (with a best fit of $t_0 \sim 15 \times 10^9 yr$) we must say that $H_0^{-1} \gtrsim 17 \times 10^9 yr$. Thus, $H_0 \lesssim 60km/sec/Mpc$, or $h_{1/2} \lesssim 1.2$. From the number of neutrinos and photons in the Universe, we know that the most massive neutrino is bounded by (see ref. 18 and references therein)

$$m_\nu \lesssim (25eV)\Omega h_{1/2}^2 \lesssim 35eV.$$

It is curious that the requirement that we want the neutrinos to give us the large-scale structure, $\lambda_J \sim 40Mpc$, or $M_J \sim 10^{16} M_\odot$, also gives us $m_\nu \sim 30eV$, a mass about what is necessary to get $\Omega \sim 1$. Also, we have a lower bound from the nucleosynthesis argument[26] that the number of neutrino species with $m_\nu \lesssim 10MeV$ is three or at most four. Since the sum of all neutrino masses cannot exceed the $35eV$ limit mentioned above, and since the lowest mass for the most massive one occurs when they are all equal, then if $N_\nu \leq 4$,

$$m_\nu \gtrsim 9eV.$$

The first scale to be able to condense and thus have their density grow will be the horizon scale when the neutrinos become non-relativistic, which is M_J. However, in the string option, loops of string will exist down to scales of galaxy size (scales smaller than galaxy size gravitationally radiate away[31]). So as the neutrinos become non-relativistic they can be trapped on smaller scales. The baryons will not be able to begin clustering until after recombination. However, the slow-moving baryons will rapidly fall on to the pre-existing loops of string plus neutrinos. Thus, galaxies will be able to form shortly after recombination, and well before $z \sim 1$.

9. PROBLEMS WITH STRINGS?

Unfortunately, just after matter domination the bulk of the neutrinos will still have relatively high velocities so their Jean's mass, while dropping, will not be low enough for most neutrinos to cluster on the galaxy size loops. Even after recombination the characteristics Jean's mass for the bulk of the neutrinos will still be much larger than galaxy size, so there will be a relatively smooth background of neutrinos which will slow the rate of growth of baryons falling onto the loops of string. Thus, strings plus neutrinos do not grow galaxies as rapidly as strings plus cold matter; however, strings definitely help the neutrino picture along. The quantitative question of whether the neutrino-string picture can form rapidly enough remains to be worked out in detail, since quick and dirty calculations indicate that the results are marginal[33]. With neutrinos, the dimensionless string tension 6μ needs to be higher than for strings with cold matter where $6\mu \sim 10^{-6}$. Unfortunately, it cannot be arbitrarily raised since high values ($\geq 10^{-5}$) cause problems in microwave anisotropy and in radiating too much energy at the time of nucleosynthesis, thus running into the equivalent of the neutrino country bound[34].

Also, it is not clear how the combination of ν's and strings deals with the very large scale structure. While strings by themselves give the scale- free correlation function out through the scales of Abell clusters[28], if neutrino pancaking is too strong, it could mess this up. On the other hand, string perturbations existing on scales smaller than $\sim$ 40Mpc may prevent pancaking from ocurring at all. Horizon length strings at matter-radiation equality will produce large scale adiabatic fluctations that could induce pancake formation in the neutrinos, going non-linear at redshift $z \sim 1$. However, the strength of the fluctuations relative to the normal string fluctuations needs to be checked to see which, if any, dominates.

If they really do not go non-linear until $z \sim 1$, they might not mess up the more rapidly forming galaxy and cluster scale fluctuations, so the smaller scale correlation functions might be retained while the neutrino pancake collapse might induce the very large scale velocity field and pancakes, filaments, and voids. Obviously the whole combined picture needs to be examined in much greater detail to see if it really can retain the best features of both models, rather than the two components destroying each others better features.

Because the string picture looks like the current front runner, people have begun looking at it in far greater detail, to see if it really can yield the observable universe. In particular, Peebles has privately circulated a "screed", stating possible problems. At a workshop held at the Aspen Center for Physics, these problems were examined and possible ways out were found. Let us now summarize the Peebles problems and possible solutions.

Problems not previously mentioned:

1. Strings produce loops following a power spectrum $\sim M^{-5/2}$, whereas galaxies are observed from their light to have a much flatter spectrum, up to $\sim 10^{12} M_{\odot}$ and then exponential fall off. Thus, at first glance, it appears that strings give too many small *and* large galaxies if their spectrum is normalized to fit the L^* galaxies at $\sim 10^{12} M_{\odot}$.
2. Strings are small relative to their separation distances. Thus, collapse onto static strings appears unlikely to give large quadropole moments, and thus tidal interactions will not produce the angular momentum observed in galaxies.
3. With strings as seeds, both cold and hot dark matter will cluster on small scales so that Ω measured for clusters should be a good estimate of Ω_{total} which would yield ~ 0.2, not 1. Biased suppression of galaxy formation with strings as seeds is evn more ad hoc than normal cold-matter biasing, so is not a convenient escape.

The possible solutions to these problems are:

1. Excess amounts of small strings forming galaxies can be supressed in a variety of ways.
   a. For larger 6μ, such as in the neutrino models, gravitational radiation eliminates the excess low mass loops.
   b. Vilenkin[35] has shown that global strings rather than gauge strings radiate Goldstone bosons in addition to gravitational radiation. Thus few mass global strings would also not be a problem.
   c. Strings do not radiate symmetrically. The differential radiation for small strings results in a rocket effect[36] which supresses their ability to acrete.
   d. More fragmentation of the small loops which form early could lower their abundance as the smaller are radiated away.

   The excess amounts of large loops may be a more complex problem and more work

needs to be done here. Possible solutions include:

a. Finite velocity may affect accretion.

b. Fragmentation of large loops will reduce their numbers.

c. Big loops may yield CD galaxies at centers of clusters with velocity curves rising as $r^{1/4}$ rather than normal flat rotation curves.

2. Angular momentum may be formed by tidal interactions because accretion is not spherical but sausage-like, due to the finite velocity of loops. Distances moved are comparable to separations so quadropole moments will be approximately large.
3. The solution to the Ω problem requires that somehow clusters don't sample a standard segment of the universe. One way to accomplish this would be if galaxies correlated more with clusters than randomly. Such could occur if large, cluster-producing strings fragment to produce smaller galaxy-producing strings, and the resultant small strings didn't get too far from the clusters. Clearly, this does occur to some degree; however, can it quantitatively yield a factor of three of more enhancement in Ω between its cluster measured value and the true value remains to be shown. The dynamical range of string simulations has not yet enabled such quantitative tests between small and large loops. Note that if galaxy strings are strongly correlated with clusters, then many regions in space will be without loops of strings, and so will not form galaxies even though they have baryons and either hot or cold dark matter.

Another possible problem is that, while the string scenario may naturally yield $D \sim 1$, it does not so naturally give $D = 1.2$. Fine tuning[39] of string parameters may enable such variation on the scale of the galaxy–galaxy correlation function, or some modification of the criteria for the formation of light-emitting regions around the strings may be necessary.

In this regard it should be remembered that because of possible systematic errors, not everyone agrees that 1.2 is significantly different from 1.0, even for the galaxy–galaxy correlation function, which is the best determined[38]. The uncertainties in the exponent of the cluster–cluster correlation functions are *far* larger, thus problems in trying to explain variations from $D = 1$ fractals are not serious at the present time. With strings there is the additional problem of tuning the primordial phase transition so as to inflate first, and then produce strings[39]. While not impossible, this is constraining.

10. EXPLOSIVE GALAXY FORMATION

The second way to get neutrinos to work involves explosive galaxy formation. Here we need initial seeds to lead to condensations which produce massive baryonic objects which explode. As mentioned before, such a model does not naturally give us $40 Mpc$ structure. If we use neutrinos then the seeds must be in a form which does not get free-streamed away by the relativistic neutrinos. Strings don't work well here because the string scales that might lead to rapidly evolving baryonic objects are radiated away gravitationally. Thus, the seeds must come in some other isothermal-like form. Perhaps the best option would be condensates from the quark–hadron transition, either planetary mass black holes[40] or Witten nuggets[41]. Both have formation problems[42] and the latter have survival problems[43] also. If such objects could form and survive, they do lead naturally[44] to very massive ($\sim 1000M_\odot$) baryonic objects which would explode on rapid timescales. Another option is cold dark matter clumps, in which case small strings work as seeds, but the large scale problems are aggravated.

The scale affected by explosions of single galaxy size[45] is at most a few Mpc; however, it has been shown[30] that at sufficiently high densities and high trigger rates, the explosions can percolate at least out to scales of a few 10's of Mpc. The fractal dimension of such percolated ensembles is quite sensitive to parameter assumptions and usually varies with scale, thus showing that it is not a true scale-free fractal. If it is made to fit the small scale (few Mpc) with $D \sim 1$ it is usually larger ($D \sim 2$) on scales of $\gtrsim 10Mpc$. Since, as mentioned above, the exponent of the cluster–cluster correlation function is not, at present, well determined, such models cannot be ruled out. With such explosions percolating within ν-pancakes, we might naturally have their pattern superimposed on the $\sim 40Mpc$ neutrino scale. In addition, although percolated explosions will initially be highly non-spherical, their shape will evolve towards sphericity with the smaller axes catching up in length to the largest one. In order for large-scale percolation to occur, several generations[21] of explosions must occur; however, cooling arguments and time to initial explosions, plus the need for condensed objects by $z \sim 4$ and the need to hide from present observers, the radiation produced by the explosions, severely restrict the possibility of such percolation and thus quite a bit of fine tuning is required to escape the constraints.

11. CONCLUSION

Thus, while we cannot explicitly rule out this latter case, unless some new physics can be developed to show how the fine-tuned parameters are natural for other reasons, we must lean towards the string option as the present frontrunner. Strings, of course, would have other observational consequences[32] like gravitational double lensing of distant objects and shifts in the $3°$ background across such a line of lenses, and a background of gravitational radiation from the evaporation of small-scale strings which might affect the millisecond pulsar. Thus, observations should eventually be able to confirm or deny this frontrunner. Table 1 gives a summary of current proposed models and their ability to solve the problems. Note that the location of dark baryons may eventually be detectable and a discriminator of models. No model is yet a clear winner. Some require more calculations to see if they can be made to work. Others require some key bit of observational data to be proven wrong.

In summary, we have come full circle and once again massive neutrinos are looking good. However, with them comes the need for galaxy and structure formation triggered by something other than random phase adiabatic fluctuations. The non-random phase fractal initial conditions such as produced by strings[46] or fractal generating explosions[16,30] seem to be the way to go. It is comforting that the exotica of cosmic strings do seem to be a natural consequence[47] of the current, in vogue, superstring Theories of Everything (T.O.E.).

Acknowledgements to co-workers J. Charlton, K. Olive, A. Melott, G. Steigman, A. Szalay, and M. Turner are gratefully given. I also acknowledge many useful discussions with N. Turok and P.J.E. Peebles. This work was supported in part by NSF AST 85-15447, and by DOE DE-FG02-85ER40234 at the University of Chicago, and was prepared at the Aspen Center for Physics.

12. REFERENCES

1. deLapparant, V., Geller, M. and Huchra, J. 1986, Center for Astrophysics preprint
2. Koo, D. and Kron, R. 1986, in preparation.
3. Kirschner, R., Oemler, G., Schecter, P., and Shectman, S. 1982, *Ap.J.* **248**, L57.
4. Faber, S., Aaronson, M., Lynden-Bell, D. 1986, *Proc. of Hawaii Symposium on Large-Scale Structure.*
5. Bahcall, N. and Soniera, R. 1983, *Ap.J.* **270**, 20; Klypin and Khlopov 1983, *Soviet Astron. Lett.* **9**, 41.
6. Szalay, A. and Schramm, D. 1985, *Nature* **314**, 718.
7. Gott, J.R., Gunn, J., Schramm, D.N., and Tinsley, B.M. 1974, *Ap.J.* **194**, 543.
8. Freese, K. and Schramm, D. 1984, *Nucl. Physics* **B233**, 167.
9. Yang, J., Turner, M., Steigman, G., Schramm, D., and Olive, K. 1984 *Ap.J.* **281**, 493.
10. Steigman, G., Schramm, D.N., and Gunn, J.E. 1977, *Phys.Lett.* **B66**, 502.
11. Cline, D. 1986, The 6th Proton–Anti-proton Conference, Aachen, West Germany, review talk.
12. Vittorio, N. and Silk, J. 1984, *Ap.J.* **L39**.
13. Bond, J., Efstathiou, G., and Silk, J. 1980, *Phys.Rev.Lett.* **45**, 1980.
14. Rowan-Robinson, M. 1986, in The Proc. ESO/CERN Symposium on Cosmology.
15. Loh, E. and Spillar, E. 1986, Princeton University preprint
16. Ostriker, J. and Cowie, L. 1980, *Ap.J.* **243**, L127.
17. Olive, K. and Schramm, D.N. 1986 *Comments on Nuclear and Particle Physics*, in press.
18. Schramm, D. and Steigman, G. 1981, *Ap.J.* **243**, 1.
19. Lubimov, A. 1988, in this volume.
20. Frenk, C., White, S., and Davis, M. 1983, *Ap.J.* **271**, 417.
21. Davis, M. 1986 *Proc. 1984 Inner Space/Outer Space*, University of Chicago Press.
22. Melott, A. 1986 *Proc. 1984 Inner Space/Outer Space*, University of Chicago Press.
23. Blumenthal, G., Faber, S., Primack, J., and Rees, M. 1984, *Nature* **311**, 517.
24. Melott, A., Einasto, J., Saar, E., Suisalu, I., Klypin, A., and Shandarin, S. 1983, *Phys.Rev.Lett* **51**, 935.
25. Efstathiou, G., Frenk, C., White, S., and Davis, M. 1985 *Ap.J.Suppl.* **57**, 241.
26. Schramm, D. 1985, *Proc. 1984 Rome Conf. on Microwave Background.*
27. Bardeen, J., Bond, J., Kaiser, N., and Szalay, A. 1985, submitted to *Ap.J.*.
28. Turok, N. 1985, U.C. Santa Barbara preprint
29. Melott, A. 1986, Univ. of Chicago preprint.

30. Charlton, J. and Schramm, D. 1986, submitted to *Ap.J.*.
31. Vilenkin, A. 1985, *Physics Reports* **121**, 1.
32. Vittorio, N. and Schramm, D. 1985, *Comments on Nuclear and Particle Physics* **15**, 1.
33. Turok, D.N. and Schramm, D.N. 1986, in preparation
34. Bennet, D. 1986, SLAC preprint
35. Vilenkin, A. 1986, Tufts Univesity preprint
36. Rashiputi 1986, preprint
37. Pagels, H. 1986, Rockefeller University preprint.
38. Peebles, P.J.E. 1981, *The Large Scale Structure of the Universe*, Princeton University Press.
39. Olive, K. and Seckel, D. 1986, FNAL preprint.
40. Crawford, M. and Schramm, D. 1982, *Nature* **298**, 538.
41. Witten, E. 1984, *Phys.Rev.* **D30**, 272.
42. Applegate, J. and Hogan, C. 1985, *Phys.Rev.* **D31**, 3037.
43. Alcock, C. and Farhi, J. 1985, MIT preprint.
44. Freese, K., Price, R., and Schramm, D. 1983, *Ap.J.* **275**, 405.
45. Vishniac, E., Ostriker, J., and Bertschinger, E. 1985, Princeton University preprint.
46. Turok, N. and Schramm, D. 1984, *Nature* **312**, 598.
47. Witten, E. 1985, *Physics Letters* **B153**, 243.

Table I: Models and Problems

	Hot & Adiabatic	Cold & Adiabatic	Cold & Strings	Hot & Strings	Cold & Explosions	Hot & Explosions
$\Omega = 1$ with $\Omega_{cluster} \sim 0.2$	o.k.	requires ad hoc biasing	requires large cluster–galaxy correlation	requires large cluster–galaxy correlation	requires special biasing	o.k.
Large Cluster–Cluster correlation function	difficult	no	o.k.	o.k. if not destroyed by pancaking	requires fine tuning	requires fine tuning
Filaments, sheets, and voids structure at ~ 40Mpc	o.k.	difficult	not easy	maybe	difficult	o.k.
Large scale high velocities	o.k.	no, worse with biasing	no	maybe	difficult	o.k.
Galaxy formation by $z \gtrsim 4$	no	o.k.	o.k.	marginal	o.k.	depends on seeds
Galaxy mass spectrum	pancake fragmentation	depends on biasing scheme	maybe	probably o.k.	maybe	maybe
Galaxy angular momentum	pancake fragmentation	o.k.	probably o.k.	probably o.k.	probably o.k.	probably o.k.
Contents of voids	mostly hot stuff	$\sim 90\%$ cold $\sim 10\%$ baryons	$\sim 90\%$ cold $\sim 10\%$ baryons	$\gtrsim 90\%$ hot $\lesssim 10\%$ baryons	$\sim 90\%$ cold $\sim 10\%$ baryons	mostly hot stuff
Halos of galaxies (including dwarfs)	mostly baryonic	$\sim 90\%$ cold $\sim 10\%$ baryonic	$\sim 90\%$ cold $\sim 10\%$ baryonic	$> 10\%$ baryonic $< 90\%$ hot	$\sim 90\%$ cold $\sim 10\%$ baryonic	mostly baryonic

COSMOLOGY AND EXTRA DIMENSIONS

Edward W. Kolb
Theoretical Astrophysics
Fermi National Accelerator Laboratory
Batavia, Illinois 60510 USA

ABSTRACT. In the past few years the search for a consistent quantum theory of gravity and the quest for a unification of gravity with other forces have led to a great deal of interest in theories with extra spatial dimensions. These extra spatial dimensions are unseen because they are compact and small, presumably with typical dimensions of the Planck length, $l_{Pl} = 1.616 \times 10^{-33}$cm. If the "internal" dimensions are static and small compared to the large "external" dimensions the only role they would play in the dynamics of the expansion of the Universe is in determining the structure of the physical laws. However, if the big bang is extrapolated back to the Planck time, then the characteristic size of *both* internal and external dimensions were the same, and the internal dimensions may have had a more direct role in the dynamics of the evolution of the Universe. This chapter presents some speculations about the role of extra dimensions in cosmology.

1. MICROPHYSICS IN EXTRA DIMENSIONS

Theories that have been formulated in extra dimensions include Kaluza-Klein theories [1], supergravity theories [1], and superstring theories [2]. The exact motivation and goals of these approaches are quite different, but for many applications to cosmology they have several common features and they will be referred to simply as theories in extra dimensions. Among the common features of theories in extra dimensions are:

• *There are large spatial dimensions and small spatial dimensions*: If some of the dimensions are compact and smaller than the three large dimensions, it is possible to dimensionally reduce the system (integrate over the extra dimensions) and obtain an "effective" 3+1-dimensional theory. Present accelerators have probed matter at distances as small as 10^{-16}cm without finding evidence of extra dimensions. This is not surprising, as the extra dimensions are expected to have a size characteristic of the Planck length. The large dimensions may also be compact. If so, their characteristic size is greater than the Hubble distance, 10^{28}cm. This disparity of about 61 orders of magnitude is somewhat striking. This disparity is

P. Galeotti and D. N. Schramm (eds.), Gauge Theory and the Early Universe, 225–256.

THEORY	α/α^0	G/G^0	G_F/G_F^0
Kaluza-Klein	$(b/b_0)^{-2}$	$(b/b_0)^{-D}$	$(b/b_0)^{-2}$
(D internal dimensions)			
Superstrings	$(b/b_0)^{-6}$	$(b/b_0)^{-6}$	$(b/b_0)^{-6}$
(6 internal dimensions)			

Table 1: Variation of fundamental constants with the size of the internal manifold

usually posed by the question "what makes the extra dimensions so small?" However, if gravity has anything to do with the size of dimensions, the only reasonable size *is* the Planck length, and a more appropriate question to ask is "what makes the observed dimensions so large?" One possible answer to the the last question is inflation. The possible connection between inflation and extra dimensions will be explored.

• *The effective low-energy theory depends upon the internal space*: In Kaluza-Klein theories the low-energy gauge group is determined by the continuous isometries of the internal manifold. In superstring theories, the structure of the internal space determines the number of generations of chiral fermions, whether there is low-energy supersymmetry, etc. If the internal space is distorted in any way the effective low-energy physics could be very different.

• *The fundamental constants we observe are not truly fundamental*: In theories with extra dimensions the truly fundamental constants are constants in the higher dimensional theory. The constants that appear in the dimensionally reduced theory are the result of integration over the extra dimensions. If the volume of the extra dimensions would change, the value of the constants we observe in the dimensionally-reduced theory would change. Exactly how they would change depends upon the theory. In Kaluza-Klein theories, gauge symmetries arise from continuous isometries in the internal manifold, while in superstring theories the gauge symmetries are part of the fundamental theory. In all theories the gravitational constant is inversely proportional to the volume of the internal manifold. In the most general case there is not a single radius in the internal manifold. However, for the sake of simplicity it will be assumed that there is a single radius, b, which characterizes the internal manifold. The b dependence of some fundamental constants are given in Table 1. In Table 1, α^0 is the present value of the fine structure constant, G^0 is the present value of the gravitational constant, G_F^0 is the present value of Fermi's constant, and b_0 is the present value of b.

• *The internal dimensions are static*: If the internal dimensions change, fundamental constants change. Limits on the time variability of the fundamental constants can be converted to limits on the time variability of the extra dimensions. Limits on time rate of change of the fine structure constant (assuming

$\lvert\dot{\alpha}/\alpha\rvert$	METHOD	$\Delta\tau$
$5\times10^{-15}\mathrm{yr}^{-1}$	$^{187}\mathrm{Re}/^{187}\mathrm{Os}$	$5\times10^{9}\mathrm{yr}$
$1\times10^{-17}\mathrm{yr}^{-1}$	Oklo reactor	$1.8\times10^{9}\mathrm{yr}$
$13\times10^{-13}h\ \mathrm{yr}^{-1}$	Radio galaxies	$2\times10^{9}h^{-1}\ \mathrm{yr}$
$2\times10^{-14}h\ \mathrm{yr}^{-1}$	QSO	$5\times10^{9}h^{-1}\ \mathrm{yr}$
$15\times10^{-15}h\ \mathrm{yr}^{-1}$	Primordial nucleosynthesis	$6.6\times10^{9}h^{-1}\ \mathrm{yr}$

Table 2: Constraints on the time variation of the fine structure constant

that the change is a power law in cosmological time) are given in Table 2. The look-back time, $\Delta\tau$, is the maximum time over which the limit may be applied. For the look-back time, an $\Omega = 1$ cosmology was assumed, i.e., a present age of $(2/3)H_0^{-1} = 6.6\times10^9h^{-1}$yr. Long look-back times are relevant if the change is not a power law in cosmological time. It is interesting to know how soon after the bang the internal space had essentially the size it has today. The limit with the longest look-back time is the limit from primordial nucleosynthesis.

Primordial nucleosynthesis is a sensitive probe of changes in α, since the neutron-proton mass difference $Q = m_n - m_p = 1.293$ MeV has an electromagnetic component. Although the details of the neutron-proton mass difference are not known, it is reasonable to assume that the electromagnetic contribution is the same size (but the opposite sign) as the entire difference. With this assumption $\alpha/\alpha^0 = Q/Q^0$, where Q^0 is the value today.

The neutron-proton ratio at freeze out given by Eq. 1.78 is $\exp(-Q/T_f)$, so n/p is very sensitive to small changes in Q. The primordial ^{4}He mass fraction as a function of b/b_0 is given in Fig. 1, assuming that α, G, and G_F depend on b/b_0 as in Table 1. The curve labeled "SS" is the superstring model ($D = 6$), and the curves marked "KK_2" and "KK_7" are Kaluza-Klein models with $D = 2$ and $D = 7$ internal dimensions. The allowed range of the primordial ^{4}He, $Y_P = X_4 = 0.24\pm0.01$. For the superstring model, the primordial helium is within acceptable limits only if at the time of primordial nucleosynthesis $1.005 \geq b/b_0 \geq 0.995$. The Kaluza-Klein models give the slightly less stringent result $1.01 \geq b/b_0 \geq 0.99$. In either case, by the time of primordial nucleosynthesis the internal dimensions had obtained a size very close to the size they have today [3].

• ***The ground state geometry does not have all the symmetries of the theory:*** It is generally assumed that the ground state geometry is of the form $M^4 \times B^D$, where M^4 is four-dimensional Minkowski space, [1] and B^D is some compact

[1]The assumption of M^4 is not quite correct in a cosmological context, and should be replaced by $R^1\times S^3$ for the closed model, $R^1\times Q^3$ for the open model.

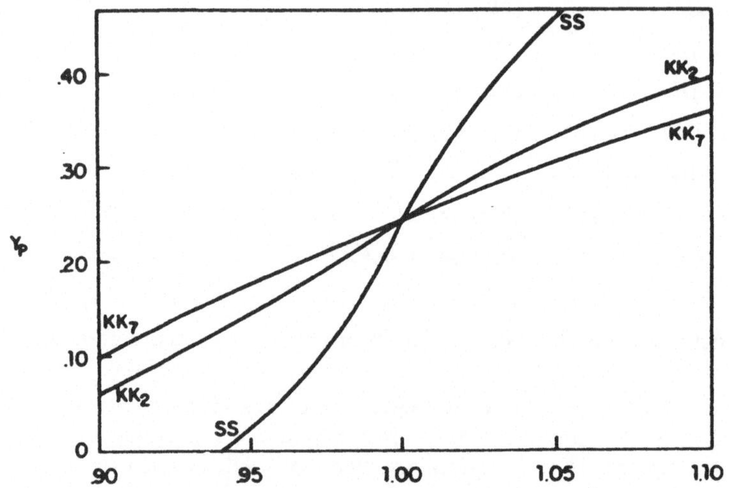

Figure 1: The primordial mass fraction as a function of b/b_0

D−dimensional space. The symmetries of the ground state are generally not as large as the symmetries of the theory, i.e., there is spontaneous symmetry breaking. One of the results of SSB is the existence of a massless (at least at the classical level) Nambu-Goldstone boson, which is sometimes called the dilaton.

• *The spectrum contains an infinite number of massive states*: If the radius of the internal space is b, then b^{-1} sets the scale for the massive states. The spectrum of the massive states depends upon the type of theory and the structure of the internal manifold. Since b is expected to be close to l_{Pl}, the massive states should have masses close to m_{Pl}.

2. STABILITY OF THE INTERNAL SPACE

All theories formulated in extra dimensions must contain some mechanism to keep the internal dimensions static. In the absence of such a mechanism, the extra dimensions would either contract or expand. The origin of the vacuum stress responsible for this is unknown. Here, some toy models are given, along with some possible cosmological effects.

In theories with extra dimensions new types of interactions may arise. For a starting point, consider the Chapline-Manton action [4], which is an $N = 1$ supergravity and an $N = 1$ super-Yang-Mills theory in 10 space-time dimensions. This theory is thought to be the field theory limit of a 10-dimensional superstring theory. It is not at all clear that the 10-dimensional field theory limit of the superstring ever makes sense. The 10-dimensional field theory description obtains only in the region between two similar energy scales. The first scale is determined by the string tension. It is the scale above which it is necessary to include the massive excitations of the string. Above this scale physics is "stringy" and any point-like field theory description is inadequate. The second scale is the compactification scale, which is determined by the radius of the internal space. At distances smaller than the compactification scale dimensional reduction no longer makes sense, the 3+1-dimensional description is inadequate, and the 10-dimensional theory must be used. The 10-dimensional field theory description makes sense at distance scales larger than the string tension scale, but smaller than the compactification scale. Since these two scales are expected to be the same order of magnitude, it is not clear if the 10-dimensional field theory description ever obtains. Nevertheless, it offers a convenient starting point for an exploration of cosmology in extra dimensions.

The Chapline-Manton Lagrangian contains the $N = 1$ supergravity multiplet $\{e^A_M;\ \psi_M;\ B_{MN};\ \lambda;\ \sigma\}$, where e^A_M is the vielbein, ψ_M is the Rarita-Schwinger field, B_{MN} is the Kalb-Ramond field, λ is the sub-gravitino, and σ is the dilaton, and the super Yang-Mills multiplet $\{G_{MN};\ \chi\}$, where G_{MN} is the Yang-Mills field strength and χ is the gluino field. The Lagrangian is [2]

$$
\begin{aligned}
e^{-1}\ \mathcal{L} &= -\frac{1}{2}R - \frac{1}{2}\bar{\psi}_M\Gamma^{MPS}D_P\psi_S - \frac{3}{4}\exp(-\sigma)H_{MNP}H^{MNP} \\
&-\frac{1}{4}\partial_M\sigma\partial^M\sigma - \frac{3\sqrt{2}}{8}\bar{\psi}_M\ \partial\!\!\!/\sigma\Gamma^M\lambda - \frac{1}{2}\bar{\lambda}\ D\!\!\!\!/\lambda \\
&+\frac{\sqrt{2}}{16}\exp(-\sigma/2)H_{MNP}\left(\bar{\psi}_Q\Gamma^{QMNPR}\psi_R + 6\bar{\psi}^M\Gamma^N\psi^P\right. \\
&\left.-\sqrt{2}\bar{\psi}_R\Gamma^{MNP}\Gamma^R\lambda\right) - \frac{1}{2}\mathrm{Tr}\bar{\chi}\ D\!\!\!\!/\chi - \frac{1}{4}\exp(-\sigma/2)\mathrm{Tr}G_{MN}G^{MN} \\
&-\frac{3}{4}\left(\mathrm{Tr}\bar{\chi}\Gamma_{MNP}\chi\right)^2 + \exp(-\sigma/2)H_{MNP}\mathrm{Tr}\bar{\chi}\Gamma^{MNP}\chi + \cdots
\end{aligned}
\tag{2.1}
$$

where $\Gamma^{MNP} = \Gamma^{[M}\Gamma^N\Gamma^{P]}$, and $H_{MNP} = \partial_{[M}B_{NP]}$. Four fermion couplings and other terms have been omitted.

The "Einstein equations" are straightforward to obtain:

$$
\begin{aligned}
R_{MN} &= \frac{9}{2}\exp(-\sigma)\left(H_{MPQ}H_N^{\ PQ} - \frac{1}{12}g_{MN}H_{PQR}H^{PQR}\right) \\
&- \exp(-\sigma/2)\left(\mathrm{Tr}G_{MP}G_N^{\ P} - \frac{1}{16}g_{MN}\mathrm{Tr}G_{PQ}G^{PQ}\right)
\end{aligned}
$$

[2]The following notation will be used: D =number of extra dimensions; $M,\ N,\ P,\ Q,\ \ldots$ run from 0 to $D+3$; $\mu,\ \nu,\ \rho,\ \ldots$ are indices in the extra dimensions; and $m,\ n,\ p,\ q,\ \ldots$ are indices in the large spatial dimensions.

$$-\frac{1}{2}\partial_M\sigma\partial^M\sigma - \frac{1}{8}\left(\mathrm{Tr}\bar{\chi}\Gamma_{PQR}\chi\right)\left(\bar{\lambda}\Gamma^{PQR}\lambda\right)g_{MN}$$

$$-\frac{3}{16}\left(\mathrm{Tr}\bar{\chi}\Gamma_{PQR}\chi\right)^2 g_{MN} + \frac{9}{2}\exp(-\sigma/2)H_M^{\ PQ}\mathrm{Tr}\bar{\chi}\Gamma_{NPQ}\chi$$

$$-\frac{3}{16}\exp(-\sigma/2)g_{MN}H_{PQR}\mathrm{Tr}\bar{\chi}\Gamma^{PQR}\chi + \cdots \qquad (2.2)$$

The task at hand is to solve Eq. 2.2 to find the equations of evolution of the scale factor(s) in the expansion of the Universe toward the quasi-static ground state of the system where there are D static dimensions and 3 dynamic dimensions expanding as in a standard FRW cosmology.

In general it is necessary to choose background field configurations. For example consider the "bosonic" parts of the equations. What are the symmetries of the metric? What are the vacuum (background) values of H_{MNP}, of G_{MN}, of $\bar{\chi}\Gamma\chi$, of $\bar{\lambda}\Gamma\lambda$, of σ? In general, many (possibly infinitely many) solutions of the field equations are expected, even if there is but one ground state that describes the microphysics of our Universe. The immediate question to ask is what picks out the ground state and what is the evolution of the Universe to this ground state? Perhaps when the true string nature of the equations are taken into consideration there will be but one possible solution to the string equations even if there are many solutions to the field theory. Perhaps something in the evolution of the Universe prefers a unique or small number of possibilities. Such questions are reminiscent of the questions considered in inflation. If the conditions in some region of the Universe are such as to enter an inflationary phase, that region of the Universe will grow relative to a region that does not undergo inflation. It is possible to imagine that the Universe starts in a state with no particular background field configuration, but in a quantum state described by a wave function Ψ that describes the probability of a given configuration, Ψ(field configurations). If in some region of the Universe the wave function is peaked about a particular configuration that will inflate some spatial dimensions, that region will grow. All that is required to produce the Universe we observe is that there is some region that will lead to three spatial dimensions inflating (and some mechanism to keep D dimensions static). It may be that the theory is unique, but the ground state is not. It may be that somewhere outside of our horizon the Universe is quite different. There may be a different number of small versus large dimensions, or the internal space may have different topological properties leading to drastically different microphysics. Before this speculation is considered, it is necessary to understand the mechanism that leads to the stabilization of the internal space. This problem will be studied by considering individual contributions to the right-hand side of Eq. 2.2.

For simplicity, the metric will be taken to have the symmetry $R^1 \times S^3 \times S^D$

$$g_{MN} = \begin{pmatrix} 1 & & \\ & -a^2(t)\tilde{g}_{mn} & \\ & & -b^2(t)\tilde{g}_{\mu\nu} \end{pmatrix} \tag{2.3}$$

where $\tilde{g}_{mn}$ is the metric for S^3 of unit radius and $a(t)$ is the actual radius, and $\tilde{g}_{\mu\nu}$ is the metric for S^D of unit radius and $b(t)$ is the actual radius. The components of the Ricci tensor are

$$\begin{aligned} -R_{00} &= 3\frac{\ddot{a}}{a} + D\frac{\ddot{b}}{b} \\ -R_{mn} &= \left[\frac{\ddot{a}}{a} + 2\frac{\dot{a}^2}{a^2} + D\frac{\dot{a}}{a}\frac{\dot{b}}{b} + \frac{2}{a^2}\right] g_{mn} \\ -R_{\mu\nu} &= \left[\frac{\ddot{b}}{b} + (D-1)\frac{\dot{b}^2}{b^2} + 3\frac{\dot{a}}{a}\frac{\dot{b}}{b} + \frac{D-1}{b^2}\right] g_{\mu\nu}. \end{aligned} \tag{2.4}$$

With the Einstein equations in the form

$$R_{MN} = 8\pi\bar{G}\left[T_{MN} - \frac{1}{D+2}g_{MN}T^P_{\ P} - \frac{1}{D+2}\frac{\Lambda}{8\pi\bar{G}}g_{MN}\right] \tag{2.5}$$

where $\bar{G}$ is the gravitational constant in $D+4$ dimensions, [3] and Λ is a possible cosmological constant in $D+4$ dimensions. All the terms on the right-hand side of Eq. 2.2 contribute to T_{MN} and Λ.

Symmetries of the stress tensor are usually chosen such that the only non-vanishing components of the stress tensor are

$$\begin{aligned} T_{00} &\equiv \rho \\ T_{mn} &\equiv -p_3 g_{mn} \\ T_{\mu\nu} &\equiv -p_D g_{\mu\nu} \end{aligned} \tag{2.6}$$

with $T^M_M = \rho - 3p_3 - Dp_D$. In terms of ρ, p_3, p_D, and $\rho_\Lambda = \Lambda/8\pi\bar{G}$ the Einstein equations are

$$\begin{aligned} 3\frac{\ddot{a}}{a} + D\frac{\ddot{b}}{b} &= -\frac{8\pi\bar{G}}{D+2}[(D+1)\rho + 3p_3 + Dp_D - \rho_\Lambda] \\ \frac{\ddot{a}}{a} + 2\frac{\dot{a}^2}{a^2} + D\frac{\dot{a}}{a}\frac{\dot{b}}{b} + \frac{2}{a^2} &= \frac{8\pi\bar{G}}{D+2}[\rho + (D-1)p_3 - Dp_D + \rho_\Lambda] \\ \frac{\ddot{b}}{b} + (D-1)\frac{\dot{b}^2}{b^2} + 3\frac{\dot{a}}{a}\frac{\dot{b}}{b} + \frac{D-1}{b^2} &= \frac{8\pi\bar{G}}{D+2}[\rho - 3p_3 + 2p_D + \rho_\Lambda]. \end{aligned} \tag{2.7}$$

[3] $\bar{G}$ is related to Newton's constant G by $\bar{G} = GV_D^0$, where V_D^0 is the volume of the internal space today.

Some possible contributions to the right hand side will be considered in turn.

•R_{MN} =NOTHING: The simplest possible form for the right hand side is zero. For the moment abandon the choice of $R^1 \times S^3 \times S^D$, and consider a $D+3$ torus for the ground state geometry. The spatial coordinates can be chosen to take the values $0 \leq x^i \leq L$, where L is a parameter with dimension of length. The general cosmological solutions of the vacuum Einstein equations are the Kasner solutions. The Kasner metric is

$$ds^2 = dt^2 - \sum_{i=1}^{D+3} \left(\frac{t}{t_0}\right)^{2p_i} (dx^i)^2. \tag{2.8}$$

The Kasner metric is a solution to the vacuum Einstein equations provided the Kasner conditions are satisfied

$$\sum_{i=1}^{D+3} p_i = \sum_{i=1}^{D+3} p_i^2 = 1. \tag{2.9}$$

In order to satisfy the Kasner conditions at least one of the p_i must be negative. It is possible to have 3 spatial dimensions expanding in an isotropic manner and D dimensions contracting in an isotropic manner by the choice [5]

$$\begin{aligned} p_1 = p_2 = p_3 &\equiv p = \frac{3 + (3D^2 + 6D)^{1/2}}{3(D+3)} \\ p_4 = \ldots = p_{3+D} &\equiv q = \frac{D - (3D^2 + 6D)^{1/2}}{D(D+3)}. \end{aligned} \tag{2.10}$$

Note that $p > 0$ and $q < 0$. With this choice the metric may be written

$$ds^2 = dt^2 - a^2(t)d\vec{x}^2 - b^2(t)d\vec{y}^2, \tag{2.11}$$

where x^i are coordinates of the 3 expanding dimensions, and y^i are coordinates of the D contracting dimensions. The two scale factors are given by $a(t) = (t/t_0)^p$, $b(t) = (t/t_0)^q$.

Somewhat more complicated classical cosmologies have been considered. The Kasner model can be regarded as an anisotropic generalization of the flat FRW cosmology, i.e., a Bianchi I cosmology. A generalization of the closed FRW model is the Bianchi IX model. The Bianchi IX vacuum solutions have the feature that the general approach to the singularity is "chaotic." [6] On approach to the initial singularity the scale factors in different spatial directions undergo a series of oscillations, contractions, and expansions. This feature is quite general, and independent of the state of the Universe after the singularity. The oscillation of the scale factors is well described by a sequence of Kasner models in which expanding and contracting dimensions are interchanged in "bounces." Such anisotropic behavior is predicted to be the general approach to the initial singularity. The question of whether such a chaotic approach to the initial singularity is present in more than three spatial dimensions has been considered. It has been shown that chaotic behavior obtains only for models with between 3 and 9 spatial dimensions [7]. The importance of this observation is clouded by the fact that at the approach to the singularity curvature may not dominate the right hand side of the Einstein equations, and near the singularity classical gravity may be a poor description.

The solutions above do not have solutions with a static internal space and if they are ever relevant, it is only for a limited time. The right-hand side must be more complicated than nothing. The next simplest thing to consider on the right-hand side is free scalar fields. Before discussing their effect on the evolution of the Universe it is necessary to discuss regularization in the background geometry.

The free energy of a non-interacting spinless boson of mass μ is given by [8]

$$F = T\frac{1}{2}\ln \mathrm{Det}\left(-\Box_{4+D} + \mu^2\right). \tag{2.12}$$

since finite temperature effects are of interest, the time is periodic with period of $1/2\pi T$, the relevant geometry is $S^1 \times S^3 \times S^D$, and the radii of the spheres are $1/2\pi T$, a, and b. The eigenvalues of $\Box$ on the compact space are discrete, and are given by the triple sum (hereafter μ will be set to zero)

$$2T^{-1}F = \sum_{r=-\infty}^{\infty}\sum_{m,n=0}^{\infty} D_{mn}\ln\Big[r^2(2\pi T)^2 + m(m+2)a^{-2} + n(n+D-1)b^{-2}\Big], \tag{2.13}$$

where D_{mn} is a factor that counts the degeneracy

$$D_{mn} = (m+1)^2(2n+D-1)(n+D-2)!/(D-1)!n!. \tag{2.14}$$

The free energy given by Eq. 2.13 is, of course, infinite. To deal with the infinities, a regularization scheme will be found to extract the relevant finite part. For the purpose of regularization, each term in the sum can be expressed as an integral using the formula [8] [4]

$$\ln X = \frac{d}{ds}X^s\,|_{s=0} = \frac{d}{ds}\left(\frac{1}{\Gamma(-s)}\int_0^\infty dt\, t^{s-1}\exp(-tX)\right)_{s=0}. \tag{2.15}$$

The finite part of the free energy is given by

$$2T^{-1}F = \frac{d}{ds}\left[\frac{1}{\Gamma(-s)}\int_0^\infty dt\, t^{-s-1}\sigma_1(4\pi^2T^2t)\sigma_3(a^{-2}t)\sigma_D(b^{-2}t)\right]_{s=0}, \tag{2.16}$$

[4]This regularization is only valid for D =odd. The D =even case will be discussed below.

where the functions σ_i are given by

$$\sigma_i(x) = \sum_{n=0}^{\infty} \frac{(2n+i-1)(n+i-1)!}{(i-1)!n!} \exp[-n(n+i-1)x]. \tag{2.17}$$

The full expression for the free energy is quite difficult to evaluate, but the free energy is simple in several limits. In the "flat-space" limit the radius of S^3 is much larger than the radius of S^D $(a \gg b)$ and $\sigma_3 \to (\sqrt{\pi}/4)a^3t^{-3/2}$. In the limit $a \gg b$ the free energy can be approximated by

$$F = \frac{\Omega_3 a^3}{b^4}\left[c_1 - c_2(bT)^4 - c_3(bT)^{D+4}\right], \tag{2.18}$$

where Ω_i is found from the volume of the i-sphere, $V_i = R^i\Omega_i$ with R the radius and $\Omega_i = (2\pi)^{(i+1)/2}/\Gamma[(i+1)/2]$. For S^3, the volume is $V_3 = R^3 2\pi^2$, and Ω_3 has the familiar form $\Omega_3 = 2\pi^2$. The term proportional to c_1 is the Casimir term (c_1 is c_N of Candelas and Weinberg [9]). The term proportional to $c_2 = \pi^2/90$ is the leading temperature-dependent term when $T \ll b^{-1}$. When $T \gg b^{-1}$, the term proportional to $c_3 = (2\varsigma(D+4)/\pi^{3/2})\Gamma[(D+4)/2]/\Gamma[(D+1)/2]$ dominates. In the "low-temperature" limit the radius of the S^1 becomes large and $\sigma_1 \to (4\pi t T^2)^{-1/2}$. In the flat-space, zero-temperature limit only the term proportional to c_1 survives.

The internal energy is given in terms of the free energy, the temperature, and the entropy

$$S = -\left[\frac{\partial F}{\partial T}\right]_{a,b}, \tag{2.19}$$

by $U = F + TS$. The thermodynamic quantities ρ, p_3, and p_D are defined in terms of the internal energy:

$$\begin{aligned} \rho &= \frac{U}{\Omega_3\Omega_D a^3 b^D} \\ p_3 &= -\frac{a}{3\Omega_3\Omega_D a^3 b^D}\left[\frac{\partial U}{\partial a}\right]_{b,S} \\ p_D &= -\frac{b}{D\Omega_3\Omega_D a^3 b^D}\left[\frac{\partial U}{\partial b}\right]_{a,S}. \end{aligned} \tag{2.20}$$

The thermodynamic quantities in zero temperature, low temperature, and high temperature limits are given in Table 3. There are several obvious limits of Table 3. In the zero-temperature or in the low-temperature limits, dimensional reduction is possible. Upon integration over the internal dimensions the effective three-dimensional energy density and pressure is obtained by multiplication by $V_D = \Omega_D b^D$. After dimensional reduction the Casimir terms are proportional to $c_1 b^{-4}$.

	Casimir $T=0$	Low Temperature $0 \le T \le b^{-1}$	High Temperature $T \gg b^{-1}$	Monopole T=0
ρ	$c_1/\Omega_D b^{4+D}$	$(\pi^2/30)T^4/\Omega_D b^{4+D}$	$(D+3)c_3 T^{D+4}/\Omega_D$	$f_0^2/2b^{2D}$
p_3	$-c_1/\Omega_D b^{4+D}$	$(\pi^2/90)T^4/\Omega_D b^{4+D}$	$c_3 T^{D+4}/\Omega_D$	$-f_0^2/2b^{2D}$
p_D	$4c_1/D\Omega_D b^{4+D}$	0	$c_3 T^{D+4}/\Omega_D$	$f_0^2/2b^{2D}$
T^M_M	0	0	0	$(4-D)f_0^2/2b^{2D}$

Table 3: Contributions to thermodynamic quantities

The low-temperature limit after dimensional reduction is $\rho = 3p_3 \to (\pi^2/30)T^4$ and $p_D = 0$, which is the expected contribution for a spinless boson in 3+1 dimensions. In the high-temperature limit dimensional reduction does not make sense.

It is possible to perform a similar analysis for particles of higher spin. The technical details are more difficult, but the physics is quite similar.

•R_{MN} =RADIATION: [10] Consider the "high-temperature" ($T \ge b^{-1}$) "flat-space" ($a \gg b$) limit with $\Lambda = 0$. In this limit T_{MN} is isotropic in the sense that $p_3 = p_D \equiv p$ (see Table 3). The Einstein equations are

$$\begin{aligned} 3\frac{\ddot{a}}{a} + D\frac{\ddot{b}}{b} &= -8\pi\bar{G}\rho \\ \frac{\ddot{a}}{a} + 2\frac{\dot{a}^2}{a^2} + D\frac{\dot{a}}{a}\frac{\dot{b}}{b} &= 8\pi\bar{G}p \\ \frac{\ddot{b}}{b} + (D-1)\frac{\dot{b}^2}{b^2} + 3\frac{\dot{a}}{a}\frac{\dot{b}}{b} + \frac{D-1}{b^2} &= 8\pi\bar{G}p. \end{aligned} \tag{2.21}$$

In keeping with the flat space assumption the $2/a^2$ term has been dropped in R_{mn}. The equation of state is $\rho = Np$, where $N \equiv D+3$. The conservation law $T^{MP}{}_{;P} = 0$ implies

$$\rho\bar{\sigma}^{N+1} = \text{constant}, \tag{2.22}$$

where $\bar{\sigma} \propto (a^3 b^D)^{1/N}$ is the mean scale factor. Since $\rho \propto T^{N+1}$, there is a conserved quantity $S_N = (\bar{\sigma}T)^N$ that is constant. This is simply the total N-dimensional entropy.

The Einstein equations (or a subset of the Einstein equations and the $T^{MP}{}_{;P} = 0$ equation) can be integrated to give $a(t)$ and $b(t)$. A typical solution is shown in Fig. 2. Both scale factors emerge from a initial singularity. The scale factor for the internal space reaches a maximum and recollapses to a second singularity. As b approaches the second singularity a is driven to infinity. The parameter x/x_s in

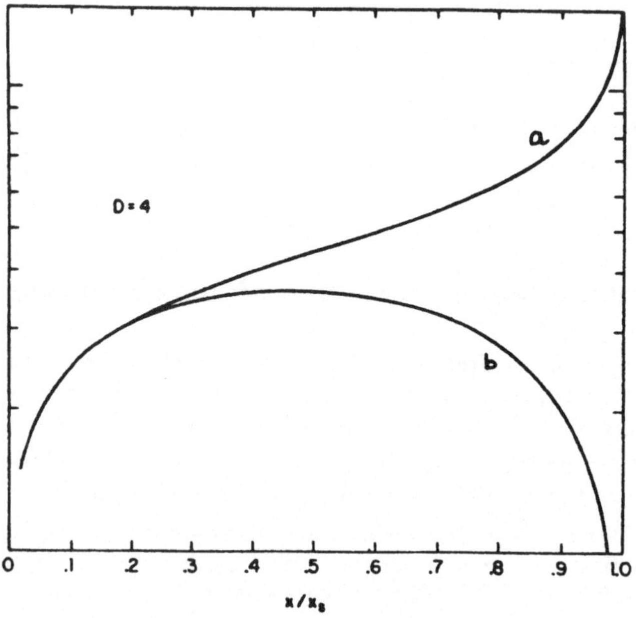

Figure 2: Evolution of the scale factors for R_{MN} =Radiation

Fig. 2 is a measure of the time in units of the time necessary to reach the second singularity.

The evolution of the temperature is shown in Fig. 3. The figure demonstrates the rather striking feature that as the second singularity is approached, the temperature increases. The expansion of a together with an increase of T seems unusual. However it is simply due to the conservation of entropy. In the region of growing T the mean volume of the Universe is actually *decreasing*, and the temperature must increase to keep S_N constant.

The assumption of the flat-space limit for S^3 can be easily justified. Imagine that the spatial geometry is $S^3 \times S^D$. If $a \geq b$ in the high-temperature region, once the maximum of b is reached, the S^3 will be inflated. The only requirement is that the curvature term, $1/a^2$, is small compared to the thermal term, $8\pi\bar{G}\rho$, at $b = b_{\text{MAX}}$.

In the approach to the second singularity the combination of expanding and contracting dimensions behaves like a Kasner model. A recurring feature in the analysis as presented in this review is that as the models become more baroque, there are limits in which the expansion can be approximated by only a part of the entire model. This is why consideration of the influence of individual terms contributing to T_{MN} is relevant.

In the period of increasing a and T, the entropy in the three expanding dimensions increases. Of course the *total* entropy is conserved, but in the approach to the second singularity entropy is squeezed out of the contracting dimensions into

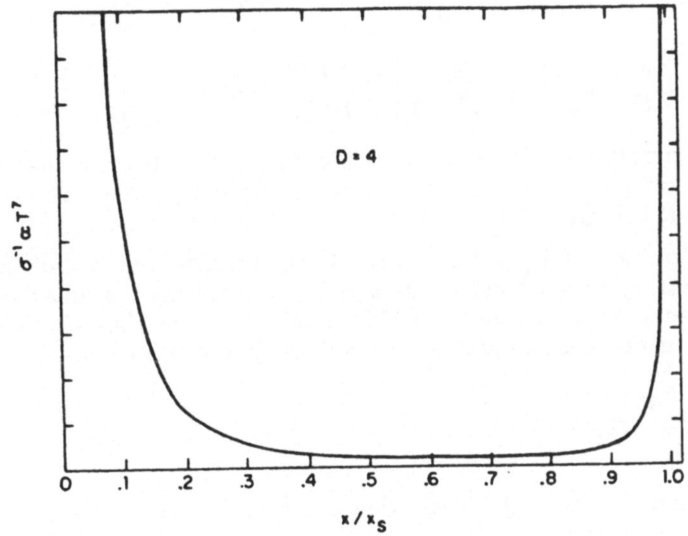

Figure 3: Evolution of the temperature for the solution of Fig.2

the expanding dimensions. The 3-entropy, S_3 will be defined as $S_3 = (d_{H3}T)^3$, where d_{H3} is the horizon distance in the 3-space

$$d_{H3} = a(t)\int_0^t dt'\, a^{-1}(t'). \tag{2.23}$$

In the approach to the second singularity, $d_{H3} \to \infty$ and $T \to \infty$, so $S_3 \to \infty$.

Before the second singularity is reached, two things must happen. First, there must be some mechanism to stabilize the internal dimensions. The other thing that must happen is that the high-temperature assumption will break down. The decrease of b outpaces the increase in T and eventually the assumption $T \geq b^{-1}$ will fail. When this occurs it is necessary to use the "low-temperature" form of the free energy and the only dynamical effect of the extra dimensions is the change in G. The increase in S_3 shuts off at this time. The conditions necessary to generate a significant amount of entropy in the three expanding dimensions have been studied. It is impossible to create an enormous amount of entropy without either very special initial conditions or extrapolating the solutions beyond the point where the high-temperature assumption breaks down.

•R_{MN} =Casimir + Λ [9]: The combination of Casimir forces plus a cosmological constant can lead to a classically stable ground state. With ρ, p_3, and p_D from Table 3, the Einstein equations in Eq. 2.7 becomes

$$3\frac{\ddot{a}}{a} + D\frac{\ddot{b}}{b} = -\frac{8\pi\bar{G}}{D+2}\left[\frac{(D+2)c_1}{\Omega_D}b^{-4-D} - \rho_\Lambda\right]$$

$$\frac{\ddot{a}}{a} + 2\frac{\dot{a}^2}{a^2} + D\frac{\dot{a}\,\dot{b}}{a\,b} + \frac{2}{a^2} = -\frac{8\pi\bar{G}}{D+2}\left[\frac{(D+2)c_1}{\Omega_D}b^{-4-D} - \rho_\Lambda\right]$$

$$\frac{\ddot{b}}{b} + (D-1)\frac{\dot{b}^2}{b^2} + 3\frac{\dot{a}\,\dot{b}}{a\,b} = \frac{8\pi\bar{G}}{D+2}\left[\frac{4(D+2)c_1}{D\Omega_D}b^{-4-D} + \rho_\Lambda\right] - \frac{D-1}{b^2}. \quad (2.24)$$

Note that the curvature of S^3 has been neglected $(1/a^2 \to 0)$, and that the curvature term for S^D $((D-1)/b^2)$ has been moved to the right hand side of the $\mu\nu$ equation where it belongs.

The search for static solutions involves setting the left-hand side of the equations to zero. Setting the left-hand side to zero involves setting the time derivatives of *both* a and b equal to zero. The value of b for this static solution will be denoted as b_0. The first or the second equation determines b_0 in terms of ρ_Λ

$$b_0^{-4-D} = \frac{\Omega_D}{(D+2)c_1}\rho_\Lambda. \quad (2.25)$$

Remembering that $\bar{G} = GV_D$ the $\ddot{b}$ equation can then be used to determine b_0 in terms of the Planck length

$$b_0^2 = \frac{8\pi c_1(4+D)}{D(D-1)}l_{Pl}^2. \quad (2.26)$$

It is useful to rewrite the equations once again, this time in terms of b_0

$$3\frac{\ddot{a}}{a} + D\frac{\ddot{b}}{b} = -(D-1)b_0^{-2}\left[\frac{D}{4+D}\left(\frac{b_0}{b}\right)^{4+D} - \frac{D}{4+D}\right]$$

$$\frac{\ddot{a}}{a} + 2\frac{\dot{a}^2}{a^2} + D\frac{\dot{a}\,\dot{b}}{a\,b} + \frac{2}{a^2} = -(D-1)b_0^{-2}\left[\frac{D}{4+D}\left(\frac{b_0}{b}\right)^{4+D} - \frac{D}{4+D}\right]$$

$$\frac{\ddot{b}}{b} + (D-1)\frac{\dot{b}^2}{b^2} + 3\frac{\dot{a}\,\dot{b}}{a\,b} = (D-1)b_0^{-2}\left[\frac{4}{4+D}\left(\frac{b_0}{b}\right)^{4+D} + \frac{D}{4+D} - \left(\frac{b_0}{b}\right)^2\right]. \quad (2.27)$$

Of course at $b = b_0$ the right-hand sides of the equations vanish.

In general there may be other interesting solutions to the system of equations. For instance in the limit where a and b *both* go to infinity, then the right-hand sides of all the equations approach a constant given by

$$H^2 = \frac{D(D-1)}{4+D}b_0^{-2}. \quad (2.28)$$

In this limit the solution to the system is $a(t) = b(t) = \exp(\pm Ht/\sqrt{3})$. This solution describes exponentially growing scale factors for both S^3 and S^D.

The static minimum $b = b_0$ is stable against small perturbations, since $\delta b(t) = b(t) - b_0$ has no exponentially growing modes. However the existence of the exponentially growing solution for a and b implies that if b is ever large, it would grow without limit. This suggests that the static minimum is not stable against arbitrarily large dilatations. This point will be discussed in detail shortly.

In order to search for other solutions, and to study the semiclassical instability in compactification, the radius of the extra dimension will be expressed as a scalar field in a potential in four dimensions. The equation for $\ddot{b}$ looks like the equation of motion for a scalar field if the $\dot{b}^2$ term is neglected on the right hand side, and the left hand side is regarded as $\partial V(b)/\partial b$. The correct function of b to regard as the scalar field is determined by the kinetic part of the action. The kinetic part of the gravitational action is

$$S_k = -\frac{1}{16\pi\bar{G}} \int d^{4+D}x\sqrt{-g_{4+D}}R_k, \tag{2.29}$$

where R_k is the part of the Ricci scalar containing time derivatives of b:

$$R_k = -D\left[2\frac{\ddot{b}}{b} + (D-1)\left(\frac{\dot{b}}{b}\right)^2 + 6\frac{\dot{a}}{a}\frac{\dot{b}}{b}\right]. \tag{2.30}$$

Upon integration by parts and integration over the internal space the kinetic part of the action becomes

$$S_k = -D(D-1)\frac{m_{Pl}^2}{16\pi}\int d^4x\sqrt{-g_4}\left(\frac{b}{b_0}\right)^{D-2}\left(\frac{\dot{b}}{b_0}\right)^2. \tag{2.31}$$

If a scalar field ϕ is defined as

$$\phi(b) = \left[\frac{D-1}{2\pi D}\right]^{1/2}\left(\frac{b}{b_0}\right)^{D/2} m_{Pl} \tag{2.32}$$

it will have a canonical kinetic term. With this definition of ϕ the $\ddot{b}$ equation becomes

$$\ddot{\phi} + 3\frac{\dot{a}}{a}\dot{\phi} + \frac{\dot{\phi}^2}{\phi} = -\frac{dV}{d\phi}, \tag{2.33}$$

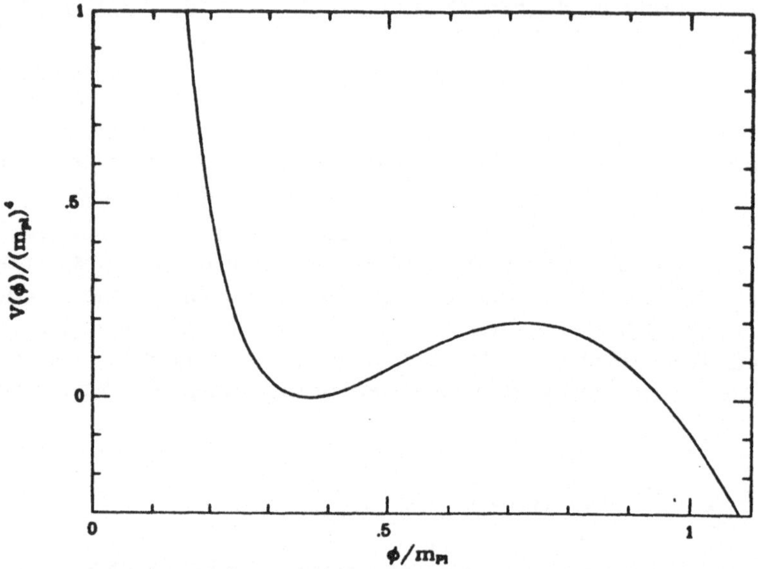

Figure 4: The potential for Casimir+Λ

where $dV/d\phi$ is the right hand side of Eq. 2.27 with the substitution of $\phi(b)$ for b. The potential is found by integrating $dV/d\phi$:

$$\begin{aligned} V(\phi) &= \left(\frac{D(D-1)}{8\pi(D+4)}\right)^2 \frac{(D-1)}{c_1} m_{Pl}^4 \left\{ \left(\frac{\phi}{\phi_0}\right)^{-8/D} - \left(\frac{\phi}{\phi_0}\right)^2 \right. \\ &+ \left. \frac{D+4}{D-2}\left[\left(\frac{\phi}{\phi_0}\right)^{2(D-2)/D} - 1\right]\right\}, \end{aligned} \tag{2.34}$$

where $\phi_0 = \phi(b_0)$ is the value of ϕ at the static minimum, $\phi_0 = [(D-1)/2\pi D]^{1/2} m_{Pl}$. There is an integration constant from integrating $dV/d\phi$ to find $V(\phi)$. The integration constant has been chosen to give $V(\phi_0) = 0$. A graph of $V(\phi)$ is given in Fig. 4 for $D = 7$ and $c_1 = 1$.

The figure illustrates several interesting features. The first feature is that the static minimum is perturbatively stable, but for ϕ greater than some value the potential is unstable. There is also a maximum to $V(\phi)$ that corresponds to $dV/d\phi = 0$ that corresponds to a solution with b static, but a expanding exponentially. A discussion of the semiclassical instability of the static solution will be discussed shortly.

•R_{MN} =Monopole+Λ [11]: The previous model used quantum effects from the Casimir effect to stabilize the extra dimensions against the cosmological constant. It is also possible to balance the effects of a classical field against the cosmological constant. Consider the Einstein-Maxwell theory in six space-time dimensions. The action for the model is given by

$$S = -\frac{1}{16\pi\bar{G}} \int d^6x\sqrt{-g_6}\left[R + \frac{1}{4}F_{MN}F^{MN} + 2\Lambda\right]. \tag{2.35}$$

The effect of the Maxwell field in the Einstein equations will through its contribution to the stress tensor

$$T_{MN} = F_{MQ}F_N^Q - \frac{1}{4}g_{MN}F_{PQ}F^{PQ}. \tag{2.36}$$

The ground state geometry will be assumed to be $R^1 \times S^3 \times S^2$, where as before $a \gg b$. The monopole ansatz has vanishing components of F_{MN} except for indices in the internal space:

$$F_{\mu\nu} = \sqrt{-g_2}\varepsilon_{\mu\nu}f(t), \tag{2.37}$$

where $f(t)$ is a function of time and g_2 is the determinant of the S^2 metric. This ansatz, of course, satisfies the field equations for F_{MN}. The Bianchi identities can be used to express $f(t)$ in terms of the S^2 radius, $f(t) = f_0/b(t)$, where f_0 is a constant.

With the monopole ansatz for F_{MN} the non-vanishing components of the stress tensor are

$$T_{00} = \frac{1}{2}\frac{f_0^2}{b^4}; \quad T_{mn} = -\frac{1}{2}\frac{f_0^2}{b^4}g_{mn}; \quad T_{\mu\nu} = \frac{1}{2}\frac{f_0^2}{b^4}g_{\mu\nu}. \tag{2.38}$$

The contributions of the monopole configuration to ρ, p_3, and p_2 are given in Table 3. The Einstein equations with the cosmological constant plus monopole are

$$\begin{aligned} 3\frac{\ddot{a}}{a} + 2\frac{\ddot{b}}{b} &= -2\pi\bar{G}\left[\frac{f_0^2}{b^4} - \rho_\Lambda\right] \\ \frac{\ddot{a}}{a} + 2\frac{\dot{a}^2}{a^2} + 2\frac{\dot{a}\,\dot{b}}{a\,b} + \frac{2}{a^2} &= -2\pi\bar{G}\left[\frac{f_0^2}{b^4} - \rho_\Lambda\right] \\ \frac{\ddot{b}}{b} + \frac{\dot{b}^2}{b^2} + 3\frac{\dot{a}\,\dot{b}}{a\,b} &= 2\pi\bar{G}\left[3\frac{f_0^2}{b^4} + \rho_\Lambda\right] - \frac{1}{b^2}. \end{aligned} \tag{2.39}$$

The static solution in terms of f_0 is

$$\rho_\Lambda = \frac{f_0^2}{b_0^4} \qquad b_0^2 = 8\pi\bar{G}f_0^2. \tag{2.40}$$

To illustrate the potential it is again useful to express the Einstein equations in terms of b_0

$$\begin{aligned}
3\frac{\ddot{a}}{a}+2\frac{\ddot{b}}{b} &= -\frac{1}{4b_0^2}\left[\left(\frac{b_0}{b}\right)^4-1\right] \\
\frac{\ddot{a}}{a}+2\frac{\dot{a}^2}{a^2}+2\frac{\dot{a}\,\dot{b}}{a\,b}+\frac{2}{a^2} &= -\frac{1}{4b_0^2}\left[\left(\frac{b_0}{b}\right)^4-1\right] \\
\frac{\ddot{b}}{b}+\frac{\dot{b}^2}{b^2}+3\frac{\dot{a}\,\dot{b}}{a\,b} &= \frac{1}{4b_0^2}\left[3\left(\frac{b_0}{b}\right)^4+1-4\left(\frac{b_0}{b}\right)^2\right].
\end{aligned} \tag{2.41}$$

In addition to the static solution at $b = b_0$, there is a quasi-static solution at $b = \sqrt{3}b_0$ where b is static, but a increases exponentially $a = a_0\exp(Ht)$, where $H = \sqrt{2}/3b_0$. Finally, there is the solution as *both* a and $b \to \infty$ where both scale factors increase exponentially with rate $H = 1/2\sqrt{5}b_0$.

By the same methods as developed for the Casimir case, it is possible to define a scalar field and a potential for the scalar field. The potential is very similar to Fig. 4. This model is also unstable against large dilatations of the internal dimensions.

The monopole compactification was considered in $D = 2$ for simplicity. The extention to larger D will be considered in the section on inflation.

•$R_{MN} = R^2 + \Lambda$ [12]: The Casimir, monopole, and cosmological constant terms can arise in the Chapline-Manton action. Although terms such as R^2, $R_{MN}R^{MN}$, and $R_{MNPQ}R^{MNPQ}$ do not appear in the Chapline-Manton action, they are expected to be present in superstring theories, and probably all other extra-dimension theories as well. Consider the gravitational action for a theory with such terms given by

$$\begin{aligned}
S = -\frac{1}{16\pi\bar{G}}\int d^{4+D}x\sqrt{-g_{4+D}}\,[R &+ 2\Lambda + a_1R^2 + a_2R_{MN}R^{MN} \\
&+ a_3R_{MNPQ}R^{MNPQ}].
\end{aligned} \tag{2.42}$$

There is a $M^4 \times S^D$ solution if the following conditions are met:

$$\begin{aligned}
0 &< D(D-1)a_1 + (D-1)a_2 + 2a_3 \\
0 &< (D-1)a_2 + 2a_3 \\
0 &< a_3 \\
\Lambda &= \frac{1}{4}\frac{D(D-1)}{D(D-1)a_1 + (D-1)a_2 + 2a_3}.
\end{aligned} \tag{2.43}$$

At the $M^4 \times S^D$ minimum, the value of b is

$$b_0^2 = 2D(D-1)a_1 + 2(D-1)a_2 + 4a_3. \tag{2.44}$$

The potential in this case is more difficult to analyze since there are higher derivative terms in the equations of motion. Nevertheless it has been shown that there is a solution corresponding to b ~constant and a increasing exponentially. Such a solution corresponds to a local maximum in the potential as in the Casimir or monopole cases. The difference in this case is that the location of this local maximum is a function of the a_i's, and for

$$a_3 = \frac{12/D - 3}{2 - 24/D(D-1)} a_2, \tag{2.45}$$

the local maximum will be at $b = \infty$. This means that the $M^4 \times S^D$ minimum is a true global minimum and is stable against large dilatations of the internal space. For the $D = 6$ case, the ghost-free action obtains for the case $a_3 = -a_2/4$, while Eq. 2.45 gives $a_3 = -5a_2/6$. The effect of the higher derivative terms in the equations of motion for $a(t)$ and $b(t)$ have been studied in both cases.

The possibility of using this model for inflation will be discussed below.

3. SEMICLASSICAL INSTABILITY OF COMPACTIFICATION [13]

In the Casimir $+\Lambda$ case, the monopole $+\Lambda$ case, and the $R^2 + \Lambda$ case where Eq. 2.45 is not satisfied, the static solution is not the true minimum of the theory. If the radius of the extra dimensions can be treated as a scalar field, it is possible to calculate the lifetime of the Universe against the decay of the false vacuum. The Casimir case will be used as an example.

Eq. 2.34 looks like the potential for a scalar field $\phi(x,t)$. The definition of ϕ in terms of b has been done to have the proper kinetic term for ϕ. With the four-dimensional gravitational degrees of freedom treated as a classical background, the problem of calculating the lifetime of the metastable state is identical to the decay of the false vacuum. For $D = 7$, $V(\phi)$ has a local minimum at $\phi_0 \simeq 0.37m_{Pl}$, a local maximum at $\phi_m \simeq 0.725m_{Pl}$, and a point degenerate with the local minimum at $\phi_T \simeq 0.96m_{Pl}$ (see Fig. 4).

The potential can be approximated in the region $0 \leq \phi \leq \phi_T$ by (c_1 has been set to 1)

$$V(\bar{\phi}) \simeq 0.093\Lambda\bar{\phi}^2 - 0.159\Lambda\bar{\phi}^3/m_{Pl}, \tag{3.1}$$

where $\phi = \bar{\phi} + \phi_0$ has been shifted to place the metastable state at the origin. The potential has the form $V(\bar{\phi}) = M^2\bar{\phi}/2 - \delta\bar{\phi}^3/3$ for which the tunnel action has been calculated. The tunnel action is $S_E \simeq 205M^2/\delta^2$ [14], which in terms of Λ and m_{Pl} is $S_E \simeq 165m_{Pl}^2/\Lambda$.

The decay rate per unit four volume is

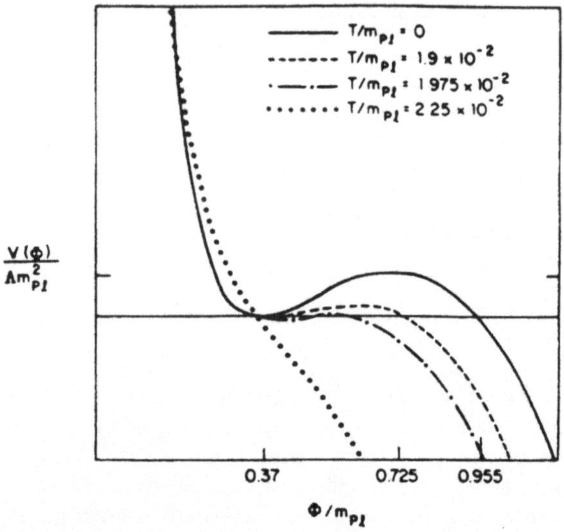

Figure 5: The temperature dependence of the Casimir potential

$$\Gamma \simeq m_{Pl}^4 \exp(-S_E), \tag{3.2}$$

where the pre-factor has been chosen as m_{Pl}^4 on dimensional grounds. In a matter-dominated Universe the probability for decay becomes of order unity in a time τ given by $\tau^4 \simeq 9\pi\Gamma/165 \simeq m_{Pl}^{-1} \exp(41 m_{Pl}^2/\Lambda)$. This is longer than the age of the Universe if $\Lambda \leq 0.3 m_{Pl}^2$.

In the Casimir case

$$\Lambda = \frac{D^2(D-1)^2(D+2)}{(D+4)^2 8\pi c_1} m_{Pl}^2 = 5.22 m_{Pl}^2/c_1 \quad (D=7). \tag{3.3}$$

In order to have the internal dimensions stay small for the age of the Universe requires $c_1 \geq 17.4$. For S^7 a single scalar field contributes $c_1 = 8.16 \times 10^{-4}$, so to satisfy the demand of longevity requires that there be more than 21,326 scalar fields. [5] Since the effective c_1's for higher-spin fields are larger, somewhat fewer are required.

There is also a finite-temperature instability present in the compactification. If the temperature-dependent terms in the free energy are included, the potential as a function of temperature has the form of Fig. 5. At high temperature the potential has no metastable state. The scalar field would not be trapped in the metastable phase if when $b \simeq b_0$ the temperature is large and temperature effects

[5] If there are N scalar fields, the effective c_1 is N times the c_1 for a single field.

are important. The temperature when $b = b_0$ depends upon the initial entropy. In a high-entropy initial condition the temperature will be large and compactification will not occur. The requirement that b should be trapped in the metastable state requires a low-entropy Universe, and the large entropy of the Universe must be created after compactification.

4. INFLATION AND EXTRA DIMENSIONS

The models of the previous section have illustrated the point that there are several mechanisms to force the internal space to be static and small. Although the mechanisms have different origins they all have in common the feature that there is a balance of forces at a particular value of $b \equiv b_0$. If $b \neq b_0$ there is an unbalanced stress in the vacuum. This unbalanced stress in the vacuum looks like a cosmological constant that can drive exponential expansion of all the dimensions, or just three dimensions. For instance in the monopole case discussed above, at $b = \sqrt{3}b_0$ there is a solution corresponding to static internal dimensions and exponentially expanding external dimensions. At $b = \sqrt{3}b_0$ the equation of motion for a is found from the (00) equation: $3\ddot{a}/a = 2/9b_0^2$, which has solution $a \propto exp(Ht)$, with $H^2 = 2/27b_0$

It is possible to imagine a scenario of new inflation where the exponential phase occurs for $b = \sqrt{3}b_0$, and is terminated when b settles to the local minimum at $b = b_0$. This is probably not a good example, because the potential is similar to the potential in Fig. 4, which is not the type of potential needed in new inflation. Even if for some unknown reason the Universe was ever in a configuration of $b = \sqrt{3}b_0$ and b static, quantum or thermal fluctuations would push b away from the unstable extremum. Even if it would roll in the correct direction toward the metastable minimum, the transition would be completed before sufficient inflation occurs.

A lesson learned from new inflation is that one should not be deterred by failure of simple models. For instance the $R^2 + \Lambda$ model is an existence proof that a model can be found. Recall that for a particular value of a_3/a_2 the potential does not turn over for large b and becomes flat. There can be a large amount of inflation as b evolves toward the ground state.

Inflation with the inflaton identified as the radius of the extra dimensions has some interesting features. In the evolution toward the ground state the radius of the extra dimensions grows, leading to an increase in the four-dimensional gravitational constant. The reheating is probably due to the change in the internal metric. For example, consider a minimally coupled scalar field χ with action

$$S \propto \int d^{D+4}x\, \chi \partial_M \left(\sqrt{-g_{4+D}}\, g^{MN} \partial_N \chi\right). \tag{4.1}$$

As b oscillates about the minimum of the potential there will be a non-zero value of $\dot{g}$ that results in an increase of χ. Although the details of the reheating remain to be worked out, the basic picture has been explored [12, 15].

All the models discussed above involve a $D+4$-dimensional cosmological constant that must be fine tuned to obtain the four-dimensional cosmological constant zero at b_0. All models (except the $R^2+\Lambda$ model with Eq. 2.45 satisfied) do not inflate and involve an unstable ground state. The introduction of the cosmological constant in the higher dimensional theory is not attractive. The fine tuning certainly must be incorrect. The unstable ground state cannot be ruled out, but seems undesirable. It would be nice if the existence of extra dimensions would lead to inflation.

Surely any realistic model should work without fine tuning of Λ. One might expect a realistic model to work for any effective value of Λ, and any change in Λ would simply lead to a change in b_0. In other words, if the vacuum energy would change, the only physical result would be a slight readjustment of b_0. This would be very attractive, since any cosmological constant produced as a result of SSB could be completely absorbed by a small change in b_0 and it would be unnecessary to fine tune Λ at high energies to account for phase transitions at low energies. Without extra dimensions there is nothing to do with the vacuum energy produced in phase transitions. Extra dimensions may provide a rug under which to sweep unwanted vacuum energy. After all, some vacuum energy is needed to keep the extra dimensions static.

The prospect of inflation from extra dimensions has not been realized in a realistic model, but there are no realistic models for compactification. In the Chapline-Manton theory there are two massless scalar fields, the dilaton and the radius of the internal dimensions. Perhaps one, or both, of these fields are the dilaton. Both fields have the promising feature that at the classical level they have flat potentials. The possibility of a unique field configuration that will lead to inflation is interesting.

The instability for large b in the Casimir and monopole models can be removed by considering combinations of the models.

•R_{MN} =ALL OF THE ABOVE [16]: Before combining the contributions it is useful to extend the analysis to products of spheres. Assume a ground state geometry of the form $R\times S^3\times\sum_{i=1}^{\alpha} S_i^d$, with metric g_{MN} =diag$(1, -a^2(t)\tilde{g}_{ij}(x), -b_1^2(t)\tilde{g}_{\mu\nu}(y), \ldots, -b_\alpha^2(t)\tilde{g}_{\rho\sigma}(y))$. The D extra dimensions are split into α d_i-spheres ($\sum d_i = D$). The stress tensor will be extended in a similar way by the definition of additional p_{di}. In the monopole and the Casimir cases, the large-b instability was caused by the presence of a cosmological constant, which was unbalanced as $b\to\infty$. For a stable ground state a cosmological constant is probably impossible. The Einstein equations without a cosmological constant are

$$
\begin{aligned}
3\frac{\ddot{a}}{a}+\sum_{i=1}^{\alpha} d_i\frac{\ddot{b}_i}{b_i} &= -\frac{8\pi\bar{G}}{D+2}\left[\rho - T^M_M\right] \\
\frac{\ddot{a}}{a}+2\frac{\dot{b}^2}{b^2}+\frac{\dot{a}^2}{a^2}\sum_{i=1}^{\alpha}\frac{\dot{b}_i}{b_i}+\frac{2}{a^2} &= \frac{8\pi\bar{G}}{D+2}\left[p_3 - T^M_M\right]
\end{aligned}
$$

$$\frac{\ddot{b}_i}{b_i} + (d_i - 1)\frac{\dot{b}_i^2}{b_i^2} + 3\frac{\dot{a}}{a}\frac{\dot{b}^i}{b^i} + d_i\frac{\dot{b}_i}{b_i}\sum_{j\neq i}\frac{\dot{b}_j}{b_j} + \frac{d_i - 1}{b_i^2} \quad = \quad \frac{8\pi\bar{G}}{D+2}\left[p_{di} - T^M_M\right]. \quad (4.2)$$

with the last equation for each internal sphere and $T^M_M = \rho - 3p_3 - \sum_{i=1}^{\alpha} d_i p_{di}$.

For forces to balance at a unique value of $b = b_0$ it is necessary to have contributions to T_{MN} that have different dependences on b. For this reason a combination of Casimir and monopole forces will be considered.

The generalization of the $D = 2$ monopole ansatz will be used. An antisymmetric tensor field of rank $d_i - 1$ has a field strength $F_{M,N,\ldots Q}$ of rank d_i and has a natural Freund-Rubin ansatz on the d_i-sphere. The stress tensor in terms of the field strength is

$$T_{MN} = F_{MP\ldots Q}F_N^{\ P\ldots Q} - \frac{1}{2d_i}g_{MN}F_{SP\ldots Q}F^{SP\ldots Q}. \quad (4.3)$$

With this assumption the monopole configuration leads to

$$\rho = -p_3 \quad = \quad \sum_{i=1}^{\alpha}\frac{1}{2}\left(\frac{f_{0i}}{b_i^{di}}\right)^2$$

$$p_{di} \quad = \quad \frac{1}{2}\left[\left(\frac{f_{0i}}{b_i^{di}}\right)^2 - \sum_{j\neq i}\left(\frac{f_{0j}}{b_j^{dj}}\right)^2\right]. \quad (4.4)$$

The generalization of the Casimir forces for products of spheres is also straightforward. The first generalization is a single sphere in even dimensions. For even dimensions there is an additional contribution to the free energy proportional to $\ln(2\pi\mu^2 b^2)$, where μ is a parameter that sets the scale of the path integral. This parameter can be set by imposing certain conditions on the effective potential. The second generalization is to products of spheres. The free energy becomes (ignoring the ln term)

$$F = \Omega_3 a^3 \sum_{i=1}^{\alpha}\frac{c_{i1}}{b_i^4}, \quad (4.5)$$

which leads to the thermodynamic quantities

$$\rho \quad - \quad -p_3 = \left(\prod_{i=1}^{\alpha}\Omega_i b_i^{di}\right)^{-1}\sum_{i=1}^{\alpha}\frac{c_{1i}}{b_i^4}$$

$$p_{di} \quad = \quad \frac{4}{d_i}\left(\prod_{i=1}^{\alpha}\Omega_i b_i^{di}\right)^{-1}\frac{c_{1i}}{b_i^4} \quad (4.6)$$

The first example of combining Casimir and monopole forces is a single internal D-sphere. Ignoring here and below the possible logarithmic dependence of the Casimir force for even dimensions, the Einstein equations are

$$
\begin{aligned}
3\frac{\ddot{a}}{a} + D\frac{\ddot{b}}{b} &= -\frac{8\pi\bar{G}}{D+2}\left[\frac{(D+2)c_1}{\Omega_D}b^{-4-D} + (D-1)f_0^2 b^{-2D}\right] \\
\frac{\ddot{a}}{a} + 2\frac{\dot{a}^2}{a^2} + D\frac{\dot{a}\,\dot{b}}{a\,b} + \frac{2}{a^2} &= -\frac{8\pi\bar{G}}{D+2}\left[\frac{(D+2)c_1}{\Omega_D}b^{-4-D} + (D-1)f_0^2 b^{-2D}\right] \\
\frac{\ddot{b}}{b} + (D-1)\frac{\dot{b}^2}{b^2} + 3\frac{\dot{a}\,\dot{b}}{a\,b} &= \frac{8\pi\bar{G}}{D+2}\left[\frac{4(D+2)c_1}{D\Omega_D}b^{-4-D} + 3f_0^2 b^{-2D}\right] \\
&\quad -\frac{D-1}{b^2}.
\end{aligned}
\tag{4.7}
$$

From the first two equations it is obvious that either c_1 or f_0^2 must be negative in order to have $\ddot{a}$ and $\ddot{b}$ vanish at b_0. The combination of the first two equations and the last equation gives

$$
b_0^2 = \frac{D(D-1)^2}{8\pi(D+2)(D-4)c_1}l_{Pl}^2; \qquad b_0^{D-6}l_{Pl}^2 = -\frac{\Omega_D 8\pi(D-4)}{D(D-1)}f_0^2. \tag{4.8}
$$

For $D < 4$, f_0^2 must be positive and c_1 must be negative. Although c_1 is positive for scalar fields on spheres, the sign of the Casimir force is notoriously slippery, and for other spins or other geometries it could easily be negative. For $D > 4$, c_1 must be positive and f_0^2 must be *negative*. Therefore this simple model is only viable for $D < 4$.

There are other problems with the model. If the potential is constructed along the lines of the previous section it is found that the static extremum is a local *maximum* of the potential. The potential is shown in Fig. 6. The point $\phi/\phi_0 = 1$ is the point where *a and b* are static. The potential becomes flat for large b, but there is a small b instability. This potential is sicker than Casimir+Λ or monopole+Λ. The same problem occurs for a product of D-spheres for the internal space.

The presence of fermion condensates in the Chapline-Manton action can cure the problem. Assume that $\mathrm{Tr}\bar{\chi}\Gamma_{MNP}\chi$ and $\bar{\lambda}\Gamma_{MNP}\lambda$ also have the Freund-Rubin form on a product of three S^3's. [6] The radius of one of the S^3's will be assumed to be much larger than the other two radii which will be assumed to be equal. If all other background fields are set to zero, a classically stable ground state with potential given by Fig. 7 is obtained. The new ingredient present in this model is that the presence of the fermion condensates change the right hand side of the Einstein equations. For the monopole+Casimir example on a single S^D, the coefficients of the monopole terms in the (00) and ($\mu\nu$) equations were fixed to be

[6] The dilaton is assumed to be a constant in space-time, $\sigma = \sigma_0$. The dilaton field equation gives $(H_{MNP})^2 = (3/2)\exp(\sigma_0/2)H_{MNP}(\mathrm{Tr}\bar{\chi}\Gamma^{MNP}\chi)$.

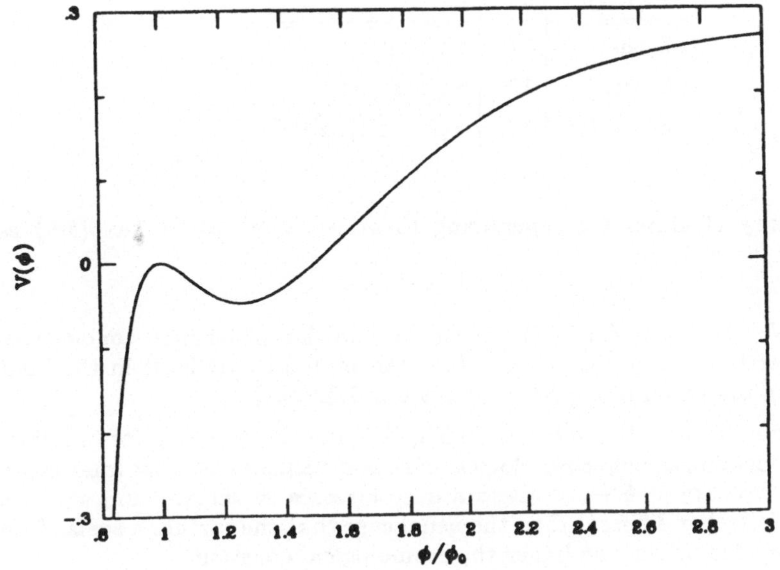

Figure 6: The potential for the Casimir + monopole case

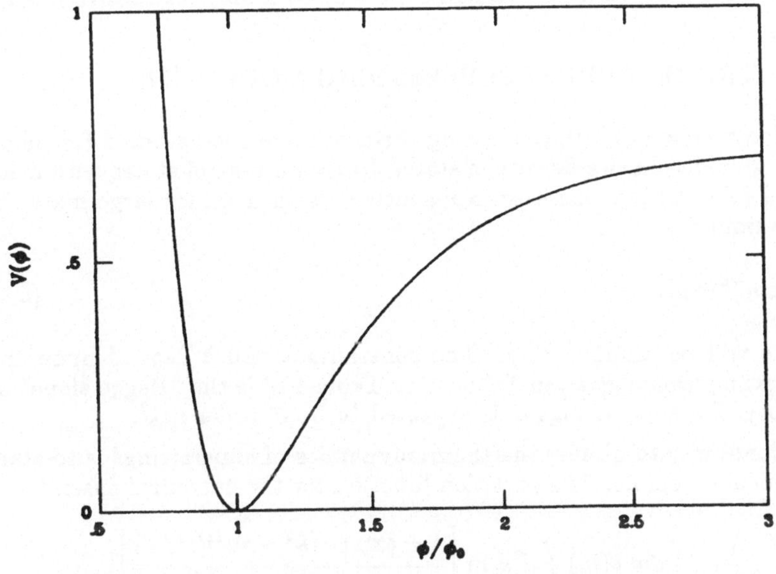

Figure 7: A possible potential for the Chapline-Manton action

THEORY	a	b
Open	9/2	$\pi\sqrt{8}(\alpha')^{1/2}$
Closed	10	$\pi\sqrt{8}(\alpha')^{1/2}$
Heterotic	10	$\pi(2+\sqrt{2})(\alpha')^{1/2}$

Table 4: Density of states for superstring theories: $\rho(m) \propto m^{-a}\exp(bm)$ as $m \to \infty$

in the ratio $(D-1)/3$ (see Eq. 4.7). With the addition of fermion condensates this is no longer true. A stable ground state can be found (at least in the limit that the radii of the two internal S^3's are not too different).

It should be noted that the potential in Fig. 7 is *not* the potential for inflation. The effective four-dimensional cosmological constant vanishes as b becomes large. This is simply because as $b \to \infty$ there are no stresses in the vacuum to drive inflation. This is rather different than the usual case that the further a scalar field is displaced from the origin, the larger the cosmological constant.

One of the lessons from new inflation is that there is a lot to be learned by models that fail. All of the models for stable extra dimensions and inflation from extra dimensions either fail or have some very undesirable features. Hopefully the lessons learned from these failures will point the way to a more attractive model.

5. LIMITING TEMPERATURE IN SUPERSTRING MODELS [17]

The thermodynamic properties of string theories have been studied for many years. All string models have a density of states $\rho(m)$=number of states with mass between m and $m+\delta m$ that increase exponentially with mass for large mass. In the large mass limit

$$\rho(m) = cm^{-a}\exp(bm). \tag{5.1}$$

The constant c will be uninteresting. The constants a and b depend upon the theory. Some examples are given in Table 4. In Table 4 α' is the "Regge slope" of the string theory. For superstrings α' is expected to be of order m_{Pl}^{-2}.

The traditional way to discuss the thermodynamics of superstrings is to start with the canonical ensemble. The partition function for the canonical ensemble is

$$\ln Z \;=\; \frac{V}{(2\pi)^9}\int dm\,\rho(m)\int d^9k\,\ln\left[\frac{1+\exp\left[-(k^2+m^2)^{1/2}/T\right]}{1-\exp\left[-(k^2+m^2)^{1/2}/T\right]}\right]$$

$$\simeq V \sum_{n=0}^{\infty} \left[\frac{1}{2n+1}\right]^5 \int_\eta dm\, m^{-a} \exp(bm) m^5 K_5[(2n+1)m/T], \tag{5.2}$$

where V is the (9-dimensional) spatial volume, η is the mass below which the exponential form of ρ is a bad approximation, and K_n is a modified Bessel function of the second kind. Using the limiting form $K_n(x) \to x^{-1/2} \exp(-x)$ the partition function may be expressed in terms of the incomplete gamma function

$$\ln Z \simeq \left(\frac{TT_0}{T_0 - T}\right)^{-a+11/2} \Gamma\left[-a + \frac{11}{2}, \eta\left(\frac{T_0 - T}{TT_0}\right)\right], \tag{5.3}$$

where $T_0 = b^{-1}$.

The partition function diverges for $T \geq T_0$. The pressure (p), average energy ($\langle E \rangle$), and specific heat (C_V) are given in terms of $\ln Z$ by

$$p = T\frac{\partial \ln Z}{\partial V}; \quad \langle E \rangle = T^2 \frac{\partial \ln Z}{\partial T}; \quad C_V = \frac{d\langle E \rangle}{dt}. \tag{5.4}$$

For $a \leq 13/2$, all diverge as $T \to \infty$. For $a > 13/2$, p and $\langle E \rangle$ approach a constant as $T \to T_0$. For $a > 15/2$, C_V also approaches a constant. If the thermodynamic quantities approach a constant as $T \to T_0$, T_0 is *not* a limiting temperature. Therefore the open string has a limiting temperature, but the closed or heterotic string does not. What is happening in this case is that the energy fluctuations are becoming so large that the thermodynamic description based upon the canonical ensemble breaks down. In this case it is more appropriate to use the microcanonical ensemble. When the microcanonical ensemble is used it is found that the most likely configuration is that one string carries almost all the energy and the remaining strings have very little energy. The specific heat in this case is *negative*.

The negative specific heat is quite interesting. A system of strings cannot come into thermal equilibrium with a heat bath. The negative specific heat also obtains for black holes. A possible connection between black holes and superstrings has been the subject of recent speculation.

6. GUT SYMMETRY BREAKING IN EXTRA DIMENSIONS

It has been shown that the phase transitions associated with spontaneous symmetry breaking have a multitude of interesting physical and cosmological effects. In theories with extra dimensions there is a new type of mechanism for symmetry breaking that does not depend upon the Higgs mechanism. The new mechanism depends upon a topological non-trivial nature of the internal space and will be referred to as topological symmetry breaking (TSB) [18].

In the absence of external sources the vacuum configuration for gauge fields is $F^a_{MN} = 0$. If the fields are defined on a topologically trivial manifold, the vanishing of F implies that $A^a_M = 0$ also. However if the manifold is not simply connected, then the vanishing of F in the vacuum does *not* imply that $A^a_M = 0$. $A^a_M \neq 0$ implies that the gauge symmetry is broken.

To determine the details of symmetry breaking the relevant quantity is the Wilson line $\vec{\mathcal{U}}$ related to the path-ordered exponential

$$\vec{\mathcal{U}} = P \exp\left(\oint_\Gamma \vec{A}_\mu dx^\mu\right) \tag{6.1}$$

where Γ represents some path in the manifold. If there are non-contractible paths in the manifold, then $\vec{\mathcal{U}} \neq 1$ and the original symmetry $\mathcal{G}$ is broken to some subgroup $\mathcal{H}$ that commutes with $\vec{\mathcal{U}}$. The Wilson lines replace adjoint Higgs fields.

This mechanism has very many interesting properties. Of interest here are the properties relevant for cosmology. The first question of interest is whether the symmetry will be restored at high temperature. Does $\vec{\mathcal{U}}$ go to unity if the system is put in a heat bath? Assuming there is a cosmological phase transition with this mechanism are topological defects (monopoles, cosmic strings, domain walls) produced in the transition? What is the dynamics of the evolution of the system to the ground state? If the system is away from the ground state at high temperature, can inflation occur in the evolution to the ground state?

Finally, in general there may be several possible ground states associated with different $\mathcal{H}$'s (including $\mathcal{H} = \mathcal{G}$). At the classical level at zero temperature they all have the same energy, namely zero. At finite temperature the state with the most massless degrees of freedom will have the lowest free energy. This will correspond to the unbroken state. As the temperature decreases a strong coupling phase will occur and massive bound states will form and the number of massless degrees of freedom in the unbroken state will fall below the number in one of the broken state. Will there follow a cascading of symmetry and does it have any physical effect. These questions are unanswered at present and are under investigation.

The Higgs mechanism and SSB has proved to be an interesting part of early Universe cosmology. It is likely that TSB will also.

7. REMNANTS

The final aspect of extra dimensions and cosmology that will be considered here is the survival of a stable massive particle somehow connected with extra dimensions. Before discussing specific particles it is useful to recall some facts about the survival of massive particles. The expansion of the Universe generally stops the annihilation of massive particles (mass M) at a temperature T_f given by

$$x_f \equiv M/T_f \sim \ln\left(m_{Pl} M \sigma_0\right), \tag{7.1}$$

where σ_0 is related to the annihilation cross section σ_A by

$$\langle |v|\sigma_A \rangle = \sigma_0 \left(\frac{M}{T}\right)^{-n}. \tag{7.2}$$

It is useful to compare the density of particles under consideration (denoted as ψ) to the entropy density. After annihilation freeze out and if entropy is conserved this ratio will be constant in the expansion. After annihilation ceases, the ratio of ψ's to entropy is given by

$$Y_\psi \sim \frac{x_f^{n+1}}{m_{Pl} M \sigma_0}. \tag{7.3}$$

In general $\sigma_0 \propto M^{-a}$. Since the effective annihilation cross section decreases with mass, the more massive a particle, the more likely it is to survive annihilation. For masses close to the Planck mass and $\sigma_0 \simeq M^{-2}$, annihilation is not effective and a particle would survive with $Y_\psi \sim 1$, i.e., about as abundant as photons. This would be a great embarrassment, since it would result in a contribution to Ω from the massive particles of about 10^{26} or so. Creation of entropy, as in inflation, could greatly reduce this number. If inflation occurs and the universe is reheated to a temperature of $T_f \ll M$, the ratio of ψ to entropy would not be determined by freeze out, but would be determined by $\exp(-M/T_R H)$. It is likely that this number is too small to be interesting today, but it is possible to imagine that M is just small enough to result in an interesting value of Y_ψ.

Here "interesting" means a value large enough to one day be detectable, but small enough not to be already ruled out. The most general limit on the abundance of massive stable particles comes from the overall mass density of the Universe. For a particle of mass M, the limit $\Omega h^2 \leq 1$ implies $Y_\psi \leq 5 \times 10^{-27}(m_{Pl}/M)$, or $n_\psi \leq 1.4 \times 10^{-23}(m_{Pl}/M)$ cm^{-3}. The most useful limit is in terms of the flux of ψ's, $F_\psi \leq 10^{-16}(m_{Pl}/M)$ cm^{-2}s^{-1}sr^{-1}. It is likely that very massive particles would be trapped in the galaxy and contribute to the mass density of the galaxy. In this case the limit is more restrictive. The relevant limit as a function of M is shown in Fig. 8. It is denoted "ρ_G."

Now consider candidates for ψ.

• *PYRGONS* [19]: In Kaluza-Klein theories there is an infinite tower of four dimensional particles corresponding to the non-zero modes of the harmonic expansions in mass eigenstates of the higher-dimensional fields. These non-zero modes are called Pyrgons.

In the five-dimensional theory the mass spectrum of the pyrgons is a series of spin-2 particles with mass $m_k = kR^{-1}$, where k is an integer and R is the radius of the internal space (in the five-dimensional theory the internal space is a circle). In the five-dimensional theory the $k = 1$ pyrgons are stable. This is because the charge operator is proportional to the mass operator. The zero modes are neutral and the $k = i$ mode has charge $e_k = i$. The kth pyrgon can decay to k number of $k = 1$ pyrgons, but the $k = 1$ pyrgons cannot decay to zero modes.

In more complicated Kaluza-Klein theories the mass spectrum is more complicated, but the general features remain, namely that there are zero modes and

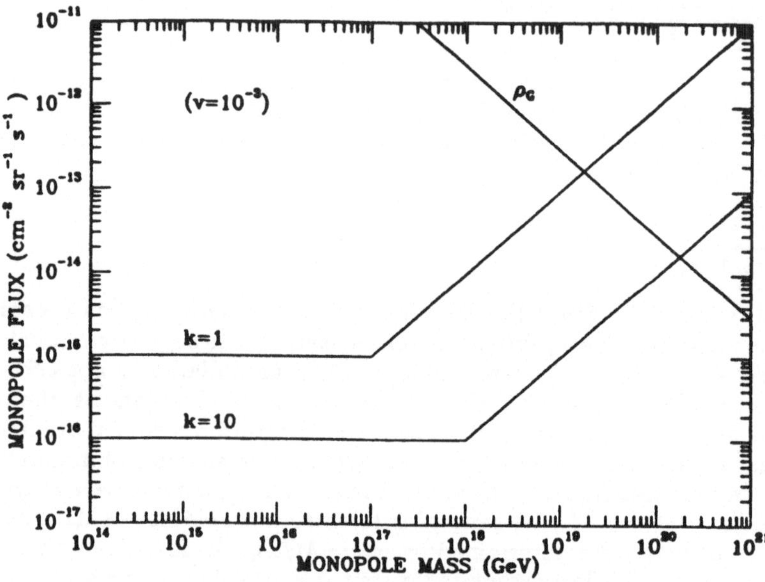

Figure 8: Flux limits as a function of mass

massive modes with mass proportional to the inverse of radii in the extra dimensions. The question of stability of the pyrgons is a more complicated one. In general there may be selection rules that prevent some massive modes from decaying. Such a selection rule is present in $N = 8$ supergravity models with an S^7 as the internal space. In general, the only reason one might imagine the pyrgons to be stable is if the pyrgon has a quantum number that is not represented by zero modes, which will be assumed to include only the observed particles. One possibility is if the pyrgon breaks the relationship of electric charge and triality. If the pyrgon is color neutral with fractional electric charge, or is fractionally charged but a color singlet it could not decay to the known particles (so long as SU_3 of color is unbroken). The second possibility is that the pyrgon has a quantum number that is not shared with any new particle.

In superstring theories the gauge symmetries arise from a different source, but there still might be excitations of the extra dimensions that are stable. There might also be excited string states that are stable. In the heterotic superstring there are 8,064 zero modes, 18,883,584 $k = 1$ modes, 6,209,272,160 $k = 2$ modes, ... (remember the increase is exponential!). Some of these massive modes might be stable.

• *MONOPOLES*: Just as GUT monopoles correspond to topological defects in the orientation of the vacuum expectation value of a Higgs field, there are magnetic monopoles in Kaluza-Klein theories that correspond to topological defects in compactification [20]. The Kaluza-Klein monopoles satisfy the Dirac quantization condition $ge = 1/2$ and have masses given by $m_M \sim m_{Pl}/e \sim 10^{20}$GeV. The cosmological production of Kaluza-Klein monopoles is uncertain because there is nothing that corresponds to a Kibble mechanism. It is unclear what the high-temperature behavior of the SSB will be [15]. In this case the SSB corresponds to the process of compactification, i.e., the symmetry breaking $\mathrm{Diff}^{D+4} \rightarrow \mathrm{Diff}^{4} \times I$ where Diff^{n} is the diffeomorphism group in n dimensions and I is the isometry group of the internal space. Since the symmetry breaking that gives rise to the Kaluza-Klein monopoles is topological in nature, the restoration of the symmetry cannot be studied by classical methods.

In theories with TSB, there are additional topologically stable excitations. There are magnetic monopoles and particles with fractional electric charge [22]. The striking feature of these particles is that the minimum magnetic charge is some integer times the Dirac quantum, $g_{\mathrm{MIN}} = k g_{\mathrm{DIRAC}}$. The minimum electric charge is also determined by the integer k, $e_{\mathrm{MIN}} = e/k$. The expected cosmological abundance of these particles has not been estimated. The present flux of the magnetic monopoles is limited by the Parker bound, which is the maximum number of monopoles that can be present without "shorting out" the galactic B-fields. The Parker limit as a function of mass and magnetic charge is shown in Fig. 8 [23]. Of course, it is always possible to avoid the Parker limit if the monopoles are abundant enough that coherent oscillations of the monopoles are the source of the galactic B-field [24].

There are perhaps other possibilities for massive stable particles. The searches for massive stable particles in cosmic rays should be pushed. The detection of any particle with mass comparable to the Planck mass would have enormous implications for both particle physics and cosmology.

ACKNOWLEDGEMENTS

This work was supported by NASA and the Department of Energy. I would like to thank Richard Holman and Frank Accetta for a careful reading of the manuscript. The section on inflation has been greatly influenced by Michael Turner's writings on the subject.

REFERENCES

1. M. J. Duff, B. E. W. Nilsson, and C. N. Pope, *Phys. Rep.* **130**, 1 (1986).

2. J. H. Schwarz, *Superstrings* (World Scientific, Singapore, 1985).

3. E. W. Kolb, M. J. Perry, T. P. Walker, *Phys. Rev. D* **33**, 869 (1986).

4. G. Chapline and N. Manton, *Phys. Lett.* **120B**, 105 (1983).

5. A. Chodos and S. Detweiler, *Phys. Rev. D* **21**, 2176 (1980).

6. J. Barrow, *Phys. Rep.* **85**, 1 (1982).

7. A. Hosoya, L. G. Jensen, J. A. Stein-Schabes, Fermilab preprint.

8. S. Randjbar-Daemi, A. Salam, and J. Strathdee, *Phys. Lett.* **135B**, 388 (1984).

9. P. Candelas and S. Weinberg, *Nucl. Phys.* **B237**, 397 (1984).

10. E. Alvarez and M. Belen-Gavela, *Phys. Rev. Lett.* **51**, 931 (1983); E. W. Kolb, D. Lindley, and D. Seckel, *Phys. Rev. D* **30**, 1205 (1984).; R. B. Abbott, S. Barr, and S. Ellis *Phys. Rev. D* **30**, 720 (1984).

11. Y. Okada, *Phys. Lett.* **150B**, 103 (1985).

12. Q. Shafi and C. Wetterich, *Phys. Lett.* **129B**, 387 (1983).

13. J. Frieman and E. W. Kolb *Phys. Rev. Lett.* **55**, 1435 (1985).

14. A. Linde, *Nucl. Phys.* **B216**, 421 (1983).

15. M. Yoshimura, *Phys. Rev. D* **30**, 344 (1984).

16. F. Accetta, M. Gleiser, R. Holman, and E. W. Kolb, *Nucl. Phys.* B276, 501 (1986).

17. M. Bowick and S. Wijewardhana, *Phys. Rev. Lett.* **54**, 2485 (1985).

18. Y. Hosotani, *Phys. Lett.* 129B, 193 (1983).

19. E. W. Kolb and R. Slansky, *Phys. Lett.* **135B**, 378 (1984).

20. R. Sorkin, *Phys. Rev. Lett.* **51**, 87 (1983); D. Gross and M. J. Perry, *Nucl. Phys.* **B226**, 29 (1983).

21. J. A. Harvey, E. W. Kolb, and M. J. Perry *Phys. Lett.* **149B**, 465 (1984).

22. X.-G. Wen and E. Witten, *Nucl. Phys.* **B261**, 651 (1985).

23. H. M. Hodges, E. W. Kolb, and M. S. Turner, Fermilab preprint.

24. I. Wasserman, S. Shapiro, and R. Farouki, *Comments on Astrophysics* **X**, 257 (1985).

ANISOTROPY MEASUREMENTS OF THE MICROWAVE BACKGROUND

P. de Bernardis, S. Masi , G. Moreno
Dipartimento di Fisica, Universita' "La Sapienza"
00184 Roma
Italy

ABSTRACT. Anisotropy measurement techniques and results are reviewed, with special attention given to experimental problems.
The cosmological relevance of the dipole anisotropy, the only anisotropy truly detected in the Cosmic Background Radiation, is discussed.

1. INTRODUCTION

Anisotropy of the Cosmic Background Radiation at 2.7 K (CBR hereafter) is a cosmological topic with a wide range of applications. In order to define anisotropy let us consider fig. 1a, where the celestial sphere is shown with two beams A and B, with beamwidth γ and angular separation θ. We define the anisotropy of CBR at angular scale θ in terms of the difference ΔI between the CBR flux $I(\alpha,\upsilon)$ measured in the two beams.

At small angular scales ($\theta<1°$) a "stochastic" approach is preferred, and the anisotropy is defined as

$$\left.\frac{\Delta I}{I}\right|_{\theta} = \frac{\sqrt{<\Delta I^2>}}{<I>} \tag{1}$$

where the brackets indicate averages over the whole celestial sphere. At large angular scales $\theta>1°$ a deterministic approach is preferred, and the CBR flux $I(\alpha, \delta)$ is expressed as a sum of spherical harmonics

$$I(\alpha, \delta) = I \sum_{l,m} a_{lm} Y_{lm} (\alpha, \delta) \tag{2}$$

The a_{lm} coefficients give the dipole, quadrupole and higher order components of the anisotropy .

P. Galeotti and D. N. Schramm (eds.), Gauge Theory and the Early Universe, 257–282.

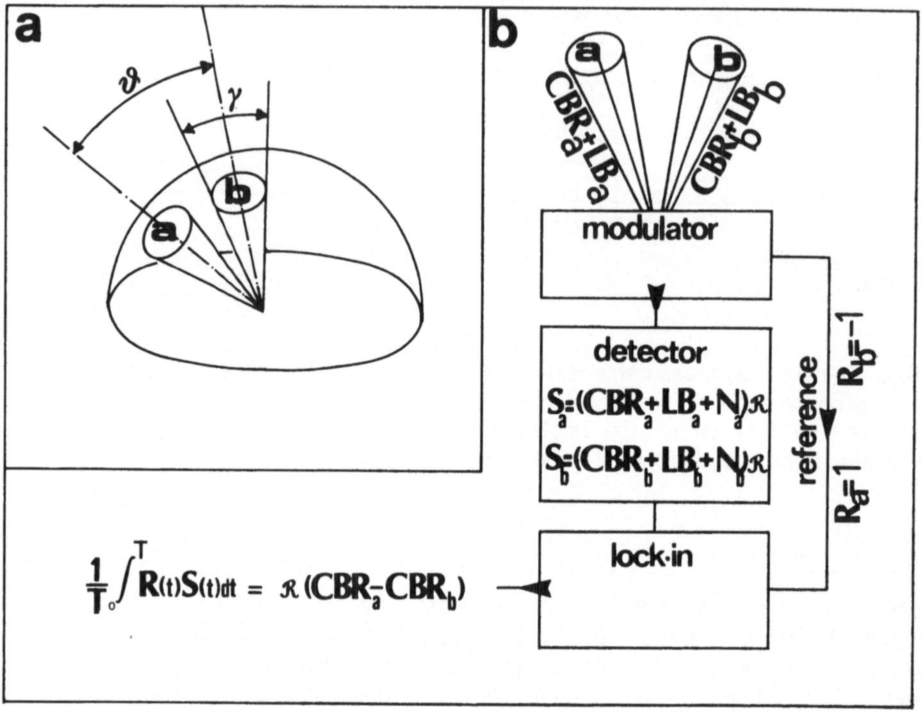

Figure 1.

a - Celestial sphere and operative definition of CBR anisotropy at angular scale θ, measuring the difference between CBR flux in the two beams a and b, with beamsize α.

b - Block diagram of differential photometer used for CBR anisotropy measurements.

Anisotropy may also be expressed in terms of temperature, recalling that for blackbody radiation

$$\frac{\Delta I}{I} = \frac{x\, e^x}{e^x - 1} \frac{\Delta T}{T} \qquad (3)$$

where $x = \frac{h\nu}{kT}$

In this paper we will discuss measurement techniques and experimental results concerning the CBR anisotropy. Let us first recall briefly the importance of these measurements in the cosmological contest.

The CBR is astonishingly isotropic: intrinsic anisotropies have not yet been discovered at a level of 1 part in 10^4 at all the angular scales.

The lack of relevant small scale anisotropies confirms the cosmological origin of the microwave background: in the case of a background originated from discrete sources, an anisotropy of the order of 100% is expected at an angular scale θ such that

$$D^3 = (c/H_0)^3 \quad \theta^2 \qquad (4)$$

where D is the mean distance between two sources ($D \simeq 1$ Mpc for galaxies) ; c/H_0 is the Hubble radius. The left side of eq. 4 is the mean volume per source, while the right side is the volume of a cone with aperture θ and height c/H_0. It emerges that $\theta \approx 1''$. Since the upper limit to CBR anisotropy is $\Delta I/I \simeq 10^{-3}$, one is forced to exclude a discrete origin of the CBR: the primaeval fireball hypothesis (Peebles, 1971) is favoured instead. However, small fluctuations are expected at all angular scales, due to the following processes:

1) The formation of structures in the universe should have left an imprint in the last scattering surface. Density fluctuations should have produced temperature fluctuations in the CBR at angular scales depending on their spectrum ($\delta\rho/\rho(M)$). The physical mechanisms responsible are listed below:
   a) In the case of "adiabatic" fluctuations, photon and barion densities are coupled and $\Delta T/T = (1/3)(\delta\rho/\rho)$.
   b) Thomson scattering couples matter and radiation: large scale matter motions produce an anisotropy $\Delta T/T = (\upsilon/c)\cos\phi$, due to the Doppler effect, where ϕ is the angle between the velocity and the line of sight.
   c) Any kind of density fluctuation produces temperature fluctuations due to gravitational redshift (Sachs and Wolfe, 1967).

However, anisotropies on scales smaller than 1' are difficult to detect: the last scattering surface has a finite thickness ($\Delta z = 100$ around $z = 1000$) containing many small bumps, which are averaged on the line of sight, smoothing the resulting anisotropy.

Morever, if a reionization occurred, Thomson scattering could have erased radiation anisotropies up to a scale of the order of a few degrees. For this reason, observation of intermediate scale anisotropies is of particular interest: at these scales fluctuations cannot be smoothed by scattering because they come from causally disconnected regions.

2) Homogeneous anisotropic models (Bianchi Universes) and inhomogeneous models imply large scale anisotropies. This topic will be discussed in more detail in section 4.2.

2. MEASUREMENT TECHNIQUES AND OBSERVATIONAL PROBLEMS

2.1 Method of Measurement

The general features of the experiments depend mainly on the angular scale being investigated. First of all the dimension of the telescopes are defined by the constraint

$$\theta > \gamma \approx \lambda/D \tag{5}$$

where λ is the wavelength of observation (which for a 2.7 K blackbody is 1 mm $< \lambda < 100$ cm) and D is the telescope diameter.

If the first inequality is not satisfied the signal is undermodulated. The equality applies for a diffraction limited antenna beamwidth. For example, in the radio region (λ = 1 cm) an angular resolution $\gamma = 1'$ requires a 40 m diameter telescope. For this reason, large ground based radiotelescopes have been used for small angular scale anisotropy measurements. Satellites like FIRST of ESA will be used in future.

For intermediate and large angular scales, smaller size experiments are adequate. To investigate the far infrared region of the spectrum, where atmospheric transmission at ground is very low, some of these experiments were flown on aircraft, balloons or satellites.

In general, anisotropy measurements reach a much higher sensitivity compared to measurements of the CBR spectrum $E(\lambda)$. Recent experiments give in fact $\Delta E/E \approx 3\%$ (Peterson et al., 1985), while the upper limit to anisotropy is $\Delta T/T < 10^{-4}$. This is due mainly to the differential technique used in anisotropy observations. In fact, in order to compare radiation in the two fields of view A and B, an apparatus similar to the one sketched in fig.1b is used: a modulator performs the detector's beam switching between the two fields, and generates a reference square wave signal R(t) synchronized to beam switching. If the emission in the two fields is different, an AC signal S(t) is generated in the detector, synchronous to the modulation. Reference and AC signal are multiplied in the lock-in amplifier and integrated for a suitable time. The result is

$$\int S(t)\,R(t)\,dt = \mathcal{R} < CBR_A - CBR_B + LB_A - LB_B >_T + (1/T)\int_0^T R(t)\,N(t)dt \qquad (6)$$

where the constant $\mathcal{R}$ is the system responsivity which is assumed to be the same in the two beams. CBR and LB are CBR and local background emissions, N(t) is the detector noise, and the brackets indicate the average over a time T. Since R(t) and N(t) are uncorrelated, the noise integral can be reduced, increasing the integration time. If the local (instrumental, atmospheric etc) background is the same in the two beams, this technique allows us to measure the difference in CBR emission between beams A and B. Repeating the measurements in many field pairs, an estimate of the CBR anisotropy can be made.

To optimize measurements it is necessary to:

1) Reduce noise in phase with modulation. This generates a non-zero contribution to the noise integral of eq. (6). It can be due to noise of the modulator or to the microphonic noise of the detector.
2) Reduce the intrinsic noise of the detector in order to allow short integration times T. This is possible only by reducing the temperature of the detector. Cryogenic detectors, both coherent (GaAs Schottky diode mixer at 4.2K, at 3 mm wavelength N ≈ 12 mK/$\sqrt{}$Hz (Epstein, 1983)) and thermal (^{3}He bolometer at 0.3 K, 2~2.5 mm wavelength range, N ≈ 5m K/$\sqrt{}$Hz (Mason et al., 1986) have been developed.
3) Have a local background similar in the two beams. This is made difficult by ground radiation, especially for large radiotelescopes which cannot be shielded: ground radiation reaches the receiver through antenna sidelobes. For this reason the telescope must be held fixed with respect to the ground, in a position maximizing sidelobe rejection.
4) Avoid asymmetries in the two beams: they can be introduced by the modulator or by the optical system. For example, a blade chopper which alternately reflects one beam or transmits the other creates a large offset due to blade emission in the reflected beam (Lubin et al., 1983); this can change slowly due to temperature variations of the system. Two slightly different antennas can also produce an offset.

2.2 Local Background

Three local backgrounds are always present: atmospheric emission, galactic emission from the interstallar medium, and discrete source background.

Atmospheric emission is due mainly to vibrational and rotational lines of O_2, H_2O and O_3. Ground based observations can be performed in the following windows (fig. 2a):

200 --> 300 GHz (1 --> 1.4 mm)
130 --> 180 GHz (1.8 --> 2.5 mm)
70 --> 110 GHz (2.7 --> 4 mm)
<30 GHz (>10 mm)

Effects of atmospheric fluctuations (quantum fluctuations and fluctuations due to density, pressure, and temperature of the atmospheric gases) must be minimized and possibly evaluated by performing observations at high elevations ($a > 80°$), by monitoring the atmospheric emission at proper wavelengths, by using high mountain observatories with low water vapour column density, and by performing observations in clear and cold atmospheric conditions. At balloon altitudes, atmospheric emission is roughly of the same order of magnitude of CBR emission (fig. 2b).

Modulation must be performed at a fixed elevation a. In fact, atmospheric emission outside saturated lines follows a cosec law

$$I(a) = I_Z / \sin a \tag{7}$$

if the modulation is tilted with respect to the horizon of an angle $\Delta\beta$, an offset is generated

$$\Delta I = I_Z \frac{\cos a}{\sin^2 a} \theta \, \Delta\beta \tag{8}$$

which can saturate the detector dynamics if $\Delta\beta$ is not small enough, or can introduce spurious signals if $\Delta\beta$ or θ are not constant.

Sources of galactic emission are HI thermal emission and syncroton emission in the centimetric region of the spectrum, and thermal emission from galactic dust in the millimetric region (fig. 3). The galactic background is a serious problem for large scale anisotropy measurements. In order to reduce galactic contamination one can choose for observations, a) a convenient range of wavelengths (a minimum in galactic emission is present at $2 \text{ mm} < \lambda < 7 \text{ mm}$ (Fixsen et al., 1983)) or, b) areas in the sky at high galactic latitudes.

The ultimate limit to searches for small scale anisotropies of the CBR is set by fluctuations due to the detection of extragalactic, discrete sources in the fields of view. Calculations carried out by Danese et al. (1982) show that this limit is much more severe in the centimetric than in the millimetric region: at angular scales of 1' they estimate $\Delta T/T \approx 7.10^{-5}$ at $\lambda = 2.8$ cm and $\Delta T/T \approx 6.10^{-6}$ at $\lambda = 0.9$ mm.

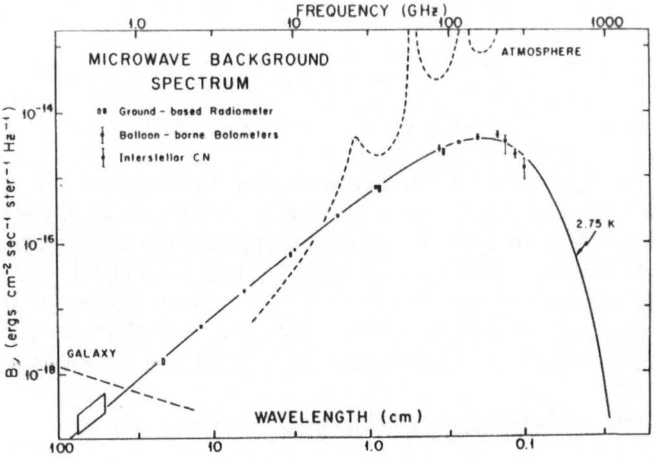

(a)

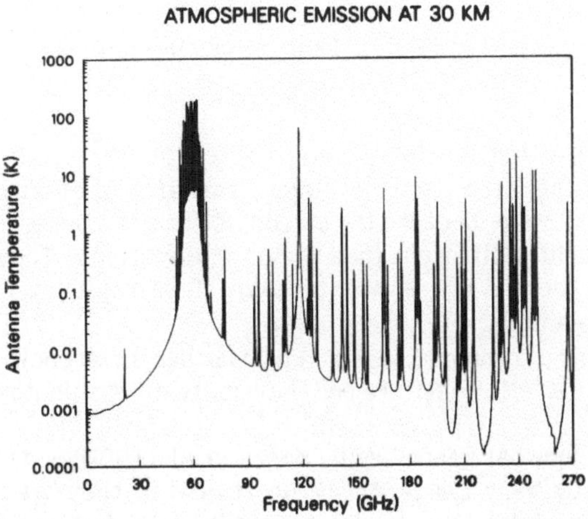

(b)

Calculated atmospheric emission at ground (a) and at a height of 30 km (b) (from Lubin and Neto (1985)).

3. SMALL AND INTERMEDIATE SCALE ANISOTROPY MEASUREMENTS

3.1 Small Scales ($\theta < 1°$)

A typical apparatus used for these measurements consists of a telescope with a nutating secondary subreflector, used as a beam-switch.

To explore different regions of the sky, most experiments use the so-called drift-scan technique: the telescope is pointed at high elevations ahead of the region to be observed, and then maintained fixed. Earth rotation allows us to scan areas at different right ascensions and constant declinations. This technique minimizes both atmospheric fluctuations and ground emission from the sidelobes. Uson and Wilkinson (1984) have used an improved version of this technique with the 43 m NRAO telescope at Green Bank. They worked at 19.5 GHz, with a 1.5' beamwidth and a modulation amplitude of 4.5'.

For each field they used two reference fields and measured both

$$\delta T_{ON} = T_{field} - T_{Ref\ 1} \quad \text{and}$$
$$\delta T_{OFF} = T_{field} - T_{Ref\ 2}$$

The anisotropy was estimated combining the two measurements:

$$\Delta T_{field} = \frac{\delta T_{ON} - \delta T_{OFF}}{2} = T_{field} - \frac{T_{Ref-1} + T_{Ref-2}}{2}$$

Advantages of this method are that the long (5 min) integration time for each δT measurement increases the sensitivity. Moreover, the effect of atmospheric background can be better accounted for because the candidate field is compared with two reference areas located at slightly different (and varying) elevations. They observed 12 fields, finding an upper limit $\Delta T/T < 4.8.10^{-5}$, which is the best fine scale anisotropy upper limit available to date.

An even more sophisticated technique based on third order differencing has been successfully developed at the Battelle-Institute and University of Washington (Seattle) (Radford S.J.(1986)).

A rather different technique has been used by Knoke et al. (1984) and Fomalont et al. (1984), using the NRAO Very Large Array: in this way they were able to investigate very small angular scales ($\theta < 1'$), which cannot be studied with a single dish telescope. A number of selected results on small scale anisotropy is shown in table I.

3.2 S-Z Effect

The Sunyaev-Zeldovich (S-Z) effect is due to hot plasmas existing in clusters of galaxies. Hot electrons are expected to scatter CBR photons to higher frequencies (inverse compton effect). This should produce a weak spot in the CBR in

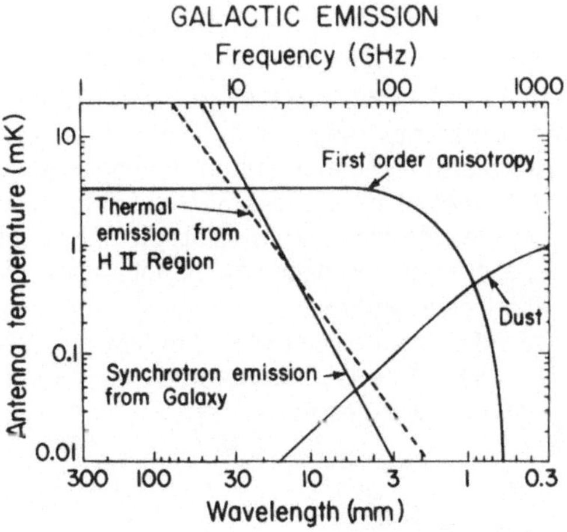

Figure 3.
Galactic background at high galactic latitudes compared to the dipole anisotropy of CBR (from Lubin et al. (1983)).

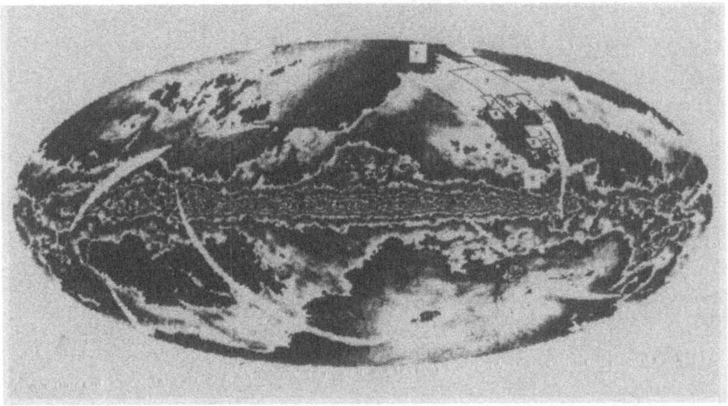

Figure 4.
Aitoff projection of IRAS all-sky map at 100 μm in a logarithmic repetitive grey scale. The map is in galactic coordinates with l = −180° on the right and l = +180° on the left. Virgo cluster (v) and Hydra cluster (H) positions are shown. The sinusoidal clear band is zodiacal thermal emission. The higher squares are the dipole anisotropy detections of experiments listed in table III; the lower squares are the directions of local group motion.

corrispondence with cluster directions: a temperature decrease is expected in the radio region ($\lambda > 1.4$ mm, maximum negative effect at $\lambda = 1.8$ mm), while a temperature increase is expected in the far IR ($\lambda < 1.4$ mm, maximum positive effect at $\lambda = 0.8$ mm) - (Sunyaev and Zeldovich, 1972). Cavaliere et al. (1979) have shown that combined CBR and X-ray flux measurements can provide an absolute determination of clusters of galaxy distances: a powerful method avoiding the uncertain cosmological distance ladder! However, the expected in-cluster out-of-cluster anisotropy is proportional to the cluster's Compton scattering optical depth, which is estimated to be less than 5.10^{-4}. For this reason, and for the presence of the local effects described in section 2, results so far obtained are contradictory.

Birkinshaw et al. (1984) found evidence for the S-Z effect at 20 GHz for three clusters (0016+16, Abell 665, Abell 2218). However, recent observations by Radford et al. (1986) failed to confirm these results.

3.3 Intermediate Scales ($\theta \approx$ some degrees)

These measurements have been performed by means of ground based, balloon and satellite borne observations.

Due to the large fields of view used, ground based observations are strongly limited by atmospheric noise also in the centimetric region: Mandolesi et al. (1985) at $\lambda = 3$ cm have found $\Delta T/T < 5.6.10^{-4}$ at $\theta = 2° / 3°$ using two gemini experiments in Bologna (Italy) and in Tromso (Norway). Better results have been reported from balloon observations: Melchiorri et al. (1981) used a liquid helium cooled Ge bolometer at wavelengths between 700 and 2000 μm, with a 5° field of view and 6° modulation, accomplished by a wobbling mirror. The most important bias affecting these short wavelength observations arises from galactic dust emission. To correct this effect, the neutral hydrogen column density was used as a dust indicator. After subtraction of a small dust contribution to the signal at high galactic latitudes, an upper limit $\Delta T/T < 4.10^{-5}$ at $\theta = 6°$ was found. Higher upper limits at the same angular scale have been found by Fixsen et al. (1983) at $\lambda = 12$ mm, by Lubin et al. (1983) at $\lambda = 3$ mm, and by the Russian "Relict" satellite (Strukov and Skulachev, 1984) at $\lambda = 8$ mm. The results are listed in Table 2.

4. LARGE SCALE ANISOTROPY

Since the late sixties (Peebles and Wilkinson 1968, Stewart and Sciama 1968) the fact that the microwave background last scattering surface provides a very distant "Machian" reference frame has been established.

Our motion would produce in the observed radiation $I(\upsilon)$, a dipole-like anisotropy, due to the Doppler effect, with amplitude

TABLE 1: SMALL SCALE ANISOTROPY MEASUREMENTS

Source	Frequency (Wavelength)	Angular Scale	λT/T (upper limits)	Experimental Apparatus
Lasenby & Davies (1983)	4.9 GHz (6cm)	10" 30" 1"	3×10^{-4} 4.6×10^{-4} 2.3×10^{-3}	Mk II telescope (125 ft x 83 ft)
Caderni et al., (1977)	240 GHz (1.2mm)	25"	1.2×10^{-4}	150cm flux collector with a germanium bolometer as detector (Testa Grigia)
Parijskij et al., (1977) Berlin et al., (1983)	4.2 GHz (7cm)	45"-9"	10^{-5}	RATAN-600 telescope
Partridge (1980)	31 GHz (9mm)	3.6" 7"	2×10^{-4} 8×10^{-5}	NRAO-11m telescope
Uson & Wilkinson (1984)	19.5 GHz (1.5cm)	4.5"	4.5×10^{-5}	NRAO-11m telescope
Fomalont, Kellermann & Wall (1984)	4.9 GHz (6cm)	18" 30" 1"	1×10^{-3} 8×10^{-4} 5×10^{-4}	Very Large Array
Knoke, Partridge et al., (1984)	4.9 GHz (6cm)	6" 18"	3.2×10^{-3} 1.2×10^{-3}	Very Large Array

TABLE 2: INTERMEDIATE SCALE ANISOTROPY

Source	Frequency (Wavelength)	Angular Scale	Anisotropy Upper Limits	Experimental Apparatus
Mandolesi et al. (1985)	90 GHz (3 cm)	2° 5°	$\frac{\Delta T}{T} < 5.6.10^{-4}$ $\frac{\Delta T}{T} < 6.8.10^{-4}$	Ground based 1 m O antennas + classical radiometer
Strukov and Skulacev (1983)	37 GHz (8.1 mm)	5°	$\frac{\Delta T}{T} < 4.10^{-4}$	RELIC satellite + classical radiometer
Melchiorri et al. (1981)	100-600 GHz (500-3000 μm)	6°	$\frac{\Delta T}{T} < 4.10^{-5}$	Balloon borne differential bolometric photometer
Fixsen et al. (1983)	24.5 GHz (12.2 mm)	10°-180°	$<\Delta T_1 \Delta T_2> \leq 0.01 (mK)^2$	Balloon borne cryogenic radiometer

$$\Delta I(\upsilon) = \beta(3 - \alpha(\upsilon))\, I(\upsilon)$$

$$\alpha = \frac{d\ln I(\upsilon)}{d\ln\upsilon} \qquad (9)$$

$$\beta = \frac{v}{c} \ll 1$$

and with the same direction of our velocity $\vec{v}$ (Mosengheil 1907, Lubin 1982, Danese and De Zotti 1981). The velocity of the sun with respect to the local group of galaxies is mainly due to the rotation of our galaxy and is well known from optical studies (Schmidt 1965) to be

$$v_{\odot LG} = 250 \text{ km/s} \quad \text{toward} \quad l = 90° \quad \text{and} \quad b = 0°$$

Consequently, a dipole anisotropy $\Delta I/I \approx 10^{-3}$ was expected in the microwave background.

Further motivation to search for large scale anisotropies came from homogeneous anisotropic models of the universe (MacCallum 1979), in which quadrupole and higher order harmonic anisotropy patterns are expected (Collins and Hawking 1973, Fabbri 1980).

4.1 The Measurements

Early measurements gave conflicting results (Conklin 1970, Henry 1971), with strong quadrupole detection due to galactic and atmospheric contamination. The first unambiguous detection of the dipole anisotropy was obtained in the radio region by two groups in Berkeley and Princeton (Corey and Wilkinson, 1976: Cheng et al., 1979) (Smoot, Gorenstein and Muller, 1977) and in the far infrared by an Italian group (Fabbri et al., 1980).

Selected recent results are summarized in table III. Experiments prior to 1983 suffered from limited sky coverage. For this reason it was difficult to separate local (i.e. galactic) from cosmological contributions in the spherical harmonic components of the detected anisotropies.

The standard data analysis was made using the Smoot angle representation of the data: in the anisotropy definition (see section 1) we considered two lines of sight $\vec{n}_1$ and $\vec{n}_2$ for each signal. However, the expected dipole anisotropy depends only on the angle θ between the dipole axis $\vec{D}$ ($\vec{D}//\vec{v}$) and the modulation direction $\vec{M} = \vec{n}_2 - \vec{n}_1$: the dipole anisotropy is proportional to

TABLE 3

SELECTED MICROWAVE ANISOTROPY MEASUREMENTS ** LARGE SCALE **

REFERENCE	EXPERIMENT	WL (mm)	F (GHz)	FOW°	MOD°	modulation system	detectors	Tx	Ty	Tz	l°	b°	DT(mK)	v(km/s) sun	v(km/s) local group	l° local group	b° local group	QUADRUPOLE (mK)
Gorenstein and Smoot 1981	airborne	8.9	33	7	60	2 horns+ dicke switch	conventional radiometer	/	/	/	228±14	66±5	3.6±0.4	400±45	568±40	255±5	40±3	L.T. 1 mK
Boughn et al 1981	balloon	6.4	46	7	90	2 horns+ dicke switch	conventional radiometer	-3.68 ±0.30	0.43 ±0.20	-0.77 ±0.31	276±4	47±5	3.78±0.30	420±30	661±30	273±3	28±3	-0.54±0.14 mK
Fabbri et al 1980	balloon	0.5-3	90-600	5	6	wobbling mirror	4-He bolometer	-2.9 ±0.4	-0.7 ±0.5	/	260±17	58±8	$2.9^{+1.3}_{-0.6}$	320^{+150}_{-60}	540^{+140}_{-60}	266±7	30±6	0.9±0.4 mK 0.7±0.2 mK
Lubin et al 1983	balloon	3.3	90	7	90	rotating mirror	cooled radiometer	-2.90 ±0.14	+0.51 ±0.08	-0.27 ±0.08	266±2	50±1	3.46±0.17	384±19	620±20	268±2	28±1	L.T. 0.2 mK
Fixen et al 1983	balloon	12	24.5	7	90	2 horns+ dicke switch	maser radiometer	-3.07 ±0.17	+0.67 ±0.09	-0.45 ±0.09	266±2	47±2	3.18±0.17	353±22	598±20	268±2	25±2	L.T. 0.15 mK
Halpern et al 1985	balloon	1.2-3	90-260	17	90	2 horns + 2 rotating mirrors	bolometers	-2.81 ±0.36	+0.25 ±0.17	-0.89 ±0.22	281±4	42±3	2.96±0.42	328±47	584±45	275±2	22±2	//
Strukov et al 1984	RELICT satellite	8	37	5.8	spin	1 main ant.+ 1 rotating horn+dicke switch	conventional radiometer	/	/	/	253±5	47±5	2.5±0.3	280±40	530±35	265±3	23±3	L.T. 0.1 mK

$$\cos \theta = \cos (\vec{M}, \vec{D})$$

For this reason anisotropy data with the same Smoot angle θ were averaged together. In fig. 6 a Smoot representation of the data obtained by de Bernardis et al. (1984) is given as an example. A general agreement on dipole measurements has now been achieved both in the radio and infrared regions: the dipole anisotropy is firmly established and has been mapped at 0.08 cm $< \lambda <$ 3 cm.

The weighted mean of the results given in table III gives

$$\Delta T_{CBR} = (3.27 \pm 0.14)\text{mK} \rightarrow v_{\odot} = (363 \pm 15) \left(\frac{2.7\ K}{T_{CBR}}\right) \frac{km}{s}$$

toward l = $(267 \pm 3)°$ and b = $(49.3 \pm 1.5)°$

The dipole apex lies in a rather clean sky zone of high latitude, in the Leo constellation. In fig. 4 the measurements of dipole direction are plotted on a repetitive grey scale representation of IRAS 100 µm data (from IRAS "low resolution all sky maps" Beichman et al., 1984).

Some zodiacal dust emission can be seen in the dipole maximum region.

Galactic and zodiacal dust emission are the main problems in the search for higher harmonic components, especially in the far infrared.

In fig. 5 estimates of the high latitude galactic background are shown: dust emission, when extrapolated at 1 mm, can produce a signal of the same order of magnitude as the dipole anisotropy, with a quadrupolar symmetry which can be fitted by a quadrupole component if the sky coverage of the experiment is not complete.

The quadrupole detection claimed by Boughn et al. (1981) at 6.4 mm was due to ground radiation leakage and microphonics of the radiometer synchronized to gondola rotation (Fixsen et al., 1983).

The infrared quadrupole detected by Fabbri et al. (1980) at 1 mm was also probably not of cosmological origin (Ceccarelli et al., 1982). In fact, a second improved differential photometer was flown with a dust-emission monitoring channel added: strong galactic emission was found even at high galactic latitudes (de Bernardis et al., 1984). After subtracting a 1/senb galactic emission model, and averaging in the Smoot angle, no quadrupole anisotropy was found (in the dipole direction) at a 0.2 mK level (Masi, 1982). The data are shown in fig. 6.

The clearest evidence for dipole anisotropy is found in the map produced by Lubin et al. (1985) at 3 mm (fig. 7). The residuals from the map, after removing the best fit dipole, show no presence of galactic emission; thus confirming that at 3 mm the local background contributions are reduced to a minimum. A similar map has been produced by Wilkinson's group (Fixsen et al., 1983) at 1.2 cm.

Maps at wavelengths near the peak of the CBR are not available at present, but would be of great interest for future research. Careful galactic emission corrections will be needed in order to extract true cosmological information. The IRAS 100 µm survey can provide a starting data-base, but cannot be used to model emission from dust at temperatures lowe r than 20K: a submillimeter survey must be used for this

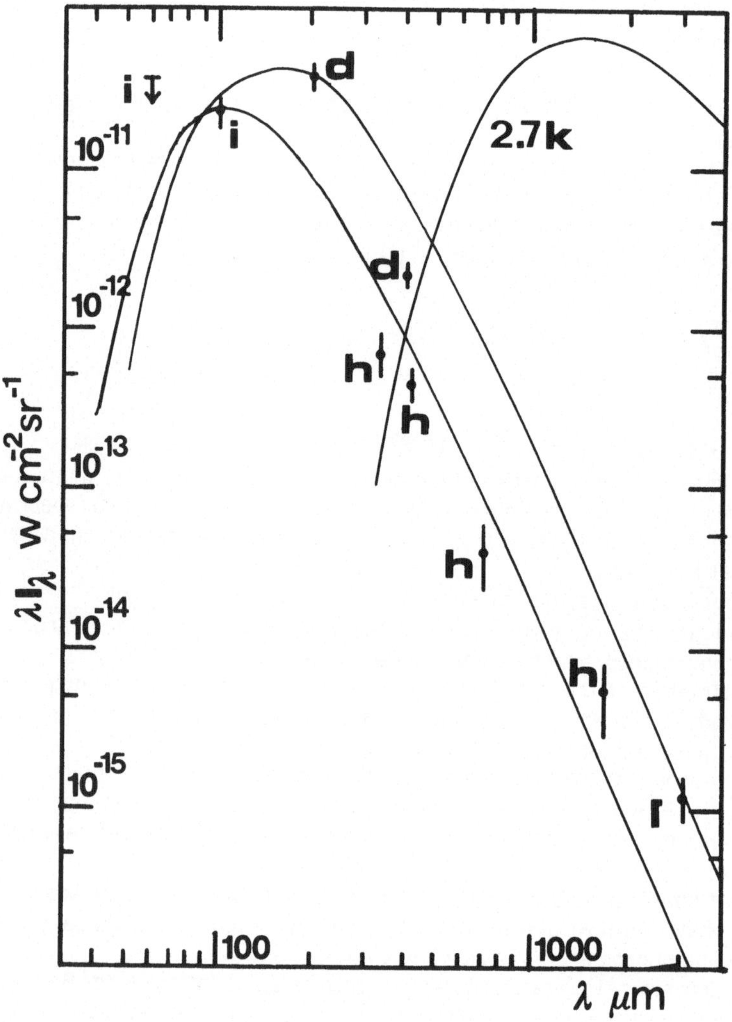

Figure 5

Estimates and measurements of the galactic background at the north pole. I) are IRAS data at 60 and 100 μm; d) are data from de Bernardis et al. (1984); h) are data from Halpern et al. (1985); l) is the value quoted by Lubin et al. (1983). Curve a) is a grey body at T = 22 K, $\xi \sim 1/\lambda^2$ arbitrarily normalized in order to fit the IRAS 100 μm datum; b) is a grey body at T = 17 K, $\xi \sim 1/\lambda^{1.8}$ normalized in order to fit the 200 μm datum.

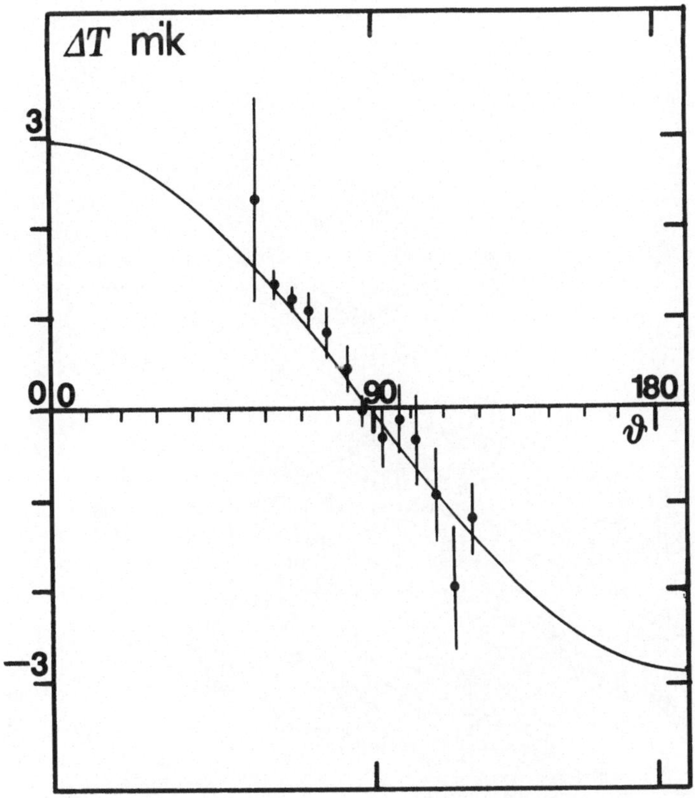

Figure 6

Smoot representation of data from the 1980 balloon flight survey at 440 μm (de Bernardis et al. 1984). A galactic component has been removed from the data before averaging in the Smoot angle. The best fit dipole is shown as a continuous line (Masi, 1982).

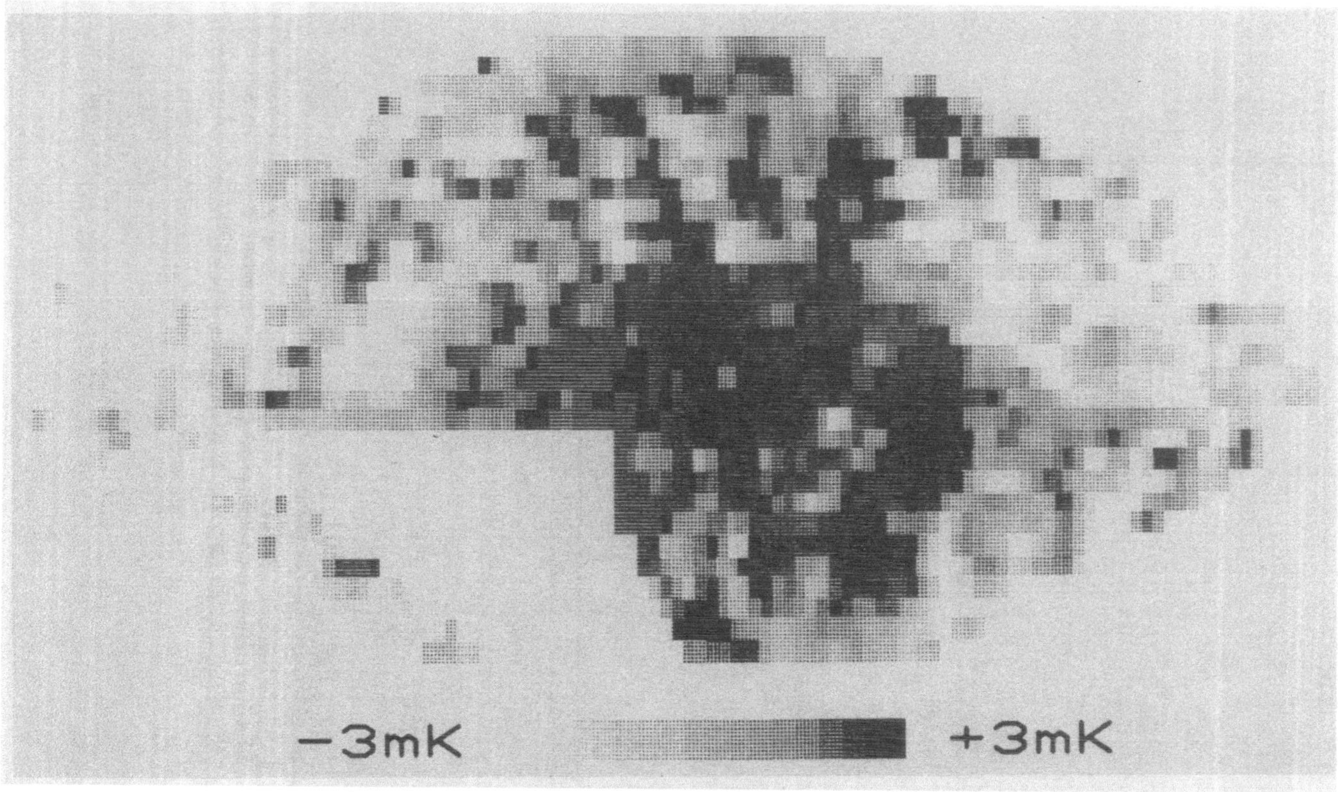

Figure 7

Map of the sky at 3 mm in celestial coordinates α, δ. $\alpha = 24^h$ at the left, and 0^h at the right. $\delta = +77°$ at the top, and $-68°$ at the bottom. Pixel size is 3° x 5° (from Lubin et al., 1985)

(Halpern et al., 1985). Such an experiment is now being developed by our group (Masi et al., 1986).

In the following we will discuss only the implications of Dipole anisotropy - the only anisotropy truly detected in the CBR until now.

4.2 Dipole Anisotropy: Cosmological Connections

Fig. 8 shows a flow-chart of what one can infer from dipole and quadrupole anisotropy measurements.

Let us start with the dipole anisotropy amplitude, and suppose that measurements at many wavelengths are available. If we assume that the CBR spectrum is a pure blackbody we have

$$\alpha = \frac{d\ln I}{d\ln \nu} = 3 - \frac{xe^x}{e^x-1} \quad : \quad x = \frac{h\nu}{kT_{CBR}} \tag{10}$$

$$\therefore \quad \Delta I(\nu) = \beta \frac{xe^x}{e^x-1} I(\nu, T_{CBR}) \quad ; \quad \beta = v/c$$

A measurement of the dipole anisotropy ΔI at frequency ν defines a relation between β and T_{CBR}. Combining two or more measurements at different frequencies one can evaluate both β and T_{CBR}. Halpern et al. (1985) have carried out an analysis of this kind, concluding that the measurements of Lubin et al. (1983), Fixsen et al. (1983), and Halpern et al. (1985) are consistent if $T_{CBR} = (2.7 \pm 0.2)$K: the dipole anisotropy measurements can be used to evaluate the temperature of the microwave background at the same level of accuracy of direct flux measurements, at least in the millimetric region.

Deviations from a pure blackbody spectrum are expected, due to various mechanisms:

1) Energy injection in the primaevel fireball at $Z>10^3$ (Zeldovich, 1969)
2) Dust emission in the population II model (Negroponte et al., 1981)
3) H and He line emission during the recombination (Lyubarsky and Sunyaev, 1983)
4) Paraphoton model of the microwave background (Georgi Ginsparg Glashow, 1982)

The expected dipole anisotropy as a function of the frequency can be computed applying formula (9) to the spectral behaviour $I(\nu)$ expected from each model. A best fit procedure to the dipole data allows us to test the models.

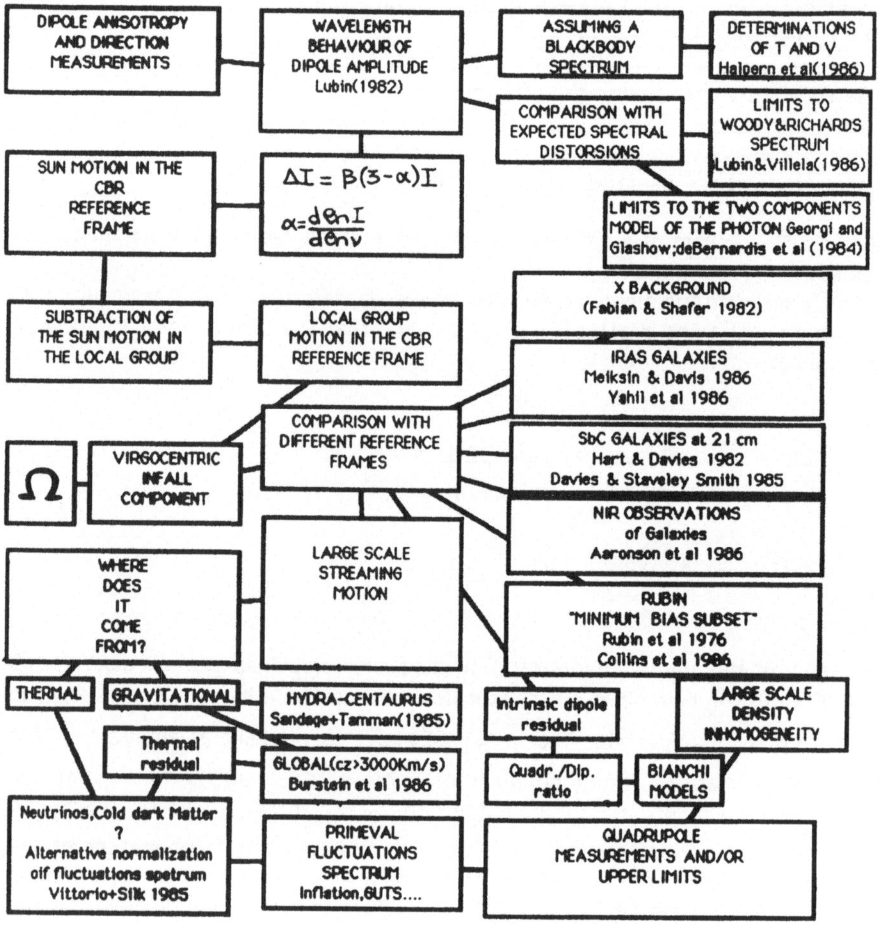

Figure 8

Large scale anisotropy measurements: cosmological connections.

For mechanism 1), dipole measurements are no longer comparable in precision to direct spectral measurements in the Raileigh-Jeans region (De Amici et al., 1985). Models 2) and 4) have lost part of their appeal after the new millimetric spectrum measurements by Peterson et al. (1985), which do not confirm the millimetric excess detected by Woody and Richards (1981). However, for model 4), a strong upper limit to the paraphoton mass has been derived using dipole anisotropy data in the radio region (de Bernardis et al., 1985). Mechanism 3) gives effects in the unobservable Wien region of the 2.7K spectrum ($\lambda \sim 600$ μm).

The direction and amplitude of the dipole anisotropy define the motion of the sun in the CBR reference frame - ($v_\odot$). Since the motion $v_{\odot LG}$ of the sun in the Local Group is well known, the Local Group motion in the CBR reference frame can be found from each dipole measurement:

$$\vec{v}_{LG} = \vec{v}_\odot - \vec{v}_{\odot LG}$$

The weighted mean of the Local Group motion determination is

$$\vec{v}_{LG} = (603 \pm 14)\text{km/s} \quad \text{toward} \quad l = 269° \pm 2° \quad b = 27° \pm 2°$$

This is a sky region far from the most obvious attractor, the Virgo Cluster which, in fact, lies at a distance of 48° (l = 284°, b = 74°), as shown in fig 2. Davies et al. (1980), Gunn (1977), and Peebles (1976) have shown that the virgocentric infall velocity v_p, if due to Virgo gravitational attraction, is related to the density parameter Ω_o by

$$\Omega_o = \left[\frac{3v_p}{\delta(v_p + v_c)} \right]^{1.5}$$

where δ is the matter mean overdensity with respect to the mean density of the universe, averaged over the sphere centered in the Virgo Cluster, and with radius equal to our distance to Virgo. Current values are $\delta = 2.8 \pm 0.5$ (Tamman and Sandage (1985) and $v_c = (967 \pm 53)$km/s (Kraan and Korteweg, 1981). If one assumes that the Virgocentric component of the Local Group motion is due to Virgo attraction, one has

$$v_p - v_{LG} \cos 48° \simeq 400 \text{ km/s} \quad \longrightarrow \quad \Omega_o \simeq 0.15$$

However, a comparison to different reference frames is needed in order to test this assumption. For example, Tamman and Sandage (1985) have studied the virgocentric infall using 8 clusters and groups of galaxies: they found $v_p = (200 \pm 50)$km/s --> $\Omega_o \simeq 0.08$. According to them, the Local Group velocity could be the sum of a virgocentric component plus a motion of all the local supercluster toward the Hydra-Centaurus supercluster. However, Burstein et al. (1986) have recently found evidence of a 700 km/s motion of the Hydra complex relative to the CBR, as a part of a relatively coherent streaming motion extending over at least $40h^{-1}$ Mpc.

Table IV summarizes the Local Group motion determinations with respect to different reference frames.

If Davies et al. (1985) and Aaronson et al.'s (1986) observations correctly describe the Local Group motion, then the intrinsic dipole component of the CBR is lower than the difference

$$\left|\vec{v}_{LG} - \vec{v}_{LG\ Davies}\right| = (90 \pm 65)\ \mathrm{km/s} \;\longrightarrow\; \left.\frac{\Delta I}{T}\right|_{\substack{INTRINSIC\\DIPOLE}} = (3 \pm 2)\,.10^{-4}.$$

This result sets a strong upper limit to the amplitude of large scale fluctuations in the early universe (Peebles, 1981). Moreover, homogeneous anisotropic models predicting a quadrupole. anisotropy of the same order of magnitude of the intrinsic dipole are still consistent with observations.

On the other hand, Collins et al.'s (1986) analysis would suggest a genuine discrepancy between the local group motion relative to their sample and that relative to the CBR. A large scale streaming motion would be present with

$$\left|\vec{v}_{LG} - \vec{v}_{LG\ Collins}\right| = (970 \pm 300)\ \mathrm{km/s}.$$

This large peculiar velocity would have important consequences for the choice of the particles dominating the galaxies' dynamics (Vittorio and Silk, 1985; Kaiser 1983; see also Nicola Vittorio's paper in these proceedings).

A related topic is that of IRAS "Galaxies". The IRAS point source data base (Beichman et al., 1984) provides nearly complete sky coverage (only galactic latitudes $|b|<10°$ must be excluded, due to confusion). This sample is unaffected by reddening at 60 μm wavelength, and consists of a large number of galaxies (~20.000 with high quality flux)(Yahil et al., 1986) providing the opportunity of extensive statistical studies. However, two main problems limit these studies: a) the greater part of IRAS "galaxies" has no optical counterparts , their redshift and distance being unknown; b) galactic cirrus (patchy, low temperature galactic dust clouds detected by IRAS for the first time) can simulate "false" galaxies.

Due to problem a), the Hubble flow anisotropy cannot be studied directly using IRAS galaxies. However, since both gravitation and radiation flux follow the $1/r^2$ law, the smoothed surface brightness due to galaxies can be used to map the local gravitational field, if one assumes that IRAS galaxies are good tracers of matter distribution. Lawrence et al. (1986) have studied the redshift distribution of a subsample of IRAS galaxies, finding all their optical identifications and measuring their redshifts. They found a median distance of 90 h^{-1} Mpc. IRAS galaxies provide an unbiased survey of cosmological structures on scales as large as 200 Mpc: more than 10 times the Local Supercluster dimensions. Yahil et al. (1986) found a dipole anisotropy of the smoothed surface brightness of these galaxies in a direction only 26° ± 10° away from the Local Group motion with respect to the CBR. Meiskin and Davies

TABLE 4: LOCAL GROUP MOTION

Reference	OBJECT	DISTANCE	V(km/s) L.G.	1°	b°	NOTES
Weighted mean	CBR last scattering surface	3000h Mpc	603±14	269±2	27±2	Machian reference frame
Shafer + Fabian (1983)	X back-ground sources	???	750±200	306±40	18±40	marginally significant but consis-tent with the CBR dipole
Yahil et al;Meiskin and Davies (1986)	IRAS 60 microns galaxies (10**4)	100h Mpc	5-10% enhancement integrated flux	248±9	40±8	consistent with CBR
Aaronson et al. (1986)	10 galaxy clusters	40-100h Mpc	IN PRESS			consistent with CBR
Davies et al.(1985) Hart + Davies (1982)	100-500 Sbc galaxies observed at 21 cm	10-55h Mpc	484±55	258±15	42±8	consistent with CBR
Rubin et al. (1976)	96 gal. minimum bias subset	10-100h Mpc	586±200	202±19	-11±17	75 deg away from CBR!
Collins Joseph Robertson (1986)	same set as Rubin with NIR and IRAS observations	"	621±300 846±334 680±330 662±220	186±30 214±40 184±35 190±20	-3±29 -22±34 -36±30 -6±20	(NIR) (NIR+Tully Fisher) (NIR+color magn.) (IRAS)
Burstein et al. (1986)		$50h^{-1}$ Mpc	700	IN PRESS		coherent streaming motion

(1986) found independently similar results fitting a dipole component to the number density of IRAS galaxies. One possible interpretation is that IRAS galaxies trace the large scale density inhomogeneity responsible for the Local Group motion. However, only a complete identification program will substantiate these results. As an independent check, a 5% density inhomogeneity at a distance of about 100 Mpc would generate a quadrupole anisotropy in the CBR at a level near the present threshold of detection: its observation would give full autoconsistency to this scenario.

REFERENCES

- Aaronson M. et al.:*Ap. J.* in press (1986)
- Beichman C.A., Neugebauer G., Habing H.J., Clegg P.E., Chester T.J., *IRAS Explanatory Supplement,* JPL D-1855, (1984)
- Berlin A.B., Bulaenko E.V., Vitkovsky V.V., Kononov V.K., Parijskij Yu. N., Petrov Z.E.: in *Early Evolution of the Universe and its Present Structure 104 IAU Sump.,* Abell G.O. and Chincarini G. Eds, Reidel, pg. 121, (1982)
- Birkinshaw M., Gull S.F., Hardbeck H.: *Nature* **309**, 34, (1983)
- Boughn S.P., Cheng E.S., Wilkinson D.T.: *Ap. J. Lett.* **243**, L113, (1981)
- Burstein et al.: *Ap. J.* in press (1986)
- Caderni N., De Cosmo V., Fabbri R., Melchiorri B., Melchiorri F., Natale V.: *Phys. Rev. D.* **16**, 2424, (1977)
- Cavaliere A., Danese L., De Zotti G.: *Astron. Astrophys.* **75**, 322, (1979)
- Ceccarelli C., Dall'Oglio G., Melchiorri B., Melchiorri F., Pietranera L.: *Ap. J.* **260**, 484, (1982)
- Cheng, Saulson P.R., Wilkinson D.T., Corey B.E.: *Ap. J. Lett.* **232**, L139, (1979)
- Collins C.B. and Hawking S.W.: *MNRAS* **162**, 307, (1973)
- Collins C.A., Joseph R.D., Robertson N.A.: *Nature* **320**, 506, (1986)
- Conklin E.K.: *Nature* **222**, 971, (1969)
- Corey F., Wilkinson D.T.: *Bull. A.A.S.* **8**, 351, (1976)
- Danese L., De Zotti G.: *Astron. and Ap.* **94**, L33, (1981)
- Danese L., De Zotti G.,Mandolesi N.: in *The Birth of the Universe*, J. Audouze, J. TranThanh Van eds, Ed. Frontieres, Moriond, pg. 205, (1982)
- Davies M. Tonry J., Huchra J., Lathan D.W.: *Ap. J. Lett.* **238**, L113, (1980)
- Davies R.D. and Stavely Smith L.: *Proc. of ESO Workshop on the Virgo Cluster,* Richter O.G. and Binggeli B. eds, pg. 391, ESO (1985)
- de Bernardis P., Masi S., Melchiorri F., Moleti A.: *Ap. J.* **284**, L21, (1984)
- de Bernardis P., Masi S., Melchiorri F., Melchiorri B., Moreno G.: *Ap. J.* **278**, 150, (1984)
- Epstein G.L.: *Ph.D Thesis* , University of California, Berkeley (1983)
- Fabbri R.: *Phys. Lett.* **79A**, 21, (1980)
- Fabbri R., Guidi I., Melchiorri F., Natale V.: *Phys. Rev. Lett.* **44**, 23, (1980)
- Fixsen D.J., Cheng E.S., Wilkinson D.T.: *Phys. Rev. Lett.* **50**, 620, (1983)

- Fomalont E.B., Kellermann K.I., Wall J.V., *Ap. J.* **277**, L23, (1984)
- Georgi H., Ginsparg P., Glashow S.L.: *Nature* **306**, 765, (1983)
- Gorenstien M.V. and Smoot G.F.: *Ap. J.* **244**, 361 (1981)
- Gunn J.: *Ap. J.* **218**, 592, (1977)
- Halpern M., Weiss R., Benford R.: in *The Cosmic Background Radiation and Fundamental Physics - Conference Proceedings of the Italian Physical Society,* **Vol.1**, pg. 83, Melchiorri F. and Maiani L. editors (1985)
- Hart L., Davies R.D.: *Nature* **297**, 191, (1982)
- Henry P.S.: *Nature* **231**, 516, (1971)
- Kaiser N.: *Ap. J. Lett.* **273**, L17, (1983)
- Knoke J.E., Partridge R.B., Ratner M.I., Shapiro I.I.: *Ap. J.* **284**, 479, (1984)
- Kraan-Korteweg R.C.: *Astron. and Ap.* **104**, 280, (1981)
- Lasenby A.N., Davies R.D.: *MNRAS* **203**, 1137, (1983)
- Lawrence A., Walker D., Rowan-Robinson M., Leech K.J., Penston M.V.: *MNRAS* , 687, (1986)
- Lubin P., Villela T., Epstein G., Smoot G.: *Phys. Rev. Lett.* **50**, 616, (1983)
- Lubin P.M.: *Int. School of Physics "E. Fermi", 86ò Course,* Melchiorri F. and Ruffini R. eds (1986)
- Lubin P.M. and Neto T.V.: in *The Cosmic Background Radiation and Fundamental Physics - Conference Proceedings of the Italian Phys. Soc.*, pg. 65, Melchiorri F. and Maiani L. eds., (1985)
- Lubin P.M., Villela T., Epstein G., Smoot G.: *Ap. J.* **298**, L1 (1985).
- Lyubarsky Y.E. and Sunyaev R.A.: *Astron. Astrophys.* **123**, 171, (1983)
- MacCallum M.A.N.: in *General Relativity, an Einstein Centenary Survey* (Hawking S.W. and Israel W. eds), Cambridge University Press, 533, (1979)
- Mandolesi N., Calzolari P., Cortiglioni S., Morigi G., Delpino F., Sironi G., Inzani P., De Amici G., Solheim J.E., Berger L., Partridge R.B., Martenis P.L. Sangree C.H., Harvey R.C., *Nature* in press (1986)
- Mosengheil K.: *Ann. Phys.* **21**, 867, (1907)
- Masi S.: 'Tesi di Laurea in Fisica', Rome University (1982)
- Masi S., Andreani P., Ballarin A., Dall'Oglio G., de Bernardis P., Giovannozzi E., Moleti A., Nisini B., Piccirillo L., Salama A.: *11ò Int. Conference on Infrared and Millimetric Waves - Tirrenia -* (1986)
- Mason C., Ceccarelli C., Dall'Oglio G., Ferri G., Masi S., Radford S.J.E.: *IR Phys.* in press (1986)
- Meikin A., Davies M.: *Astron. J.* **91**, 191, (1986)
- Melchiorri F., Olivo Melchiorri B., Ceccarelli C., Pietranera L.: *Ap. J.* **250**, L1, (1981)
- Negroponte J., Rowan-Robinson M., Silk J.: *Ap. J.* **248**, 18, (1981)
- Parijskij Yu. N., Petrov Z.E., Cherkov L.N.: *Sov. Astron. Lett.* **3**, 263, (1977)
- Partridge R.B.: *Ap. J.* **235**, 681, (1980)
- Peebles P.J.E., Wilkinson D.T.:*Phys. Rev. Lett.* **18**, 557, (1968)
- Peebles P.J.E.: *Physical Cosmology,* Princeton University Press (1971)
- Peebles P.J.E.: *Ap. J.* **205**, 318, (1976)

- Peebles P.J.E.: *Ap. J.* **205**, 318, (1976)

- Peterson J.B., Richards P.L., Timusk T.: *Phys. Rev. Lett.*, **55**, 332, (1985)

- Peterson J.B., Richards P.L., Timusk T.: *Phys. Rev. Lett.* **55**, 332, (1985)

- Radford S.J.E., Boynton P.E., Ulich B.L., Partridge R.B., Schommer R.A., Stark A.A., Wilson R.W., Murray S.S.: *Ap. J.* **300**, 159, (1986)
- Radford S.J.E.: *Ph.D. Thesis* University of Seattle (1986)
- Rubin V.C., Thonnard N., Ford W.K., Roberts M.S.: *Astron. J.* **81**, 719, (1976)
- Sachs R.K., Wolfe A.M.: *Ap. J.* **147**, 73, (1967)
- Shafer R.A., Fabian A.C.: in *Early Evolution of the Universe and its present structure,* Abell G.O. and Chincarini G. eds, pg. 333, (1983)
- Schmidt M.: in *stars and Stellar Systems,* **Vol. 5** 'Galactic Structure', The University of Chicago Press, 513, (1965)
- Smoot G.F., Gorenstein M.V., Muller R.A.: *Phys. Rev. Lett.* **39**, 898, (1977)
- Smoot G.F., De Amici G., Friedman S.D., Witebsky G., Mandolesi N., Partridge B.R., Sironi G., Danese L., De Zotti G.: *Phys. Rev. Lett.* **51**, 1099, (1983)
- Stewart J.M. and Sciama D.W.: *Nature,* **216**, 748, (1968)
- Strukov I.A., Skulachev D.P.: *Sov. Astron. Lett.* **10**, 1, (1984)
- Sunyaev R.A., Zeldovich Ya. B.: *Comments Ap. Space Phys.* **4**, 173, (1972)
- Tamman G.A., Sandage A.: *Ap. J.* **294**, 81, (1985)
- Uson J.M., Wilkinson D.T.: *Ap. J.* **283**, 471, (1984)
- Vittorio N., Silk J.: *Ap. J.* **293**, L1, (1985)
- Wilkinson D.T.: *Science,* in press (1986)
- Woody D.P., Richards P.L.: *Ap. J.* **248**, 18, (1981)
- Yahil A., Walker D., Rowan-Robinson M.: *Ap. J. Lett.* 301, L1, (1986)
- Zeldovich Ya.: *Soviet Physics JETP* **4**, 730, (1969)

STRANGE MATTER IN THE UNIVERSE*

Angela V. Olinto**
Center for Theoretical Physics
Laboratory for Nuclear Science and Department of Physics
Massachusetts Institute of Technology
Cambridge, Massachusetts 02139 U.S.A.

1. INTRODUCTION

In the course of this lecture, I would like to introduce the strange matter hypothesis and some of its consequences.

In 1984, E. Witten conjectured that quark matter with a large fraction of strange quarks is absolutely stable.[1] Imagine squeezing A nucleons together to densities much higher than nuclear density, *i.e.* let them overlap. It has long been thought that a first-order phase transition to two-flavor quark matter would occur where individual nucleons lose their identity and the system looks like a bag of $3A$ quarks (baryon number A). Up and down quarks can convert into strange quarks via weak interactions; if the mass of the strange quark, m_s, is less than the typical Fermi energy of the system, μ ($\mu \sim 313\,\mathrm{MeV}$), the equilibrium configuration will have a fraction of strange quarks. If $m_s < \mu$ the strange quark Fermi sea can be filled from m_s up to a final Fermi level $\mu' < \mu$ (it introduces more energy states between m_s and μ) and the total energy of the system will be lower. The fact that heavier quarks have masses much greater than μ rules out the possibility of "charming matter" and so on.

At zero pressure, the energy per baryon number of nuclear matter, $E/A\big|_{NM}$, is lower than that of two-flavor quark matter, $E/A\big|_{2QM}$. As long as $m_s < \mu$ we see that the energy per baryon number of three-flavor quark matter is lower than that of two-flavor, *i.e.* $E/A\big|_{3QM} = E/A\big|_{2QM} - \Delta$, $\Delta > 0$. If Δ is big enough maybe $E/A\big|_{3QM} < E/A\big|_{NM}$ at zero pressure! This is the strange matter hypothesis. Three-flavor quark matter is absolutely stable and it is the true ground state of hadronic matter. Stable three-flavor quark matter is called strange matter.[2]

---

* This work is supported in part by funds provided by the U. S. Department of Energy (D.O.E.) under contract #DE-AC02-76ER03069.
** CNPq Fellow

P. Galeotti and D. N. Schramm (eds.), Gauge Theory and the Early Universe, 283–293.

Nuclear matter does not convert spontaneously into strange matter due to the very high order of weak interaction processes needed to convert simultaneously approximately a third of the quarks into strange quarks.

The phenomenology of strange matter was studied by E. Farhi and R. Jaffe.[2] They used QCD perturbation theory up to first order in α_c and the bag model, characterizing the region where the quarks live by an energy density B. A final decision on the stability of strange matter could not be reached given the uncertainties on the values of α_c, m_s, and B. Reasonable windows on the values of these parameters are consistent with the stability of strange matter in bulk. Surface effects limit the stability of low baryon number strange matter lumps (for example "strange matter" is unstable for baryon number one, $\Lambda \rightarrow p\pi^-, n\pi^0 \cdots$). Gravitational collapse limits the other end, so they found strange matter could be stable for $10^2 \lesssim A \lesssim 10^{57}$ (for reasonable choices of α_c, m_s, and B).

Witten also identified two scenarios for the production of strange matter. First, he proposed that strange matter may have been produced as the universe cooled through the QCD phase transition at temperatures $\sim 100 - 200$ MeV. These primordial strange matter nuggets could then account for the dark matter in the universe. E. Farhi and C. Alcock[3] showed that all strange matter produced this way evaporates completely into protons and neutrons as the universe cools from 50 to 1 MeV.

Second, he pointed out that strange matter can be produced in supernovae especially during core bounce when densities much higher than nuclear density are reached. Remnants of such events should then be called strange stars instead of neutron stars.

2. STRANGE STARS

To see what a strange star looks like, we need to know the equation of state for strange matter.

Strange matter can be thought of as a Fermi gas of up, down, and strange quarks and some electrons (to guarantee charge neutrality). The masses of up, down, and electrons may be neglected given that the relevant scale is the strong interactions scale, which is around $m_{\text{proton}}/3 \sim 313\,\text{MeV}$. The strange quark has mass between 50 and 300 MeV. Chemical equilibrium is achieved by the weak processes:

$$d \rightarrow u + e + \bar{\nu}_e \quad u + e \rightarrow d + \nu_e$$
$$s \rightarrow u + e + \bar{\nu}_e \quad u + e \rightarrow s + \nu_e$$
$$d + u \leftrightarrow s + u$$

The neutrinos have a very long mean free path and therefore escape without interacting. Their chemical potential can be set to zero; equilibrium then reads:

$$\mu_d = \mu_u + \mu_e$$
$$\mu_s = \mu_d \quad (\equiv \mu)$$

Charge neutrality gives:

$$\frac{2}{3}n_u - \frac{1}{3}n_d - \frac{1}{3}n_s - n_e = 0$$

where μ_i and n_i are the chemical potential and number density of particle i, respectively. This set of equations leaves only one free parameter, say μ.

The number densities and other relevant thermodynamical functions can be determined by the thermodynamical potentials Ω_i. Expressions from Ω_i up to first order in α_c can be found in Ref. [2] and from them:

$$n_i = -\frac{\partial \Omega_i}{\partial \mu_i} \quad , \quad n_B = \frac{1}{3}(n_u + n_d + n_s)$$

$$\rho = \sum_i (\Omega_i + \mu_i n_i) + B$$

$$P = n_B \frac{\partial \rho}{\partial n_B} - \rho = n_B(\mu_u + \mu_d + \mu_s) - \rho$$

$$G = \frac{P + \rho}{n_B} = \frac{\partial \rho}{\partial n_B} = \mu_u + \mu_d + \mu_s$$

where n_B is the baryon number density, ρ is the energy density, P the pressure, and G the Gibbs potential per particle. These equations together with the equilibrium and charge neutrality conditions completely describe cold strange matter as a function of μ only. The equation of state is found to be independent of the number of flavors; therefore, in the limits $m_s \to 0$ (three flavor quark matter) and $m_s \to \mu$ (two flavor quark matter) it is simply:

$$P = \frac{1}{3}(\rho - 4B)$$

while between the limits this is still a good approximation (up to $\sim 4\%$).

2.1. Global Properties[4,5]

With the above equation of state one can integrate the Oppenheimer-Volkoff equations of hydrostatic equilibrium and find the global properties of strange stars.

Fig. 1 shows the density throughout the star. The variation is modest, starting at $\rho_0 = 4B$ [$\sim 4 \times 10^{14}\,\mathrm{g/cm^3}$ for $B = (145\,\mathrm{MeV})^4$] at zero pressure up to a maximum of $4.8\,\rho_0$, in contrast with neutron stars that have many orders of magnitude variations. The total mass versus the central density is plotted in Fig. 2. The vertical asymptote represents the region where gravity is irrelevant and densities are simply $\rho_0(= 4B)$. As M grows larger, gravity becomes important and central density rises, reaching a maximum at $4.8\,\rho_0$ for $M \simeq 2\,M_\odot$. Fig. 3 shows the total mass versus total baryon number. Here, the dependence on m_s

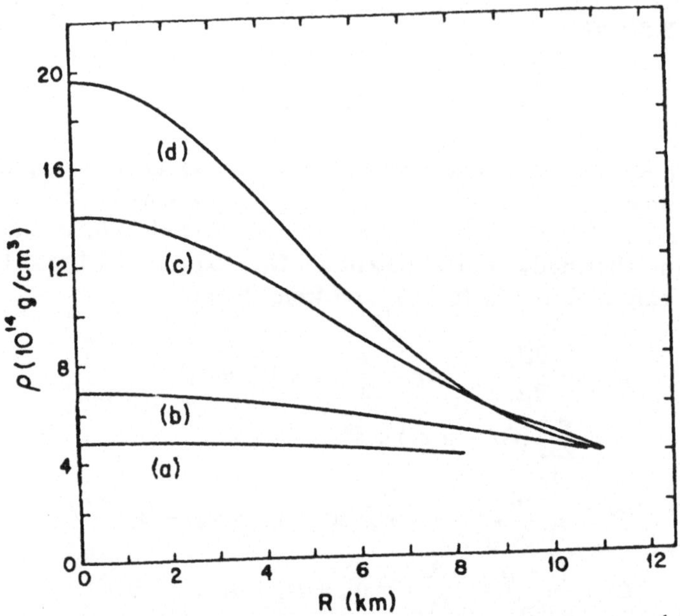

Fig. 1: Density versus radius for strange stars of mass: (a) $0.53\,M_\odot$; (b) $1.4\,M_\odot$; (c) $1.95\,M_\odot$; (d) $2\,M_\odot$.

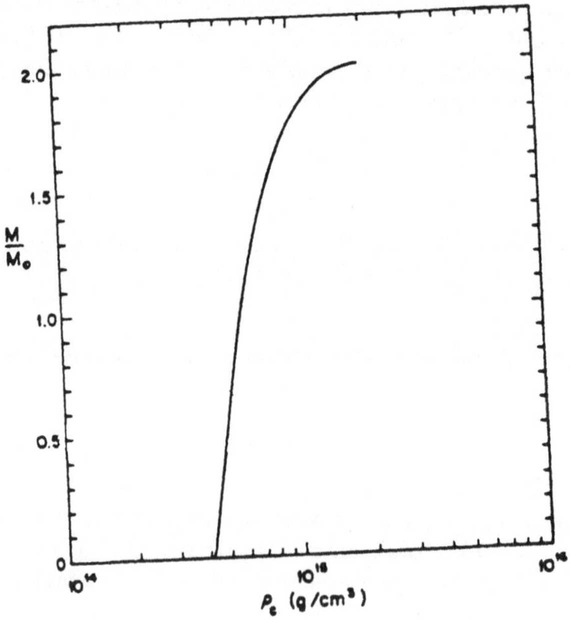

Fig. 2: Total mass (M) versus central density (ρ_c) for stable strange stars.

and α_c is significant; two limiting cases are shown: $m_s = 100\,\text{MeV}$, $\alpha_c = 0.1$ and $m_s = 300\,\text{MeV}$, $\alpha_c = 0.6$. The binding energy of the star relative to H is $(n_B m_H - M)$. Fig. 4 shows the moment of inertia versus the mass. Finally, the mass-radius relation for strange stars is plotted against that of neutron stars with different equations of state in Fig. 5. For low masses $\rho \simeq 4B$ and $M \propto R^3$, while for $M \gtrsim 1\,M_\odot$ it deviates from R^3 due to gravitational effects, becoming two-valued up to the collapse limit. Neutron stars have a minimum mass because nuclear matter is not stable at low densities; there is no minimum mass for strange stars. This qualitative difference suggests an important astrophysical distinction. However, all masses estimated so far (from observed binary systems) are around $1.4\,M_\odot$ where no obvious distinction can be made.

2.2. Strange Stars' Surfaces[5]

2.2.1. Bare.

A very interesting possibility is that of a strange star with bare surface. Strange matter can terminate abruptly (from $4 \times 10^{14}\,\text{g/cm}^3$ to zero in a few fermis). The quarks are confined by the strong interactions; the electrons, bound only electrostatically, will extend over a bigger region, and form a cloud around the strange matter core.

The electron distribution can be determined using a Thomas-Fermi model. Assuming a uniform charge distribution up to the boundary of strange matter ($z = 0$), Poisson's equations look like:

$$\frac{d^2V}{dz^2} = \begin{cases} \dfrac{4\alpha}{3\pi}\left(V^3 - V_q^3\right) & z \leq 0 \\ \dfrac{4\alpha}{3\pi} V^3 & z > 0 \end{cases}$$

Here, the Fermi momentum of the electron (p_e) is equal to the electric potential (V) because the chemical potential at infinity $\mu_\infty = -V + p_e$ is zero (*i.e.*, $n_e = p_e^3/3\pi^2 = \frac{V^3}{3\pi^2}$). The boundary conditions are $V \to 0$ as $z \to \infty$ and $V \to V_q$ as $z \to \infty$ (V_q is the potential inside the quark matter). Integration gives us $V(0) = \frac{3}{4}V_q$, the height of the Coulomb barrier. Fig. 6a shows a plot of $V(z)$ for $V_q = 20\,\text{MeV}$; the distribution of electrons extends several hundred fermis.

The integrity of such a surface is greater than that of any other astrophysical object; therefore, the Eddington limit to the luminosity of a self-gravitating object is irrelevant.

This bare surface does not radiate as a black body as one might expect. The dispersion relation associated with the propagation of a photon in strange matter resembles the familiar plasma dispersion relation: $\omega^2 = \omega_p^2 + k^2$, where the "plasma" frequency can be written $\omega_p^2 = \frac{8\pi\alpha}{3}\frac{n_u^3}{\rho_u}$, which is typically ~ 19 MeV.

Therefore, photons with energies $\lesssim 19\,\text{MeV}$ will be reflected. Bare strange stars are silver stars, like mirrors.

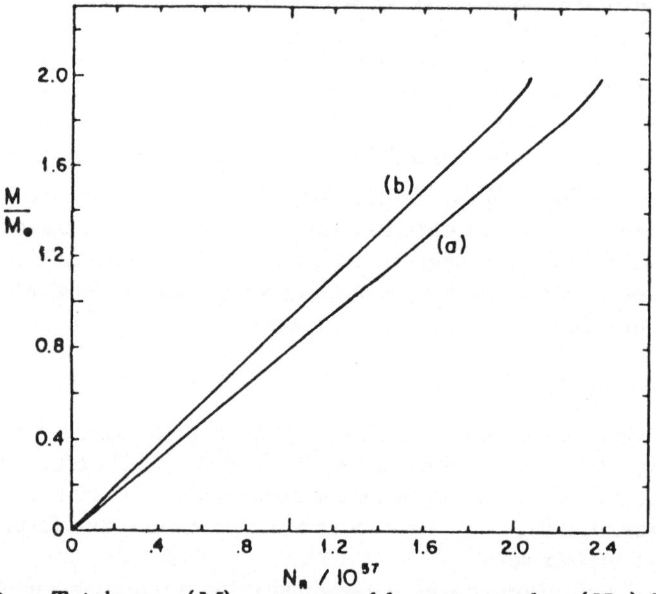

Fig. 3: Total mass (M) versus total baryon number (N_B) for strange stars for the cases: (a) $m_s = 100\,\text{MeV}$, $\alpha_c = 0.1$; (b) $m_s = 300\,\text{MeV}$, $\alpha_c = 0.6$.

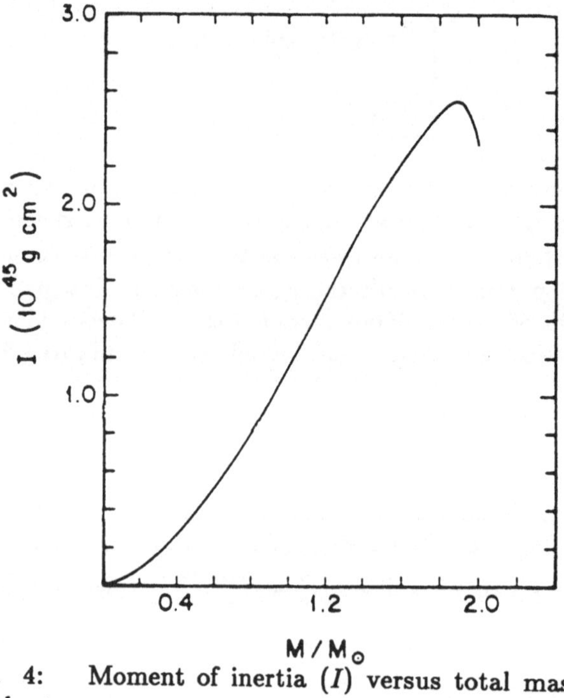

Fig. 4: Moment of inertia (I) versus total mass (M) for stable strange stars.

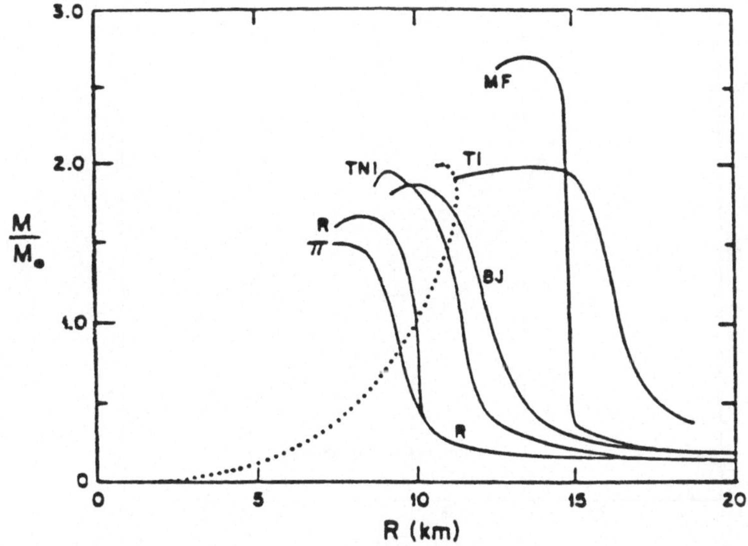

Fig. 5: The dashed line is the strange star mass versus radius relation. The solid lines represent mass versus radius relations for conventional neutron stars assuming different equations of state. These curves are discussed in *Black Holes, White Dwarfs and Neutron Stars*, by S. L. Shapiro and S. A. Teukolsky (John Wiley & Sons, Inc., Reading, MA, 1983).

2.2.2. Thin Crusts. Now suppose we allow some normal material to accrete onto the strange matter surface. How much crust can the star support?

If a nucleon "touches" a strange matter surface, it is converted into strange matter. Free neutrons will "fall right in" gravitationally and be "swallowed", while protons and heavier ions might be supported by the Coulomb barrier at the surface. The maximal crust is, therefore, the outermost layer of a neutron star, where the densities are lower than the neutron drip point (4×10^{11} g/cm^3), *i.e.* there are no free neutrons.

Even so, there must exist a gap between the strange matter surface and the crustal material for it not to be absorbed. To find the probability of a bottom-most ion tunneling the Coulomb barrier, we again integrate a Thomas-Fermi

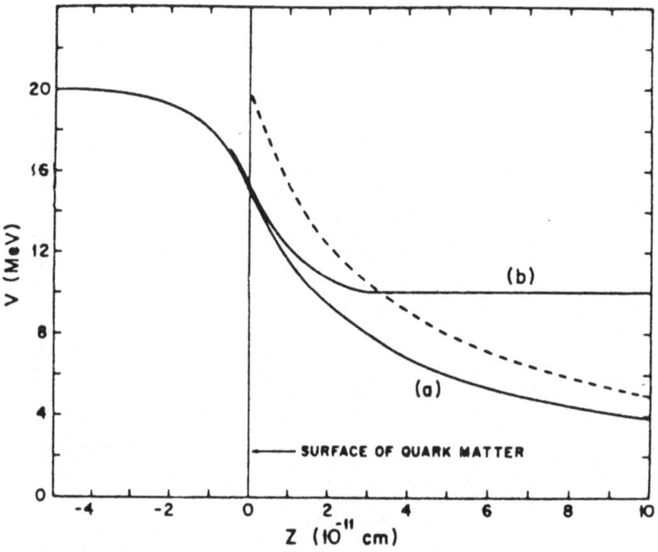

Fig. 6: Electrostatic potential (V) versus height (Z) near the surface of the strange star for two cases with $V_q = 20\,\mathrm{MeV}$: (a) $V_c = 0$; (b) $V_c = 10\,\mathrm{MeV}$. The dashed line shows V_c versus gap width (Z_G). The vertical line at $z = 0$ represents the surface of the quark matter.

model modified by assuming that the crustal material starts at $z = z_G$:

$$\frac{d^2V}{dz^2} = \begin{cases} \frac{4\alpha}{3\pi}\left(V^3 - V_q^3\right) & z \leq 0 \\ \frac{4\alpha}{3\pi}V^3 & 0 \leq z \leq z_G \\ \frac{4\alpha}{3\pi}\left(V^3 - V_c^3\right) & z > z_G \end{cases}$$

Now $V \to V_q$ as $z \to -\infty$ and $V \to V_c$ as $z \to \infty$. Integration for $V_q = 20\,\mathrm{MeV}$ and $V_c = 10\,\mathrm{MeV}$ gives a gap size $z_G \simeq 330\,\mathrm{fm}$ (Fig. 6b).

Using a WKB approximation for the transmission coefficient and the frequency of "strikes" in, say, 10^{10} years, we find a transmission probability of $\sim 10^{-104}$ for $V_q = 20\,\mathrm{MeV}$ and $V_c = 18\,\mathrm{MeV}$. This illustrates how a modest gap of a few hundred fermis is enough to support a crust, preventing strong interactions between the ions and strange matter.

The surface of a strange star might be bare or have a thin crust. A decision between the two possibilities depends on the history of the star's formation that is ultimately related to details in supernovae that are not fully understood.

2.3. Cooling of Strange Stars[5]

Observations of supernova remnants seem to contradict standard neutron star cooling curves, which cool too slowly. Strange matter is a more effective emitter of neutrinos; as it cools faster, strange stars are consistent with the observations. Standard neutron star calculations are above some of the upper limits on surface temperatures given by the Einstein Observatory. Strange stars' cooling curves are below those limits for a wide range of α_c and m_s (emissivities depend strongly on choices of m_s and α_c). More exotic models of neutron stars (pion condensate or quark matter core) are also consistent with the faster cooling observations, making it hard to distinguish between the two types of stars.

3. "STRANGE" CONVERSION[5]

Now suppose strange matter is the true ground state of hadronic matter. Is every neutron star a strange star?

Unfortunately, no conversion route has been found sufficient to prove that every neutron star converts. Some possible routes are the following:

- Two-flavor quark matter converts directly into strange matter. Therefore, if a neutron star core is made of two-flavor quark matter, it will convert into strange matter which would then eat its way up to the neutron-drip line. Unfortunately, the composition of a neutron star core is still an open question.

- At neutron stars' densities, there is a finite density of Λ's. If enough Λ's agglomerate in a region, they will convert to strange matter provided that the baryon number of the agglomerate is high enough for stability. This strange matter lump would again grow, converting the whole star but the thin crust. The problem here is determining the minimum baryon number stable strange matter lump; the uncertainties of m_s, α_c, and B preclude any definite conclusion.

- Neutron stars are born hot. Heated neutrons can form lumps of unstable strange matter (baryon number less than the minimum for stability) that serve as intermediate higher-energy states for the formation of a stable lump (analogous to a chemical burning). This route also depends on uncertain parameters that define the form of this "activation barrier" and the masses of these low baryon number strange lumps.

- Very high energy cosmic neutrinos can deposit as much as 10^{12} MeV on a quark. A hot quark-gluon region will then be formed containing $s\bar{s}$ pairs. If the fluctuations are such that when $T \sim 200$ MeV (QCD phase transition) there is an enhancement of s in a region of this bag, this region will go into strange matter while $\bar{s}$ enhanced region will go into K^+'s and K^0's. It is difficult to formulate this process quantitatively.

- A small lump of strange matter falling into a neutron star will convert the whole star as long as the lump is big enough to penetrate the thin crust. This route cannot lead us to any general conclusion since the same difficulties face the creation of the lump.
- Finally, a strange star can be formed during core bounce in a supernova. The densities are high enough to reach a two-flavor quark matter phase, Λ's agglomerates, or intermediate strange matter lumps. If the conversion is fast enough the explosion might blow away all the normal material and give birth to a bare strange star.

4. "STRANGE" EVENTS

So far, no analysis has been able to decide on the stability of strange matter. An experimental effort has been made in heavy ion collisions without a decisive result.[6] Because of this uncertainty, attention should be paid to events that present difficulties for standard models of matter.

Witten[1] suggested that strange matter could explain the Centauros event in which a cosmic ray primary "fragmented into hundreds of baryons and almost nothing else" in Earth's atmosphere. Baym, Kolb, McLerran, Walker and Jaffe[5] suggested that the source of the high-energy cosmic rays arriving from Cygnus X-3 (we will hear more about Cygnus X-3 later in the school) is a strange star. I will tell you about another "strange" event that we have been working on.[8]

On March 5, 1979, nine interplanetary spacecraft observed the most intense burst of non-solar high energy photons ever recorded. The separation of these spacecraft ($\sim$ 600 lt-sec) gave a very precise angular determination of the origin of the burst. Within the $1' \times 2'$ error box one can identify the 16,000-year-old supernova remnant N49 in the neighboring galaxy of the Large Magellanic Cloud. The burst had an amazingly fast rise time ($< 1\,\text{ms}$) and a single very intense peak of $\sim 10^{-3}\,\text{erg/cm}^2\text{-s}$ that corresponds to $\sim 10^{45}\,\text{erg/s}$ if we assume isotropic radiation from N49 (55 kpc away). This luminosity is way above the Eddington limit ($\sim 10^{38}\,\text{erg/s}$ for a $1\,M_\odot$ object). Following the peak, there was an oscillatory phase, two orders of magnitude less intense; it had a clear period of $8.00 \pm 0.05\,\text{sec}$, and decayed in a few minutes. The spectrum had a 420 keV feature traditionally associated with a neutron star redshifted e^+e^- annihilation peak. There were three recurrent less intense events. So far, no model has been successful in reproducing the data.

We propose that a strange matter lump hit a strange star, and produced this gamma-ray burst. Normal matter lumps are destroyed by tidal forces in the vicinity of such compact objects, while strange matter lumps are at most distorted, and therefore consistent with a very short rise time.

A $10^{-8}\,M_\odot$ strange lump, originating far from the star with zero total energy, arrives with enough kinetic energy (velocity $\sim 0.6\,c$) to account for the energetics of the event. Such a lump passed right through the crust and came to rest when it reached the strange matter core. Most of its kinetic energy was transformed locally into heat that was radiated and blew open a crater in the crust. This radiation corresponds to super-Eddington luminosities that were ultimately observed in our solar system (super-Eddington luminosities can be attained in bare strange matter surfaces). The energy released as this crater refilled was responsible for the longer duration, lower intensity phase. The 420 keV feature is the strange star redshifted e^+e^- annihilation line. The eight-second period is due to the rotation of the star, but an eight-second strange (or neutron) star is expected to be much older than 16,000 years. We suggest that N49 has a strange star remnant in a binary system with an older, eight-second-period strange star. The origin of the lump can then be associated with the N49 supernova explosion that might have ejected some lumps of strange matter after core bounce.

5. CONCLUSION

The interesting possibility that the ground state of matter may have been overlooked (instead of ^{56}Fe it might be strange matter) is still an open question. It is possible that this form of matter is the result of a supernova, that neutron stars are really strange stars, and that a class of gamma-ray events is associated with these stars. The decision on the stability of strange matter will most likely come from observations, and it seems important to keep in mind the unusual features of such matter when dealing with events that are not consistent with more standard models of matter.

ACKNOWLEDGEMENTS

The original work presented here has been done in collaboration with E. Farhi and C. Alcock. I would like to thank the organizers of the "Ettore Majorana" International School of Particle Astrophysics, E. Farhi, C. Alcock, V. Jenkins, and S. Gardner.

REFERENCES

1. E. Witten, *Phys. Rev. D* **30**, 272 (1984).
2. E. Farhi and R. L. Jaffe, *Phys. Rev. D* **30**, 2379 (1984).
3. C. Alcock and E. Farhi, *Phys. Rev. D.* **32**, 1273 (1985).
4. P. Haensel, J. L. Zdunik and R. Schaeffer, *Astron. Ap.* **160**, 121 (1986).
5. C. Alcock, E. Farhi and A. Olinto, *Ap. J.*, to appear November 1 (1986).
6. E. Farhi and R. L. Jaffe, *Phys. Rev. D.* **32**, 2452 (1985).
7. G. Baym, R. L. Jaffe, E. W. Kolb, L. McLerran and T. P. Walker, *Phys. Lett.* **160B**, 181 (1985).
8. C. Alcock, E. Farhi and A. V. Olinto, MIT preprint CTP #1378, July 1986.

AN EXPERIMENTAL SEARCH FOR GALACTIC AXIONS

Joseph T. Rogers
Department of Physics and Astronomy
University of Rochester
Rochester, NY 14627
U.S.A.

ABSTRACT. Several groups in the United States and elsewhere have proposed building detectors sensitive to axions of the Dine-Fischler-Srednicki (DFS) type that may have condensed into the galactic halo. These detectors are based on the principle, due to Sikivie [1], of conversion of axions to microwave photons in a magnetic field. We describe the device built by a Rochester-Brookhaven-Fermilab collaboration*, its sensitivity, and the motivation from astrophysics for attempting to observe axions.

1. Introduction

The axion was hypothesized out of a need to solve the "strong CP problem". The QCD lagrangian contains a term

$$\frac{\theta}{32\pi^2} G_{\mu\nu}\tilde{G}_{\mu\nu} \tag{1}$$

θ is an angle associated with non-perturbative QCD effects ("instanton" effects), which determines the degree of parity violation in strong interactions. Since QCD is invariant under charge conjugation, θ also determines the amount of CP violation. Yet the lack of a measurable neutron electric dipole moment places a severe limit on the degree of CP violation in strong processes, and constrains θ to be less than 10^{-8}. This seems unnatural for a parameter that is free to take on any value from 0 to 2π.

A way to naturally produce a small value of θ is to introduce a scalar field (or fields) which couples weakly to the fermions. The lagrangian now has an effective potential which depends on θ. The U(1) symmetry in θ (the Peccei-Quinn symmetry)[2] is dynamically broken; θ can evolve in time. We also get a new particle, a light pseudoscalar, the axion [3,4].

As the temperature in the early universe drops below f_a, the Peccei-Quinn symmetry breaking scale, θ takes on a definite value, likely to be of order unity, in a volume within the particle horizon. At this temperature, instanton effects are unimportant, so the effect-

P. Galeotti and D. N. Schramm (eds.), Gauge Theory and the Early Universe, 295–300.

ive potential is flat. At a temperature of about 150 MeV (the QCD scale factor), these effects create a potential with a minimum at $\theta = 0$, and coherent oscillations of θ start, corresponding to the acquisition of mass by a zero-momentum axion field. The expansion of the universe damps these oscillations, bringing θ close to zero, but the present energy density in the axion field oscillations is calculable [5,6,7,8]:

$$\rho_a = \rho_{crit}\left(\frac{N}{6}\right)^{5/6} \theta_i \left(\frac{f_a}{10^{12}\text{GeV}}\right)^{7/6} \tag{2}$$

N is the number of minima in the effective potential, which is of order 6, and θ_i is the initial value of θ.

The axion model proposed by Dine, Fischler, and Srednicki [9] has only one important parameter, f_a, which determines the axion's coupling to matter

$$L_{int} = i\left(\frac{a}{f_a}\right)[x_u m_u \bar{u}\gamma_5 u + x_d m_d \bar{d}\gamma_5 d + \ldots] \tag{3a}$$

$$x_u + x_d = 1 \tag{3b}$$

and its mass, related to the neutral pion mass

$$m_a = \frac{\sqrt{z}}{1+z}\,\frac{f_\pi m_\pi}{f_a}\,N \tag{4a}$$

$$z = \frac{m_u}{m_d} \tag{4b}$$

Yet f_a is unconstrained by theory.

Accelerator experiments [10] exclude axions with $f_a < 100$ GeV (fig. 1). Axions with 100 GeV $< f_a < 6 \times 10^9$ GeV would provide a major source of energy loss in stars and are inconsistent with our knowledge of stellar evolution [11,12]. The requirement that the axion density be less than the critical density of the universe constrains f_a to be less than 10^{12} GeV [5,6,7,8]. These bounds require the axion mass to lie between 10^{-3} eV and 10^{-5} eV.

Because of the abundance of axions implied by the lower limit on f_a, and their creation in a cold, zero-momentum state, it is possible that they compose the unseen massive halo inferred from the rotation curve of the galaxy. Using a simple isothermal model for this dark matter, Turner [8] finds its local density to be 5×10^{-25} g/cm . Axions condensing into the galaxies would acquire their virial velocities, of around 10^{-3} c. We attempt to detect these relatively dense galactic axions.

2. Detection

The fundamental interaction of axions with matter is through the axion-fermion-fermion vertex shown in fig. 2a. The fermions, being charged, can in turn couple to photons (fig. 2b). The effective coupling (fig. 2c) of the axion to the electromagnetic field is

$$L_{int} = -\left(\frac{N}{6}\right)\left(\frac{e^2}{2\pi^2}\right)\left(\frac{\xi}{f_a}\right)\mathbf{E}\cdot\mathbf{B}\ a \qquad (5a)$$

$$\xi = 4 - \frac{4+z}{1+z} \qquad (5b)$$

Because the decay of the axion into two photons occurs extremely slowly, we use a strong static magnetic field $\mathbf{B}_0$ to provide a virtual photon and detect the conversion of the axion to a second photon. The energy of this photon corresponds to microwave frequencies, 1 to 300 GHz, so we trap it in a microwave cavity within a magnet, and detect it in a microwave receiver coupled to the cavity. The cavity mode is chosen to make the integral of the lagrangian density (5a) over the cavity volume non-zero. Because the axion field is coherent over many meters, and a uniform magnetic field is the most practical to produce, we are restricted to the TM_{0m0} modes of a cylindrical cavity, which have the electric field parallel to the cavity axis (fig. 3). The presence of the cavity resonance enhances the conversion rate proportionally to the cavity's electrical Q. The microwave power available from the cavity from conversion is [13,14,15]

$$P = (2.2\times10^{-23}\,\mathrm{Watt})\left(\frac{\langle\rho_a\rangle}{5\times10^{-25}\mathrm{g/cm^3}}\right)\left(\frac{m_a c^2}{10^{-5}\mathrm{eV}}\right)\left(\frac{B_0}{10\mathrm{T}}\right)^2\left(\frac{V}{10^4\mathrm{cm^3}}\right)\left(\frac{Q}{10^5}\right)G^2 \qquad (6)$$

G is a normalization factor for the $\mathbf{E}$ field over the cavity volume V.

3. Experimental Technique

The Rochester-Brookhaven-Fermilab detector uses a 6.6 tesla superconducting solenoid magnet. The 4.2 kelvin liquid helium bath bath used to cool the magnet also cools the copper cavity, improving its Q and decreasing its thermal excitation. The cavity is tuned in frequency under microprocessor control by the insertion of a sapphire rod, chosen for its high dielectric constant and low loss tangent. For scanning wider frequency ranges, cavities of different diameters are substituted. A cryogenic GaAs FET preamplifier is coupled magnetically to the cavity and forms the low-noise first stage of the microwave receiver (fig.4). The microwave signal is mixed down in two steps, using computer-controlled oscillators, before entering a 64 channel spectrum analyzer. Some pre-averaging of the spectra to decrease the fluctuations in the cavity and amplifier noise is done by the microprocessor; further averaging and analysis occurs off-line.

Our experimental program is to cover the frequency range 1 to 10 GHz in two years with the greatest possible sensitivity. The axions' linewidth, determined by their velocity distribution in the galaxy, is estimated to be one part in 9 x 10^6 of their frequency [15]; the cavity thermal noise and amplifier noise in the same bandwidth, and the number of spectrum averages we are able to make to smooth that noise, limit our sensitivity. Table 1 illustrates the density of standard DFS axions that could be observed with the present apparatus at a 3 standard deviation cofidence level, and with an improved device that could be constructed in the following years. The improvements would enable us to see the galactic dark matter, if made of axions, and to determine the axion's contribution to the density of the universe.

Table 1: Current experimental parameters of the Rochester-Brookhaven-Fermilab axion experiment and possible improvements

Parameter	Current value	Improved value
B_o (tesla)	6.6	15
V (cm^3)	1.0×10^4	2.5×10^4
Q_o	3.6×10^5	3.6×10^5
G	.35	.35
T(noise) (kelvin)	11.1	5
$\frac{df}{dt}$ (Hz/sec)	140	140
detectable ρ_a (g/cm^3)	1.4×10^{-23}	4.8×10^{-25}

References

* The members of the collaboration are: A.C. Melissinos, S. DePanfilis, B. Moskowitz, J. Rogers, Y. Semertzidis, and W. Wuensch, University of Rochester; H. Halama and A. Prodell, Brookhaven National Laboratory; W.B. Fowler and F. Nezrick, Fermilab.

[1] P. Sikivie, Phys. Rev. Lett. 51 (1983) 1415 and 52 (1984) 695

[2] R.D. Peccei and H.R. Quinn, Phys. Rev. Lett. 38 (1977) and Phys. Rev. D 16 (1977) 1791

[3] S. Weinberg, Phys. Rev. Lett. 40 (1978) 223

[4] F. Wilczek, Phys. Rev. Lett. 40 (1978) 279

[5] J. Preskill, M.B. Wise, and F. Wilczek, Phys. Lett. 120B (1983) 127

[6] L.F. Abbott and P. Sikivie, Phys. Lett. 120B (1983) 133

[7] M. Dine and W. Fischler, Phys. Lett. 120B (1983) 137

[8] M.S. Turner, Enrico Fermi Inst. preprint no. 85-67 (1985)

[9] M. Dine, W. Fischler, and M. Srednicki, Phys. Lett. 104B (1981) 199

[10] A. Zehnder, Phys. Lett. 104B (1981) 494

[11] M. Fukugita, S. Watamura, and M. Yoshimura, Phys. Rev. Lett. 48 (1982) 1522

[12] N. Iwamoto, Phys. Rev. Lett. 53 (1984) 1198

[13] W. Wuensch, 'An Axion Search Sweep Rate Calculation' (1985) University of Rochester (unpublished)

[14] A.C. Melissinos et al., 'A Search for Galactic Axions' (1985) Brookhaven National Laboratory proposal (unpublished)

[15] L. Krauss, J. Moody, F. Wilczek, and D.E. Morris, Phys. Rev. Lett. 55 (1985) 1797

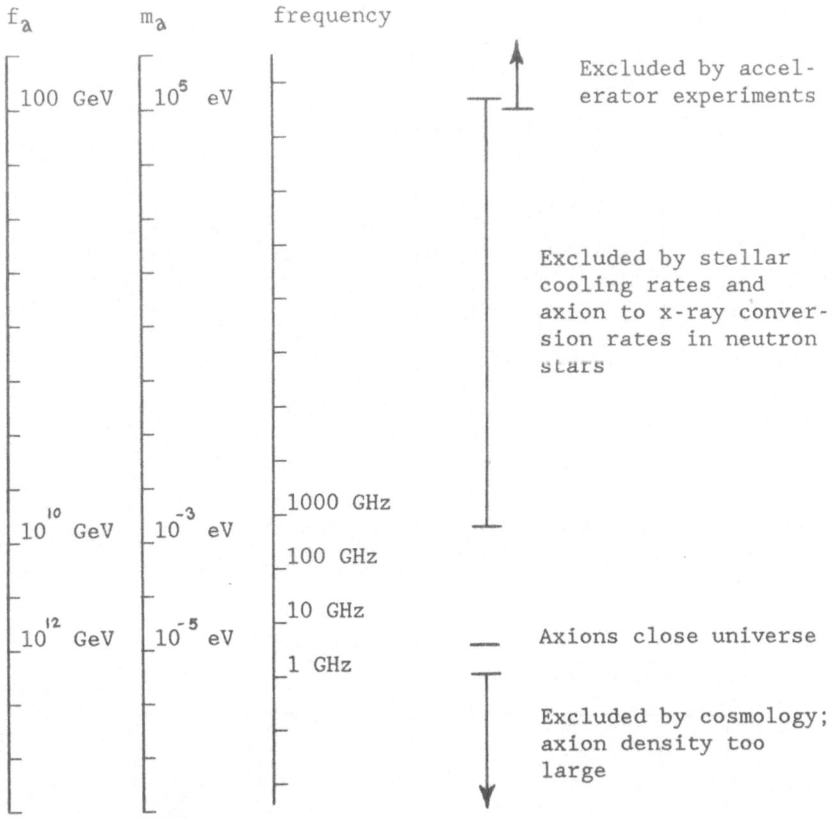

Figure 1: Values of the axion mass excluded by previous observation and analysis.

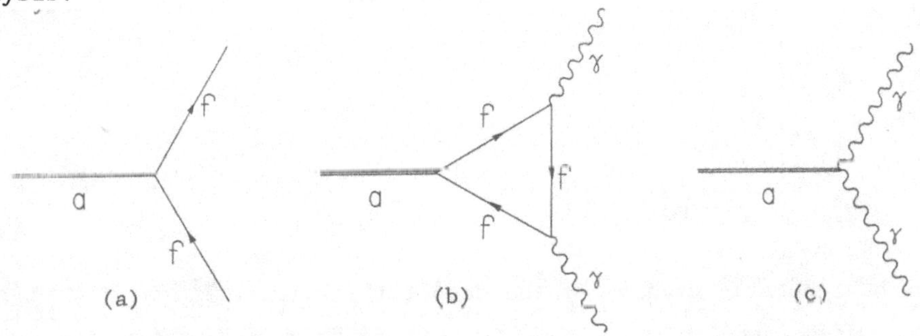

Figure 2: (a) Axion-fermion-fermion interaction; (b) Indirect coupling of axion to two photons; (c) Effective axion-photon-photon interaction.

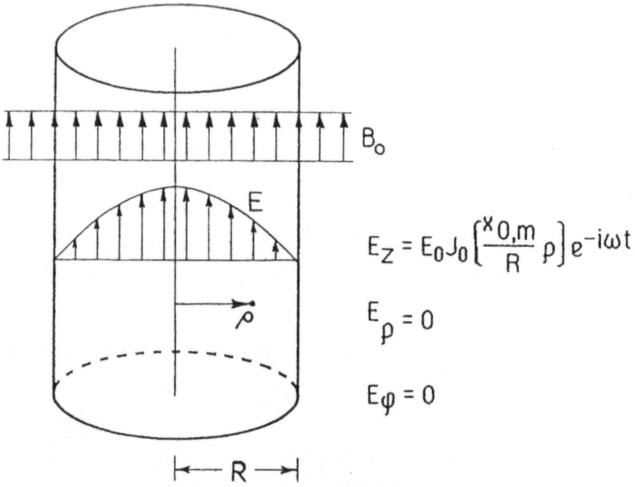

Figure 3: TM_{0m0} modes of a cylindrical cavity. Shown is the TM_{010} mode. B_0 is the static magnetic field used in the detector.

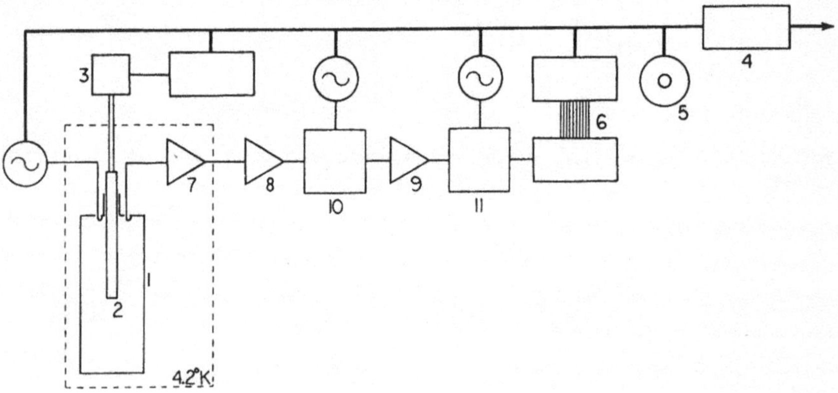

Figure 4: Schematic diagram of the detector instrumentation. (1) Microwave cavity; (2) Sapphire tuning rod; (3) Tuning rod motor and controller; (4) Microprocessor with connection to mainframe computer; (5) Disk drive; (6) 64 channel analyzer and computer interface; (7) Cryogenic preamplifier; (8) postamplifier; (9) IF amplifier; (10,11) Mixing stages with their local oscillators. The light lines represent signal paths; the dark lines are IEEE-488 digital communication lines.

R^2 Inflation

Milan Mijić
California Institute of Technology, 452–48, Pasadena, CA 91125

Abstract. I review the model of Chaotic Inflation driven by the corrections to Einstein's gravity quadratic in curvature. Initial fluctuations are discussed using particular solution of the Wheeler–De Witt equation. The coefficient of the quadratic term is restricted to be between 10^{11} and 10^{15} G.

All that I am going to talk about will be an analysis of the cosmology based on the simple Lagrangian

$$L = R + \varepsilon R^2 \tag{1}$$

Here R stands for the scalar curvature. Model is in the four dimensions, $c = \hbar = 1$, so that ε has dimension $(\text{length})^2$. A factor of $1/(16\pi G)$ should be taken to multiply whole expression when necessary.

Quite a lot of work has been done about the consequences of the higher derivative terms on the evolution of the Early Universe. Together with Mike Morris and Wai–Mo Suen we've been studying this model in the context of the Inflationary Cosmology [1]. Our approach is close in spirit to the pioneering work of Starobinsky [2], and is directly motivated by the beautiful work of Hawking and Luttrell [3].

Among the three generic types of the inflation – "old" (first order phase transitions), "new" (second order phase transitions), and chaotic, we prefer the last one for the well known reasons [4]. However, there are still three major issues which have to be understood before we can believe that Chaotic Inflation indeed occurred some time in the deep past. First, so far we have essentially only a toy model or paradigm [4,5]. The scalar field which generates inflation is introduced especially for that purpose and has nothing to do with the rest of the physics. One eventually wants inflation to occur as an unavoidable consequence in the cosmology based on some fundamental particle theory. Second, for Chaotic Inflation to take place, one needs sufficiently large fluctuations in the vacuum energy prior to the inflationary phase. How can one describe them? What is their source? What is the relevant probability distribution? In particular, is inflationary phase some typical event, or do such large fluctuations occur relatively rarely and our Universe is therefore outcome of a rare event? Finally, as in the "new" inflation, in order to have density perturbations of the proper magnitude one has to tune the available parameters – mass or self couplings of the scalar field. The complete theory should have the desired value as an input, coming from some different physics. In as much as the inflaton field is decoupled from everything else, we can say nothing about its coupling.

P. Galeotti and D. N. Schramm (eds.), Gauge Theory and the Early Universe, 301–307.

The aim of our proposal is to be a step in constructing a realistic model of Chaotic Inflation. It solves the first two problems posed above, but not yet the third. We suggest that :

(i) our Universe had in the past a Chaotic Inflationary phase based on the $R+\varepsilon R^2$ as the cosmologically relevant piece of the fundamental Lagrangian,

(ii) the initial fluctuations leading to the inflation were quantum fluctuations in the metric and the curvature, determined by the solution of the Wheeler–De Witt equation ("wave function of the Universe"[6]), and,

(iii) the parameter ε was fixed by some as yet unknown mechanism to obey $10^{11}<\varepsilon^{-1/2}<10^{13}GeV$.

This turns out to be sufficient for a consistent, successfull inflationary phase. Before I review details of the model let me comment more about the points (i) – (iii).

There are several reasons, apart from the inflation, why one may think about an R^2 term. It has been known that quantum fluctuations in curved space time induce an effective Lagrangian of the form [7,2] ,

$$\Delta L = Glog\,[\frac{\Lambda_{UV}}{\Lambda_{IR}}]R^2+\frac{R^2}{H_s^2}log\,[R/\mu] \qquad (2)$$

Here we have both IR and UV cut–offs, μ is the renormalisation mass and H_s is determined by the trace anomaly. This was the original model studied by Starobinsky. When the curvature reduces somewhat, then one has the $R+R^2$ model.

Secondly, it has been observed that gravity, with all second order invariants included, is renormalisable [8]. If the metric is restricted to be homogeneous and isotropic one has the simple Lagrangian (1). Such a theory has many nice properties [9].

Finally, it has been shown recently that the low energy limit of superstrings contains at least some quadratic terms [10]. In general, if one has in ten dimensions $(Riemann)^2+\alpha(Ricci)^2+\beta R^2$, then upon compactification to four dimensions one obtains

$$\Delta L = ((\alpha+1)/3+\beta)G\frac{V_6}{\phi}R^2 \qquad (3)$$

where V_6 is the volume of the compactified space and ϕ is the equilibrium expectation value of the dilaton field.

Thus, there are a number of reasons why we may expect R^2 to show up as we go toward higher curvatures.

Concerning point (ii) it is worth stressing that although many people believe that quantum fluctuations play a decisive role in the Very Early Universe, it is by no means established yet, that they can be described successfully in terms of the proposed wave functions [6,2]. On top of this, particular use of the wave function to be made here is even less obviously correct. We feel, however, that looking at the wave function predictions in various simple cases will help us greatly to understand better that beautiful approach.

At last, point (iii) shows that in the R^2 model we have to make one fine tunning to an accuracy of about 10^{-6}. We do not know yet what fixes $\varepsilon^{-1/2}$ to be so small in comparison to the Planck mass. Depending on what is the source of the quadratic term, the prospects to eventually fix ε are quite different. If $R+R^2$ is a fundamental theory, ε is most likely a new constant of nature. If the source is quantum fluctuations, then with the state of the art we can say nothing about the ε. But if our model is an effective theory generated by the superstrings, then the value of ε can be found once the dynamics of dilaton field and the compactification is understood. At any event, these are the issues which have to be addressed.

Now I will sketch how the model works. A detailed discussion can be found in [1]. For related and partially overlapping discussion see [11].

First we examine the classical evolution. We are interested in a regime $\varepsilon R \gg 1$, when effects of the quadratic term are important. Then, there is a simple solution,

$$H(t)=H_i-\frac{t}{36\varepsilon}, \tag{4}$$

where $H_i=H(0)$ is the (large) value of the Hubble parameter at the time when classical evolution starts to make sense. This is a superluminal and therefore inflationary expansion. It is also quasi-de Sitter as $|\dot{H}/H| \ll H$. Inflation ends when H becomes of the order of $\frac{1}{6\sqrt{6\varepsilon}}$ and oscillations in metric and curvature take over. The inflation is driven by the relaxation of the large value of initial curvature. If that initial condition is fulfilled then inflationary expansion is just a typical phase in the expansion of the Very Early Universe. The total amount of expansion is $18\varepsilon H_i^2$, and to have useful inflation we need initial value of H which is just of the order of few times $\varepsilon^{-1/2}$, which is the natural scale for the model.

There is another nice way to show existence of the inflationary phase. If one performs a conformal transformation [12]

$$\tilde{a}^2=(1+2\varepsilon R)a^2 \tag{5}$$

and defines $d\tilde{t}^2=(1+2\varepsilon R)dt^2$, (so that both the old and new line elements are in the Robertson–Walker form), then the Lagrangian (1) can be rewritten as

$$L=\tilde{R}+6\frac{\varepsilon^2}{(1+2\varepsilon R)^2}\left(-\dot{R}^2+\frac{R^2}{6\varepsilon}\right) \tag{6}$$

In this picture (we call it the conformal picture) there is Einstein's gravity built out of the conformaly transformed metric, while the physical curvature behaves as a massive scalar with peculiar nonlinear self couplings. It is clear now that R^2 inflation is just like any other – there is a potential, and there is a small parameter. Approximate classical solutions are the same as in the Chaotic Inflation driven by the massive scalar, namely,

$$\tilde{H}=(24\varepsilon)^{-1/2}, \tag{6}$$

$$R(\tilde{t})=R_i-\frac{\tilde{t}}{3\varepsilon\sqrt{6\varepsilon}}.$$

Subsequent oscillations of the geometry are found to be

$$H(t)=[\frac{3}{\omega}+\frac{3}{4}(t-t_{os})]^{-1}cos[\omega(t-t_{os})]^2 \quad (7)$$

where t_{os} is the beginning of the oscillatory phase and $\omega=(24\varepsilon)^{-1/2}$.

The time varying background leads to particle production. When the energy density of the newly produced particles exceeds the contribution of the ε terms in the equation of motion, it is the beginning of the standard Friedmann phase. One way to compute particle energy density is to evaluate perturbatively Bogoliubov coefficients between the new and the old particle modes, square them and integrate over all frequencies. Using a massless, minimally coupled scalar as a simple model for the matter fields, and assuming quick thermalisation of the particles produced, one finds that transfer of energy from geometry to the matter is rather slow process. After just a few oscillations matter can be characterized by a temperature $T_r \approx 10^{17}\sqrt{\frac{G}{\varepsilon}}GeV$, while its energy density starts to dominate at the temperature $T_F \approx 10^{17}N^{1/4}(G/\varepsilon)^{3/4}GeV$, (where N is number of particle species). While the second temperature signals the transition to the Friedmann Universe, some important processes as baryogenesis or string and monopole production take place efficiently in the range (T_r,T_F).

One can place some restrictions on ε from this, but details of these processes, due to the nonstandard space-time background, require more attention.

The major weakness of the method used in [1] is that it does not show explicitly the decay of the background solution due to particle production. Recently however, Suen developed a new method for solving the semiclassical Einstein's equations [13], and applied it to the R^2 Inflation [14]. The numerical solution shows beautifully the back reaction : there is a clear exponential decay of the oscillating background down to the radiation dominated expansion. Values of the characteristic temperatures T_r and T_F are as estimated above.

Now let me discuss the most characteristic feature of the Inflationary Cosmology – conversion of the quantum fluctuations into classical perturbations which lead to galaxy formation and (still unseen) anisotropy of the microwave background.

Recall that in all inflationary models that are considered so far, the amplitude of gravitational waves of a given wavelength is proportional to the Hubble parameter at the time when the wavelength leaves horizon. That in turn implies an upper bound of $\approx 10^{17}GeV$ for the Hubble parameter during inflation.

In the R^2 model, however, graviton has an additional coupling to the background metric given by,

$$L_{gw}=(1+2\varepsilon R)\partial h \partial h, \quad (7)$$

which changes the amplitude of the perturbations to

$$A_k=\sqrt{G/2}\frac{H(t_{hc})}{\sqrt{1+2\varepsilon R}}\approx\sqrt{\frac{G}{48\varepsilon}}. \quad (8)$$

The spectrum is scale invariant, and there is no bound on Hubble parameter, just on ε!

Another way to understand this result is to notice that conformal transformations affect backgrounds, but leave perturbations unchanged. Then one has simply $A_k = \tilde{A}_k = (G/2)^{1/2}\tilde{H} = (G/48\varepsilon)^{1/2}$, as before! Using the present bound on microwave anisotropy $\Delta T/T < 7X10^{-4}$ [15], it follows that $\varepsilon^{-1/2} < 10^{-3} m_{pl}$.

However, scalar perturbations provide a much stronger constraint on ε. To deal with them we shall use the conformal picture only. There are two major difficulties if we work with full fourth order gravity – quantization of scalar perturbations is rather difficult and a gauge invariant form of the linearized theory has not been developed yet. In the conformal picture we can use some of the standard techniques [16] built upon Bardeen's gauge invariant formalism [17], and the only difficulty comes from the nonlinearities in the "scalar field" Lagrangian for R. The result for the amplitude of the scalar perturbations is

$$A_s(k) = O(1)\sqrt{\frac{2\pi G}{\varepsilon}}\,18\varepsilon H(t_{hc}(k))^2 \tag{9}$$

This spectrum has several interesting features. It is almost scale invariant, amplitude grows slightly with the wavelength. The deviation is due to the time dependence of the Hubble parameter. Amplitude for the scalar perturbations is proportional to the amplitude for the gravitational waves times the expansion from the time of the horizon crossing until the end of the inflationary phase. Thus, scalar perturbations overpower gravitational waves by the large amount. Perturbations which are within the horizon today are generated at $H \leq 10^{-5} m_{pl}$, which is the safe domain for classical and semiclassical calculations. The bound on microwave anisotropy implies $\varepsilon^{-1/2} < 10^{13} GeV$, while the need for the galaxy formations in the adiabatic scenario requires at least $\varepsilon^{-1/2} > 10^{11} GeV$.

A parameter of the model, ε, is now constrained. What can we say about the H_i? I will report here only about the first modest step in addressing this important issue. A more complete discussion is to appear shortly [18].

We would like to consider quantum fluctuations in the metric and curvature, to see what is their typical magnitude and probability of occurrence, in order to interpret them as initial conditions for the classical evolution. For the R^2 model Wheeler–De Witt equation can be cast in the form [3],

$$[\partial^2_{yy} - \partial^2_{xx} + 4V(x,y)]\Psi(x,y) = 0$$

with,

$$x = 2a\varepsilon R \tag{10}$$

$$y = 2a(1+\varepsilon R)$$

$$V(x,y) = x^2\frac{(y-x)^2}{72\pi} - \frac{(y^2-x^2)}{4}$$

Here, a spatially closed three geometry has been considered. The space of all possible (a,R) configurations has a domain for which the embedding is euclidean and a domain for which it is lorentzian. We are interested in latter one, especially in the section where the semiclassical approximation can be applied. For this, one expects that quantum mechanical probabilities can be interpreted as a distribution of the initial conditions for the classical evolution. To evaluate Ψ one has to impose some boundary conditions. As a trial, we evaluated the wave function in WKB approximation using Vilenkin's boundary condition [2]. This picks the solution which is decreasing within euclidean domain from $a=0$ toward euclidean/lorentzian boundary, thus suggesting the interpretation that the Universe is "created by the tunneling" from the state with no classical space time and with no lorentzian signature ("nothing") [2]. Result is $\Psi(a,R)\approx a^{-2}exp[f(a,R)]$, where $f(a,R)$ is an algebraic function weakly dependent on a. This solution clearly prefers small a! One can find value of R for which the exponent is maximized and the corresponding point on the euclidean/lorentzian boundary can be taken as the most probable initial condition for the classical evolution. Thus we find

$$\begin{aligned} R_i &\geqq 4000 m_{pl}^2 \\ a_i &\geqq 0.06 l_{pl}(R_i/4000)^{-1/2} \\ H_i &\geqq 20 m_{pl}(R_i/4000)^{1/2} \end{aligned} \qquad (11)$$

As should be expected, those numbers are on the Planck scale.

Let me now summarize history of the Very Early Universe according to the R^2 model. Universe started out as nearly classical, of the finite size and with the large, but finite curvature. Initially, there was an inflationary expansion. However, it was not just a short burst. If ε is constrained to be as demanded by the (iii), inflationary expansion lasted until Hubble parameter became as low as $10^{-7}m_{pl}$. Less than hundred of the last e-foldings affected part of the Universe we see now. Perturbations generated during this phase are sufficiently weak not to be in contradiction with the present bound on the anisotropy of the microwave background, yet they were strong enough to allow for galaxy formation within adiabatic scenario. After inflationary phase metric and curvature started to oscillate, reheating the Universe to the temperature of about $10^{10}GeV$. This is a little bit low for the standard scenarios of baryogenesis, string and monopole production, however there are nonstandard scenarios too. There is also a beautiful observation [19] that phase transition might occur due to the drop in the Hawking temperature toward the end of the inflationary phase. One readily checks that remnants of the phase transition with $T_c \approx 10^{12}GeV$ might abundantly survive.

It appears therefore, that the R^2 inflation is just as good as any other model of the Chaotic Inflation. Its special virtue is that it is based on the Lagrangian we are led to think of in the context of some of the most recent and most exciting developments in the High Energy Physics and Quantum Gravity.

Acknowledgments. Work reviewed here is done in collaboration with Mike Morris and Wai-Mo Suen. I am indebted to Shahram Hamidi and Brian Warr for their kind help in editing this work. Most of all, I would like to thank John Preskill for his constant encouragement and support, and P.Galleoti and D.Schramm for organizing this most interesting school and making possible for all of us to have such a wonderful time in Erice.

References:

[1] M.Mijic, M.Morris and W.M.Suen, *Phys. Rev. D* **34**, (Oct 15) (1986).

[2] A.A.Starobinsky, *Phys. Lett.* **91B**,99,(1980); an excellent review of the Starobinsky model with many novel ideas is given in A.Vilenkin, *Phys. Rev. D* **32**,2511,(1985).

[3] S.W.Hawking and J.C.Luttrell, *Nucl. Phys. B* **247**,250,(1984).

[4] A.D.Linde, *Rep. Prog. Phys.* **47**,925,(1984).

[5] M.S.Turner, this volume.

[6] S.W.Hawking, Nucl. Phys. *B* **239**,257,(1984).

[7] N.Birrell and P.Davies, *Quantum Fields in Curved Space*, Cambridge Univ. Press, (1982).

[8] K.S.Stelle, *Phys. Rev. D* **16**,953,(1977).

[9] N.Barth and S.Christensen, *Phys. Rev. D* **28**,1866,(1985), and references there in.

[10] D.J.Gross at all., *Phys. Rev. Lett.* **54**,502,(1985).

[11] A.A.Starobinsky, **9**,579,(1983), *Pis'ma Astr. Zh.*, **9**,579,(1983); L.Kofman, A.Linde and A.A.Starobinsky *Phys. Lett.* **157***B*,361,(1985).

[12] B.Whitt, *Phys. Lett.* **145***B*,176,(1984).

[13] W.M.Suen, *Phys. Rev. Lett.* (submitted).

[14] W.M.Suen and P.Anderson, University of Florida preprint, (September, 1986).

[15] V.A.Rubakov, M.V.Sazhin and A.V.Veryaskin, *Phys. Lett.* **115***B*,189,(1982).

[16] A.Guth and S.Y.Pi, *Phys. Rev. Lett.* **49**,1110,(1982); J.M.Bardeen, P.J.Steinhardt and M.S.Turner, *Phys. Rev. D* **28**,679,(1983); R.Brandenberger and R.Kahn, *Phys. Rev. D* **29**,2172,(1984).

[17] J.M.Bardeen, *Phys. Rev. D* **22**,1882,(1980).

[18] M.Mijic, M.Morris and W.M.Suen, to appear.

[19] Q.Shafi and A.Vilenkin, *D* **29**,1870,(1984). *Phys. Rev. D* **29**,1870,(1984).

ON THE EFFECTIVE EVOLUTION EQUATION OF THE SCALAR FIELD IN THE NEW INFLATIONARY UNIVERSE

A. Ringwald
Institute for Theoretical Physics
University of Heidelberg
Philosophenweg 16
6900 Heidelberg
FRG

ABSTRACT. The one loop effective evolution equation of the scalar field driving inflation in the new inflationary universe scenario is derived using a perturbative calculation scheme proposed recently by M. Morikawa and M. Sasaki. The expectation value $\overline{\Phi}$ of the scalar field is allowed to be dependent on time. We show that for an adiabatic evolving meanfield $\overline{\Phi}(t)$ there arises the derivative of the effective potential as well as a local damping term proportional to $\dot{\overline{\Phi}}$, which has its origin in the fact that particles which will be created by the time dependence of the meanfield will decay subsequently into lighter particles.

1. INTRODUCTION

As discussed by Mike Turner in his lectures the new inflationary universe scenario /1-3/ provides a reasonable explanation of several outstanding cosmological problems /1/. Introducing a period of exponential growth of the cosmic scale factor it explains the large scale homogeneity and isotropy of the universe, and it also gives a reason why the quantity Ω (the ratio of the actual mass density to the critical mass density) is in the range of the unstable value 1. Other advantages of inflationary models include a possible explanation of fluctuations that give rise to galaxies, a dilution of magnetic monopoles and a production mechanism for essentially all the matter, energy and entropy in the observed universe.

In this talk we will be concerned with the time evolution of the scalar field which drives inflation in the new inflationary universe. There are two key ingredients in the effective evolution equation of $\overline{\Phi}$, the expectation value of the field operator ϕ , which is given by (see e.g. /4/):

P. Galeotti and D. N. Schramm (eds.), Gauge Theory and the Early Universe, 309–318.

$$\ddot{\Phi} + 3H\dot{\Phi} + \Gamma_D\dot{\Phi} + V'(\Phi) = 0 \qquad (1.1)$$

where a dot denotes d/dt and a prime $d/d\Phi$. One is the spontaneous symmetry breaking (SSB) scalar potential $V(\Phi)$ which is flat near $\Phi=0$ in order that the meanfield Φ will slowly evolve from its high temperature ground state $\Phi = 0$ to the vicinity of the zero temperature SSB ground state $\Phi = \sigma$. During this 'slow rollover' the energy density of the universe will be dominated by $V(\Phi)$ which will act like a cosmological constant in the Einstein equations leading to an exponential growth of the scale factor /2,3/, provided that $V(\sigma) = 0$, which leads to $V(0) > 0$.

The original proposal for the potential /2/ was a Coleman Weinberg one loop effective potential /5/ which is given at zero temperature and in Minkowski space by (see Fig. 1):

$$V^{(1)}_{eff}(\Phi) = B\cdot\Phi^4\left(\ln\frac{\Phi^2}{\sigma^2} - \frac{1}{2}\right) + \frac{B}{2}\sigma^4 \qquad (1.2)$$

where B is a dimensionless number depending on the coupling parameters of the specific model, and the last term ensures $V^{(1)}_{eff}(\sigma) = 0$. Coleman Weinberg (CW) potentials arise in theories, which are massless at the tree level, by taking into account radiative corrections around a constant meanfield .

After the meanfield $\Phi(t)$ has reached the region near the SSB minimum σ it will start to oscillate around σ. One expects that these oscillations will be damped by the expansion of the universe (second term in (1.1), where $H(t)$ is the expansion rate) and particle production due to the time dependence of Φ (third term in (1.1)).

The production of particles is essential because this will lead to the reheating and the entropy generation at the end of the inflationary epoch. Both, inflation and reheating, will solve together the homogeneity and flatness problems /1/.

The third term in (1.1) was simply put in by hand (see e.g. /4/). Γ_D is the total decay rate of the quasi-particle $\bar{\varphi} := \Phi - \sigma$ in its rest frame with $\text{mass}^2 = V''(\sigma)$. This can be understood in a heuristic picture if one realizes that oscillations of Φ around σ are essentially equivalent to a coherent state of the corresponding $\bar{\varphi}$ -particle with zero momentum /6/. These $\bar{\varphi}$ -particles decay into lighter particles and yield the damping term in (1.1).

This heuristic approach however has the drawback that it is not clear how (if at all) the one loop effective potential enters the effective evolution equation. The effective potential in one loop approximation is obtained by taking into account Gaussian fluctuations around a constant meanfield Φ, but we are dealing with a time-dependent mean-

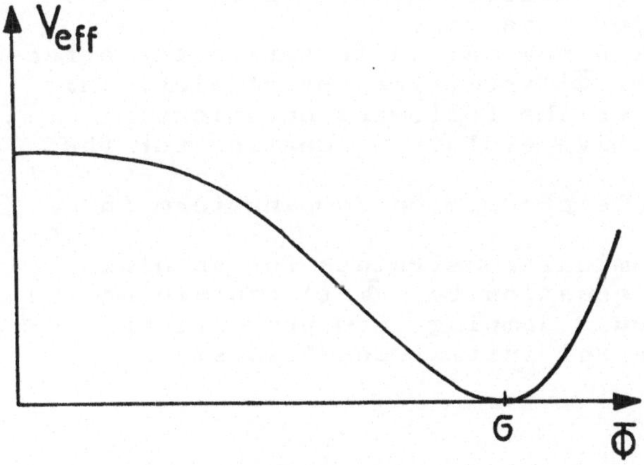

Fig.1: The Coleman Weinberg one loop effective potential

field. The other disadvantage is that one cannot decide if there is a particle production damping term far away from the minimum of the potential, because the heuristic picture yielding the damping term in (1.1) can only be applied to oscillations around σ , because only in this case the quasi-particle $\bar{\varphi}$ has a positive mass.

We therefore present a new method to derive the effective evolution equation of $\bar{\Phi}$ from first principles. This method allows one to answer the following questions:

1) What is the effective evolution equation for the meanfield ?

2) Is there a particle production damping term far away from the minimum?

3) What are the dynamical constraints for obtaining an ordinary differential equation for $\bar{\Phi}(t)$ containing the CW effective potential and a damping term proportional to $\dot{\bar{\Phi}}$?

4) What are the relevant initial conditions?

2. THE MODEL

We will work in Minkowski space because the basic mechanism is the same as in a curved space-time.

Let's start with the massless ϕ-theory given by the Lagrangian

$$\mathcal{L} = \frac{1}{2}(\partial\phi)^2 - \frac{\lambda}{4!}\phi^4 \quad . \tag{2.1}$$

Representing the Heisenberg operator $\phi(x)$ as

$$\begin{aligned} \phi(x) &= \bar{\Phi}(x) + \varphi(x) \\ \bar{\Phi}(x) &:= \langle|\phi(x)|\rangle \equiv \langle\phi(x)\rangle \end{aligned} \tag{2.2}$$

where the c-number part $\bar{\Phi}(x)$ is the expectation value of $\phi(x)$ in a state $|\rangle$ which will be specified later, one obtains the effective evolution equation for $\bar{\Phi}(x)$ in one loop approximation by taking the expectation value of the field equation following from (2.1):

$$\Box\bar{\Phi} + \frac{\delta m^2}{2}\bar{\Phi} + \frac{\lambda + \delta\lambda}{3!}\bar{\Phi}^3 + \frac{\lambda}{2}\langle\varphi^2\rangle_{ren}\bar{\Phi} = 0 \tag{2.3}$$

where λ is the renormalized coupling constant and δm^2 and $\delta\lambda$ are renormalization constants of order $\hbar$.

The operator φ satisfies in one loop approximation

$$\left(\Box_x + \frac{\lambda}{2}\bar{\Phi}^2(x)\right)\varphi(x) = 0 \quad . \tag{2.4}$$

Now let's assume that the state $|\rangle$ is homogeneous in space. In this case the meanfield $\bar{\Phi}$ will be a function of time only, i.e. $\bar{\Phi} = \bar{\Phi}(t)$. From (2.4) we see that the quantum fluctuations will have a time dependent mass. Due to this time dependence the definition of positive (negative) frequency modes of φ changes from instant to instant, i.e. even if one starts at a given time t_o with the vacuum state, this will become a many particle state for $t > t_o$. Therefore we expect that the time dependence of $\bar{\Phi}$ will lead to production of φ-particles. The decay of the produced φ-particles into lighter particles then should damp the evolution of $\bar{\Phi}(t)$. This idea, which is due to M. Morikawa and M. Sasaki /7/, is similar to the particle creation by a time varying background geometry /8/.

To evaluate $\langle\varphi^2\rangle$ we make use of the time-dependent perturbation theory. The Lagrangian corresponding to (2.4) reads:

$$\mathcal{L}(\varphi,t) = \frac{1}{2}(\partial\varphi)^2 - \frac{\lambda}{4}\bar{\Phi}^2(t)\cdot\varphi^2 \tag{2.5}$$

We split $\mathcal{L}(\varphi,t)$ in a part which depends on the mean-field at a certain time, $\bar{\Phi}(t_o) \equiv \bar{\Phi}_o$ and a remaining rest:

$$\begin{aligned}\mathcal{L}(\varphi,t) &= \mathcal{L}_o(\varphi) + \mathcal{L}_1(\varphi,t)\\ \mathcal{L}_o(\varphi) &:= \frac{1}{2}(\partial\varphi)^2 - \frac{\lambda}{4}\bar{\Phi}_o^2\cdot\varphi^2\\ \mathcal{L}_1(\varphi,t) &:= -\frac{\lambda}{4}\left(\bar{\Phi}^2(t) - \bar{\Phi}_o^2\right)\cdot\varphi^2 \quad .\end{aligned} \tag{2.6}$$

To $\mathcal{L}_o(\varphi)$ we add $\mathcal{L}_{int}$, representing couplings between φ and other particles:

$$\begin{aligned}\mathcal{L}_o^{tot} &:= \mathcal{L}_o(\varphi) + \mathcal{L}_{int}\\ \mathcal{L}^{tot} &:= \mathcal{L}(\varphi,t) + \mathcal{L}_{int} \quad .\end{aligned} \tag{2.7}$$

To be consistent one should of course take into account $\mathcal{L}_{int}$ from the beginning, in the effective evolution equation as well as in (2.6). This extension is straightforward, but let's for simplicity restrict to the above mentioned case.

Now we pass from the Heisenberg picture to the Dirac picture

$$\varphi(t) = U^+(t,t_o)\cdot\hat{\varphi}(t)\cdot U(t,t_o) \tag{2.8}$$

where the space dependence has been suppressed and the time evolution operator $U(t,t_o)$ reads:

$$U(t,t_o) = T\exp\left\{i\int_{t_o}^{t}dt'\int_{-\infty}^{+\infty}d^3x'\,\mathcal{L}_1(\hat{\varphi}(x'),t')\right\} \quad . \tag{2.9}$$

The Dirac operator $\hat{\varphi}$ evolves in time according to

$$\hat{\varphi}(t) = \exp\left\{i(t-t_o)H_o^{tot}\right\}\hat{\varphi}(t_o)\cdot\exp\left\{-i(t-t_o)H_o^{tot}\right\} \tag{2.10}$$

where H_0^{tot} is the Hamiltonian corresponding to $\mathcal{L}_0^{tot}$.

Now we are ready to calculate $\langle\varphi^2\rangle$ perturbatively in $\mathcal{L}_1(\hat{\varphi},t)$:

$$\langle\varphi^2(t)\rangle = \langle\hat{\varphi}^2\rangle + \\ + i\frac{\lambda}{4}\int_{t_0}^{t}dt'\int d^3x'\left(\Phi^2(t')-\Phi_0^2\right)\langle[\hat{\varphi}^2(x'),\hat{\varphi}^2(x)]\rangle \tag{2.11}$$

We have to specify the state $|\rangle$, with respect to which we build the expectation values. We choose it to be the vacuum at time t_0 of the quantum fluctuations φ, i.e.

$$H_0^{tot}|0\rangle = 0 \quad . \tag{2.12}$$

Expectation values with respect to this state we denote by $\langle\ldots\rangle_0$.

Using spatial translation invariance, the fact that the dynamics of $\hat{\varphi}$ is governed by H_0^{tot} and (2.12) it can be shown that for $t \geq t'$ in lowest order in λ:

$$\langle[\hat{\varphi}^2(x),\hat{\varphi}^2(x')]\rangle_0 \simeq 4\cdot i\cdot \mathrm{Im}\, G_c^{(2)}(x,x')^2 \tag{2.13}$$

where $G_c^{(2)}$ is the full connected Green function of $\hat{\varphi}$.

Inserting (2.13) in (2.11) we obtain

$$\langle\varphi^2\rangle_0 = \langle\hat{\varphi}^2\rangle_0 + \\ +\lambda\left(\Phi^2(t)-\Phi_0^2\right)\int_{t_0}^{t}dt'\int d^3x'\ \mathrm{Im}\, G_c^{(2)}(x,x')^2 \\ +\lambda\int_{t_0}^{t}dt'\int d^3x'\left(\Phi^2(t')-\Phi^2(t)\right)\ \mathrm{Im}\, G_c^{(2)}(x,x')^2 \tag{2.14}$$

The first two terms in (2.14) will lead to the derivative of the effective potential in the effective evolution equation whereas the last term will become the damping term. This term is the only one which has been calculated by Morikawa and Sasaki /7/.

The first two terms in (2.14) diverge and must be regularized. Using dimensional regularization one obtains for the sum of the first two terms in (2.14) in one loop approximation (i.e. neglecting $\mathcal{L}_{int}$) after dropping the pole term (minimal subtraction):

$$\frac{1}{(4\pi)^2}\frac{\lambda}{2}\,\Phi^2(t)\,\ln\frac{\Phi_0^2}{2\mu^2} \tag{2.15}$$

where the mass parameter μ has to be removed by imposing renormalization conditions.

After insertion of this expression into the effective evolution equation one realizes that this looks like the one loop contribution to the derivative of the CW effective potential; the only difference is the fact that the logarithm in (2.15) contains Φ_o instead of $\Phi(t)$. We will argue now under what conditions Φ_o in the logarithm can be replaced by $\Phi(t)$.

The time t_o is distinguished from later times $t > t_o$ only by the fact that the state $|0\rangle$, with respect to which the expectation values are built, is the vacuum at time t_o of the φ-sector. We therefore expect: if the state $|0\rangle$ remains the vacuum state with respect to the φ-sector even for $t > t_o$, then, concerning to the φ-sector and the calculation of the first two terms in (2.14) in one loop approximation, no time $t > t_o$ is distinguished from the other and one can replace Φ_o by $\Phi(t)$.

What is the corresponding physical situation? The explicit time dependence of the mass term in the φ-Lagrangian leads to φ-particle creation, which decay according to the couplings in $\mathcal{L}_{int}$ into lighter particles. We therefore expect that the vacuum at the time t_o will be the vacuum with respect to the φ-sector alone for all times $t \geq t_o$, if the characteristic time scale for creation of φ-particles, $|\Phi/\dot{\Phi}|$, is large in comparison with the characteristic time scale for decay of φ-particles, τ_φ, i.e.

$$|\Phi/\dot{\Phi}| \gg \tau_\varphi \tag{2.16}$$

i.e. if the meanfield is slowly varying (adiabatic). In this case we actually get the derivative of the CW effective potential in the effective evolution equation after summing the appropriate terms in (2.3) and fixing the renormalization constants by imposing Coleman Weinberg renormalization conditions /5/:

$$V_{eff}^{(1)\,'}(\Phi) = \frac{\lambda}{3!}\Phi^3(t) + \frac{\lambda^2}{64\pi^2}\Phi^2(t)\left(\ln\frac{\Phi^2(t)}{M^2} - \frac{11}{3}\right) \tag{2.17}$$

Surely, the SSB minimum of (2.17) lies outside the range of validity of the one loop approximation. But in more realistic theories, taking into account the interactions of Φ with other particles from the beginning, one gets a CW effective potential which is valid in the vicinity of the induced minimum.

The third term in (2.14) gives rise to dissipation. If the meanfield is slowly varying according to (2.16), the following factor in the integral can be expanded

$$\Phi^2(t') - \Phi^2(t) \simeq 2\cdot\Phi(t)\cdot\dot{\Phi}(t)\cdot(t'-t) \tag{2.18}$$

and one obtains /7/ for the third term in (2.14):

$$- \frac{\lambda}{128\pi} \Phi(t)\, \dot{\Phi}(t)\, \frac{\mathrm{Im}\Sigma(m^2(t))}{m^3(t)} \tag{2.19}$$

where

$$m^2(t) := \frac{\lambda}{2} \Phi^2(t) \tag{2.20}$$

and Σ is the self-energy with respect to $\mathcal{L}_{int}$. Remember that the total decay rate of a φ-particle of mass m in its rest frame and the self-energy are related by

$$\Gamma_\varphi(m) = - \frac{\mathrm{Im}\Sigma(m^2)}{m} \quad . \tag{2.21}$$

If φ decays into much lighter particles then the total decay rate will be given by (on dimensional grounds):

$$\Gamma_\varphi(m) = \tau_\varphi^{-1} \simeq \gamma^2 \cdot m = \gamma^2 \left(\frac{\lambda}{2}\right)^{1/2} \cdot \Phi \tag{2.22}$$

where γ is a dimensionless coupling constant to light particles. So finally we obtain for the third term in (2.14):

$$- \frac{1}{64\pi} \gamma^2 \left(\frac{\lambda}{2}\right)^{1/2} \dot{\Phi}(t) \quad . \tag{2.23}$$

This expression will be multiplied by $(\lambda/2)\Phi$ in the effective evolution equation (2.3).

We conclude: for a slowly varying meanfield

$$\left|\dot{\Phi}/\Phi\right| \ll \Gamma_\varphi \simeq \gamma^2 \left(\frac{\lambda}{2}\right)^{1/2} \Phi \tag{2.24}$$

the backreaction of the quantum fluctuations leads to the derivative of the CW effective potential as well as to a damping term in the effective evolution equation:

$$\ddot{\Phi} + \frac{\gamma^2 (\lambda/2)^{3/2}}{64\pi} \Phi \dot{\Phi} + V_{eff}^{(1)\prime}(\Phi) = 0 \tag{2.25}$$

where $V_{eff}^{(1)}$ is the CW effective potential in one loop approximation, given in our case by (2.17).

As a first approximation for a more detailed calculation in a Robertson Walker background geometry /9/ one has to replace $\ddot{\Phi}$ by $\ddot{\Phi} + 3H\dot{\Phi}$.

In cases in which Φ doesn't evolve slowly according to (2.24) one has to return to (2.14). In this cases it is not possible to evaluate the time integration in the third term in (2.14). After inserting (2.14) in the effective evolution equation (2.3) and renormalization one is left with a nonlocal integro-differential equation for $\Phi(t)$ - the integral in (2.14) contains the complete history of Φ from t_0 to t.

3. CONCLUSIONS

In this talk I presented a method to evaluate the one loop effective evolution equation of the scalar field in the new inflationary universe. In particular I concentrated upon the process of particle production and subsequent decay and the occurence of the derivative of the effective potential in the evolution equation of the meanfield Φ in field theories of CW type.

The effective evolution equation is given by

$$\ddot{\Phi} + 3 \cdot H \cdot \dot{\Phi} + C \cdot \Phi \cdot \dot{\Phi} + V_{eff}^{(1)\prime}(\Phi) = 0 \qquad (3.1)$$

where $V_{eff}^{(1)}$ is the CW one loop effective potential and $C \ll 1$ is a model dependent product of couplings, if following conditions hold:

a) Initial conditions:

The initial state has to be

- spatially homogeneous,
- the vacuum state with respect to all physical particles
- "localized" at $\Phi(t_0) = \Phi_0 \neq 0$ at some time t_0.

b) Dynamical constraint:

$\Phi(t)$ must vary slowly according to

$$\left| \dot{\Phi} / \Phi \right| \ll C \cdot \Phi \quad . \qquad (3.2)$$

It is plausible that the above mentioned initial conditions hold after a few e-folds in the inflationary epoch, after all physical particles have been redshifted away.

Concerning the dynamical constraint some remarks are in order. During the slow rollover the dynamical constraint will be satisfied. So we can trust (3.1) during the slow rollover phase and conclude that there will be a particle production damping term proportional to $\dot{\Phi}$ even far away from the SSB minimum σ and (3.1) describes consistently the evolution of Φ. But during the oscillations around σ :

$$\left| \dot{\Phi} / \Phi \right| \simeq \left(V_{eff}''(\sigma) \right)^{1/2} \simeq \sigma \quad . \qquad (3.3)$$

In this case the dynamical constraint (3.2) requires $C \gg 1$, which on the other hand means that perturbation theory breaks down! So we conclude that during the oscillation phase one cannot trust (3.1) and has to treat the full, nonlocal integro-differential equation for Φ .

To improve these calculations one should evaluate $\langle \varphi^2 \rangle_0$ also in curved space-time /9/. Another improvement would be to extend the calculations to non-zero temperature.

4. ACKNOWLEDGMENTS

I would like to acknowledge various instructive discussions with my supervisor M.G. Schmidt and J. Fuchs.

5. REFERENCES

/1/ A. Guth: Phys. Rev. D **23**, 347 (1981)

/2/ A. Linde: Phys. Lett. **108** B, 389 (1982)
A. Albrecht, P. Steinhardt: Phys. Rev. Lett. **48**, 1220 (1982)

/3/ A. Linde: Rep. Prog. Phys. **47**, 925 (1984)
R. Brandenberger: Rev. Mod. Phys. **57**, 1 (1985)

/4/ P. Steinhardt, M. Turner: Phys. Rev. D **10**, 2162 (1984)

/5/ S. Coleman, E. Weinberg: Phys. Rev. D **7**, 1888 (1973)

/6/ L. Abbott, E. Farhi, M. Wise: Phys. Lett. **117** B, 29 (1982)

/7/ M. Morikawa, M. Sasaki: Prog. Theor. Phys. **72**, 782 (1984)

/8/ L. Parker: Phys. Rev. **183**, 1057 (1969)

/9/ A. Ringwald: Heidelberg Preprint, in preparation

Constraints on the Geometry of the 5th dimension in cosmological solutions of Five dimensional Relativity in vacuum.

Paulo Macedo *

Dipartimento di Fisica Teorica - " Amedeo Avogadro "
Università di Torino
Cso. Massimo d'Azeglio ,46 - 10125 Torino - Italy

Keywords: Cosmology / Kaluza - Klein / Extra dimensions .

ABSTRACT.

Two vacuum solutions of 5 dimensional vacuum Einstein's Equations are obtained assuming only homogeneity and isotropy in the usual 3 space dimensions and general dependence of the metric on time t ,as well as on the 5th coordinate Ψ.

Such solutions are shown to exibit always a killing vector field in the t,Ψ submanifold as a consequence of the field equations. This justifies the usual assumption that the geometry of the 5th coordinate is a circle. It is shown they are equivalent to the Lorenz - Petzold type (ix) solutions.

* On leave from the Fac. Science of the University of Coimbra.

P. Galeotti and D. N. Schramm (eds.), Gauge Theory and the Early Universe, 319–326.

Introduction.

The original idea that the Universe could be more than 4-dimensional was due to Kaluza[1] and Klein[2] who tried to unify Gravity and Electromagnetism assuming the existence of a fifth dimension.

More recently the idea of the existence of extra dimensions became popular after it was realised that larger than 4-dimensional space could provide the right arena to develop supergravity theories[3].

The search for cosmological solutions of the n-dimensional Einstein's equations was to our knowlege pioneered by Chodos and Detweiler[4] (1980); They studied the behaviour of anisotropic Kasner type solutions in vacuum in the 5-dimensional case , reaching the conclusion that 3 of the 4 space dimensions expanded whereas 1 of them shrinked .

Bailin, Love and Vayonakis[5] (1984) , Barrow and Stein-Schabes[6] (1986), and Bose[7] (1985) found solutions with a Robertson Walker type of behaviour in the 3 usual space dimensions and a static behaviour in the other n-4 extra ones .

Lorenz - Petzold[8] (1984) found exact solutions in 5 dimensional vacuum and dust cases which also present this behaviour in the assimptotic region (for large time values).

All the above mentioned authors have implicitely assumed that the geometry of the extra dimensions is a sphere .

To our knowlege ,there has been no previous study of the local behaviour of these dimensions.

The scope of the present work is to do that for the simple case of the 5-dimensional Kaluza-Klein model in vacuum.

The formalism.

In this work , we study how the radius of the fifth dimension G , as well as the scale factor a , will vary locally with the fifth coordinate Ψ and with time t.

In order to achieve that , one assumes that the 5-dimensional Kaluza-Klein manifold has a metric of the type:

$$ds^2 = a^2[\, d\chi^2 + \sin^2\chi \,(\, d\theta^2 + \sin^2\theta \, d\phi^2)] + G^2 \, d\psi^2 - dt^2 \qquad (1)$$

where χ, θ, ϕ are the usual spherical coordinates in 3 space, ψ is the 5th coordinate, t is time .

We also assume the dependences $a = a(t,\psi)$ and $G = G(t,\psi)$.

We shall also use the following conventions:

$$\dot{a} = \frac{\partial a}{\partial t} \quad ; \quad \dot{G} = \frac{\partial G}{\partial t}$$

and

$$a' = \frac{\partial a}{\partial \psi} \quad ; \quad G' = \frac{\partial G}{\partial \psi} \qquad (2)$$

Proceding in a straitforward way ,from the metric one can calculate the Christoffel Connection $\Gamma^{A}{}_{BC}$ .The result is

$$\Gamma^{0}{}_{11} = \frac{\dot{a}}{a}\, g_{11} \qquad \Gamma^{0}{}_{44} = \frac{\dot{b}}{b}\, g_{44}$$

$$\Gamma^{i}{}_{10} = \frac{\dot{a}}{a} \qquad \Gamma^{i}{}_{14} = \frac{a'}{a} \tag{3}$$

$$\Gamma^{4}{}_{40} = \frac{\dot{b}}{b} \qquad \Gamma^{4}{}_{44} = \frac{b'}{b} \qquad \Gamma^{4}{}_{11} = -\frac{a'}{ab^{2}}\, g_{11}$$

$\Gamma^{i}{}_{jk} = \Gamma^{i}{}_{jk\,(R-W)}$ (Cristofell simbols for the 3-sphere in spherical coordinates)

$i,\ j,\ k = 1,\ldots,3$

calculating the Ricci tensor and substituting in Einstein's vacuum equations

$$R_{AB} = 0 \tag{4}$$

One obtains the following field equations:

$$\dot{G}\, a' - G\, \dot{a}' = 0 \tag{5}$$

$$G\, \ddot{a} - \frac{G'}{G^{2}}\, a' + \frac{a''}{G} - \ddot{G} a = 0 \tag{6}$$

$$a\, \ddot{G} + 3\, G\, \ddot{a} = 0 \tag{7}$$

$$\frac{(a')^{2}}{G^{2}} - \dot{a}^{2} - 1 - a\, \ddot{a} = 0 \tag{8}$$

Two cases must be considered:

a) $a'=0$ and $G'=0$

In this case one obtains the results of Lorenz-Petzold.

b) $a'=0$ and $G=0$

This case will be shown to be equivalent to the previous one since it can always reduce to it by coordinate changes.To do that one starts by dividing eq. (5) by $a'G$ .Integrating the resulting equation one obtains :

$$a' = G\, f(\Psi) \tag{9}$$

using (9) in (8) one obtains a 2nd order diferential equation in a :

$$[f(\Psi)]^2 - \dot{a}^2 - 1 + a\ddot{a} = 0 \tag{10}$$

integrating with respect to t, one gets:

$$a^2 = K_1 t^2 + K_2 t + K_3 \tag{11}$$

where K_1, K_2 and K_3 are functions only of Ψ , and K_1 is related with $f(\Psi)$ by:

$$K_1 = f^2/2 - 1 \tag{12}$$

substituting f in (9) one obtains:

$$G = \frac{1}{2\sqrt{K_1 + 1}} \cdot \frac{\partial a}{\partial \Psi} \tag{13}$$

In order to solve equation (7) ,one can use (11) and (13) ,obtaining:

$$2K_1 \frac{\partial K_3}{\partial \Psi} - K_2 \frac{\partial K_2}{\partial \Psi} + 2K_3 \frac{\partial K_1}{\partial \Psi} = 0 \tag{14}$$

which integrated becomes :

$$4K_1K_3 - (K_2)^2 = \text{const.} = c_o \tag{15}$$

Equation (6) is not independent from the others since solving it using (14) gives an identity $0 \equiv 0$.

The solution becomes therefore:

$$ds^2 = \left(k_1 t^2 + k_2 t + \frac{k_2{}^2 - c_o}{4k_1}\right)(d\chi^2 + \sin^2\chi\, d\theta^2 + \sin^2\chi \sin^2\theta\, d\phi^2) +$$

$$+ \frac{\{k_1' t^2 + k_2' t + [(k_2{}^2 - c_o)/4k_1]'\}^2}{4(k_1+1)\cdot\left(k_1 t^2 + k_2 t + \dfrac{k_2 - c_o}{4k_1}\right)} d\psi^2 - dt^2 \tag{16}$$

In order to see if it's possible to find a coordinate in which the metric will not depend on the 5th coordinate but only on t, we shall investigate if this metric admits a Killing vector field ξ of the form

$$\xi = (\xi^0, 0, 0, 0, \xi^4) \tag{17}$$

From the Killing condition

$$\pounds_\xi \, \mathbf{g} = 0 \tag{18}$$

we obtain

$$\xi_{i,j} + \xi_{j,i} - 2\, \xi_l \, \Gamma^l{}_{ij} \tag{19}$$

where

$$\xi_i = g_{ij}\, \xi^j \tag{20}$$

using the Cristofell simbols $\Gamma^A{}_{BC}$, obtained from g [eq.(3)] in (19), and making $\xi_0 = \alpha$ and $\xi_4 = \beta$, on obtains the following set of equations:

$$\dot{\alpha} = 0 \tag{21}$$

$$\alpha' - \beta' - 2\beta \frac{\dot{b}}{b} = 0 \tag{22}$$

$$\alpha \dot{a} - \beta \frac{a'}{b^2} = 0 \tag{23}$$

$$\beta' - \alpha \dot{b} b - \beta \frac{b'}{b} = 0 \tag{24}$$

As it can be easily verified, this sistem is always possible and determined and the Killing vector field is :

$$\xi = C\left(-k_1+1\ ,\ 0,\ 0,\ 0,\ \frac{1}{C}\, \frac{k_1' t^2 + k_2' t + k_3'}{\sqrt{k_1 t^2 + k_2 t + k_3}}\right) \tag{25}$$

where C is an arbitrary constant as expected.

As this point, we didn't specify the gauge choice get. We shall choose one such that the arbitrary functions $k_1 = k_1(\psi)$ and $k_2 = k_2(\psi)$ will be:

$$k_1 = c_1 = \text{const.}$$

$$k_2 = \Psi \tag{26}$$

therefore

$$k_3 = \frac{\Psi^2 - c_0}{4c_1}$$

If we make the following variable change

$$t = \frac{\sqrt{|c_0|}}{2c_1} t_1$$

$$\Psi = \sqrt{|c_0|}\,\Psi_1 \tag{27}$$

one obtains the following two solutions (corresponding to $c_2 > 0$ and $c_2 < 0$:

I)

$$ds^2 = c_1 \{[1 - (t_1 + \Psi_1)^2][d\chi^2 + \sin^2\chi\, d\theta^2 + \sin^2\chi\, \sin^2\theta\, d\phi^2] +$$

$$+ \frac{1}{1 + c_1} \frac{(t_1 + \Psi_1)^2}{1 - (t_1 + \Psi_1)^2} d\Psi_1^2\} - dt_1^2 \tag{28}$$

and

II)

$$ds^2 = c_1 \{[(t_1 + \Psi_1)^2 + 1][d\chi^2 + \sin^2\chi\, d\theta^2 + \sin^2\chi\, \sin^2\theta\, d\phi^2] +$$

$$+ \frac{1}{c_1 + 1} \frac{(t_1 + \Psi_1)^2}{(t_1 + \Psi_1)^2 + 1} d\Psi_1^2\} - dt_1^2 \tag{29}$$

At this point, looking at equations (28) and (29) one notices that:

a) Taking an observer stationary in the 5th dimension (Ψ_1 = const) the metrics seen by him are the two solutions type (ix) of Lorenz-Petzold [7] (for $\tilde{K} = 1$ and $\tilde{K} = -1$).

b) Both metrics admit a Killing vector of the form (k, 0, 0, 0, -k)

Due to this fact one can transform these solutions into diagonal metrics which components depend only on one variable u instead of two (t_1 and Ψ_1).

To ilustrate this point ,we shall take our type I metric and make the following changes of variable :

$$t_1+\Psi_1 = \tau$$

$$\Psi_1 = \psi + \tau(c_1+1) - c_1\cdot\sqrt{c_1+1}\ \text{arc th}\left(\frac{\tau}{\sqrt{c_1+1}}\right)$$

$$\tau^2 = (c_1+1) - \frac{u^2}{c_1}$$

The end result is the following corresponding metric:

$$ds^2=(c_1{}^2-u^2)(d\chi^2+ \sin^2\chi\, d\theta^2 + \sin^2\chi\, \sin^2\theta\, d\phi^2) + \frac{u^2}{(c_1{}^2-u^2)(c_1+1)}\, d\psi^2 - du^2$$

where u and are the new time and 5th coordinate respectively.

This solution has been studied in great detail by Matzner and Mezzacappa.

Conclusions

A.- One should notice that although solution I) exibits the "crack-of-doom" pathology described by Matzner and Mezzacappa[9], solution II) corresponds to a open universe ,is perfectly well behaved and could prove to be a good candidate for the universe we live in, since the radius of the 5th dimension tends assimptoticaly to a constant. This is in acordance with the fact already mentioned by several authors[10],[11] that the coupling constants did not vary within our observation limits.

This behavior does not happen in the solution mentioned by Matzner and Mezzacappa which predicts a collapse of the 5th dimension in a finite time generating a "crack of doom".

B.- As far as the geometry of the 5th dimension is concerned. one can see that in this case it can be thought as a circle since it has a constant finite radius of curvature. This is a consequence of the existence of the above mentioned Killing vector field. This justifies the use of the congruence mentioned by Matzner and Mezzacappa: in this case it corresponds to the congruence of the integral curves of the Killing vector field. This congruence is not a necessary assumption for the sake of physical understanding as sustained by these authors but a direct consequence of the field equations as we have shown.

We congecture that the existence of such Killing vector field may be connected with the homogeneity of vacuum. It is therefore an interesting open problem to study if such vector fields also appear in the presence of an inhomogeneous source. This is a problem we shall undertake for further research.

Acknowlegments

The author would like to acknowlege Prof. Tullio Regge for sugesting the problem ,enlighting discussions help and encouragement, as well as the financial support of Fundação Calouste Gulbenkian and the Italian Governmemt by providing grants to carry out this work. To Prof. M. Francaviglia, F. Caruso, I. Bediaga and M. Monteiro for helping discussions. Finaly, the University of Torino for the hospitality.

REFERENCES

1- T. Kaluza , Sitzungsber. Preuss. Akad. Wiss. Phys. Math. Kl.LIV ,966 ,(1921).

2- O. Klein , Z. Phys. 37 ,895 (1926).

3- E. Cremmer ,B. Julia and J. Sherk , Phys. Lett. 76B ,409 ,(1978).

4- A. Chodos and S. Detweiler , Phys. Rev. D 21 ,2167 ,(1980).

5- D. Bailin ,A. Love and C. Vayonakis , Phys. Lett. 142B ,344 ,(1984).

6- J. Barrow and J. Stein-Schabes , Phys. Lett. 167B ,173 ,(1986).

7- S. Bose , Phys Rev. D 31 ,1493 ,(1985).

8- D. Lorenz-Petzold ,Phys. Lett. 149B ,79 , (1984).

9- R. Matzner and A. Mezzacapa , Phys. Rev. D 32 ,3114 ,(1986).

10- F. J. Dyson , "Aspects of quantum theory" ,Eds. A. Salam and E. Wigner , Cambridge U.P. (1972).

11- H. Reeves , preprint.

RECENT DEVELOPMENTS IN QUANTUM COSMOLOGY

G.Esposito
University of Naples
Department of Physics
Mostra d'Oltremare,Padiglione 19
80125 Naples,Italy

ABSTRACT. In quantum cosmology,the functional ψ which is the probability amplitude of finding a given field configuration and a given three-metric on a compact spacelike three-surface S obeys a zero energy Schrödinger equation,the Wheeler-De Witt equation for closed universes.ψ may be given by a functional integral taken over all compact euclidean four-metrics which have a boundary at S,and over all field configurations which match the given value on S.After a re-examination of the Wheeler-De Witt equation for a closed FRW universe in which there is a massive scalar field ϕ minimally coupled to the gravitational field,I calculate the time-evolution of the cosmic scale factor and of the scalar field in the case of a $\lambda\phi^n$ theory,in order to understand when a $\lambda\phi^n$ theory is able to drive inflation.Some new topics are mentioned,too.

1. INTRODUCTION

The quantum state ψ of the universe is a functional of the three-geometry and of the matter field configuration ϕ_0 on a compact spacelike three-surface S.Hartle and Hawking(1983) have proposed that ψ is given by an euclidean functional integral taken over all field configurations which match ϕ_0 on S,and over all compact positive-definite four-metrics which induce the three-metric h_{ij} on their boundary S.

In section 2 I examine the foundations of this postulate.In section 3 I show that ψ obeys the Wheeler-De Witt equation,and I study this equation in the case of a minisuperspace model(Hawking 1983) given by a closed FRW universe in which there is a massive scalar field minimally coupled to the gravitational one and constant on the hypersurfaces of homogeneity.In section 4 I apply the formalism of sections 2 and 3 for the more general case of a $\lambda\phi^n$ theory;I calculate the time-evolution of the scalar field and of the cosmic scale factor,and I estimate the duration of the inflationary era and the value of the cosmic scale factor at the end of this era.It seems that the universe expands by a factor of order $\exp(3\ \phi_0^2/n)$ for every n,where ϕ_0 is the initial value of the scalar field.Finally,in section 5 I illustrate the advantages and the limits of the models,and some new developments.

P. Galeotti and D. N. Schramm (eds.), Gauge Theory and the Early Universe, 327–341.

2. FOUNDATIONS

In the case of quantum field theory in Minkowski space, the probability amplitude of going from a field configuration ϕ_1 at time t_1 on a spacelike surface S_1 to a field configuration ϕ_2 at time t_2 on a spacelike surface S_2 is given by:

$$<\phi_2, t_2 | \phi_1, t_1> = \int_C D[\phi] \exp(iI[\phi]) \qquad (1)$$

where $D[\phi]$ is a measure on the class C of all field configurations which match the given values on S_1 and S_2. In (1) the exponential on the right-hand side is rapidly oscillating and therefore the integral does not exist. One therefore performs a Wick rotation, by introducing the euclidean time $\tau = it$, so that:

$$<\phi_2, \tau_2 | \phi_1, \tau_1> = \int_C D[\phi] \exp(-I_E[\phi]) \qquad (2)$$

where $I_E[\phi]$ is the action of ϕ in the euclidean regime. In (2) the integrand is exponentially damped and therefore the right-hand side may be well defined.

If one tries to take into account the cases in which there is also a gravitational field, one is lead to postulate that the amplitude of going from a three-metric h_1 and a field configuration ϕ_1 on the spacelike surface S_1 to a three-metric h_2 and a field configuration ϕ_2 on the spacelike surface S_2 is given by:

$$<h_2, \phi_2 | h_1, \phi_1> = \int_C D[g_{\mu\nu}] \; D[\phi] \; \exp(-I_E[g_{\mu\nu}, \phi]) \qquad (3)$$

where C is the class of all four-metrics which induce h_1 on S_1 and h_2 on S_2 and of all matter field configurations which match the given values on S_1 and S_2. Owing to our interest for closed universes (Tipler 1986) S_1 and S_2 are assumed to be compact. Let us now suppose that S_1 shrinks to a point, and let $S_2 = S$. Then (3) becomes:

$$\int_C D[g_{\mu\nu}] \; D[\phi] \; \exp(-I_E[g_{\mu\nu}, \phi]) = \Psi(h_{ij}, \phi_o) \qquad (4)$$

which is the probability amplitude of having a given field configuration ϕ_o and a given three-metric h_{ij} on S.

We now ask: have the euclidean four-metrics which belong to the class C a peculiar property?

In the case of a cosmological constant $\Lambda > 0$, the choice of compact four-metrics is a natural one. In fact, there is a theorem due to Myers (Milnor 1962) which states that in an n-dimensional Riemannian manifold

M whose Ricci curvature is $\geqslant (n-1)/r$ (where r is a positive constant) everywhere, every geodesic whose length is $>\pi\sqrt{r}$ contains conjugate points and hence is not minimal. Furthermore, if M is assumed to be geodesically complete, the Hopf-Rinow theorem states that any two points of M can be joined by a minimal geodesic. We know also that in the geodesically complete manifolds any closed bounded set is compact. Therefore, these two theorems establish that a Riemannian geodesically complete n-dimensional manifold M whose Ricci curvature is $\geqslant (n-1)/r$ everywhere, is compact with a diameter $\leqslant\pi\sqrt{r}$.

When Λ is $\leqslant 0$ there are non compact solutions of the Einstein equations; those with greatest simmetry are flat euclidean space and euclidean anti-De Sitter space. One would have therefore to consider in the class C the following metrics (Hawking 1983, 1984):

(1) Connected asymptotically euclidean (Gibbons and Pope 1979) or anti-De Sitter (Magnon and Ashtekar 1984) metrics which have an inner boundary at S;

(2) Disconnected metrics which consist of a compact part with boundary at S and an asymptotically euclidean or anti-De Sitter metric without inner boundary.

But a calculation (Hawking 1984) shows that metrics of tipe (1) don't give the dominant contribution to the path-integral, owing to the existence of a scale transformation which contradicts a certain inequality which ought to hold. Therefore disconnected metrics give almost the same result as compact metrics, if we are concerned with observations in a finite region. But disconnected metrics seem to be suitable for scattering problems, in which one is concerned only with metrics which are connected to infinity, in order to study observables at infinity. On the other hand, in cosmology we are concerned with observables in a finite region, and we don't observe a concentration of matter in certain regions which decreases at large ditances, as it ought to occur in a scattering problem. By reason of these results we take the functional integral (4) over all compact four-metrics $g_{\mu\nu}$ which induce h_{ij} on S (Hawking 1983, 1984).

It has to be remarked that the functional ψ doesn't depend explicitly on time; this happens in closed cosmological models. For infinite worlds, on the other hand, ψ would depend also on the surrounding laboratory which determines the asymptotic coordinates (De Witt 1967). For curved spaces which are asymptotically flat this means that ψ depends explicitly on time, too (Hartle 1984).

The final test of the Hartle-Hawking's postulate is its agreement with observations (Hawking 1983, Halliwell and Hawking 1985). I now recall the calculations which show how one can interpret the functional ψ.

Let us consider the homogeneous and isotropic euclidean space with metric (Hawking 1983):

$$ds^2 = \sigma^2(d\tau^2 + H^{-2}\cos^2 H\tau\, d\Omega_3^2) = \sigma^2(d\tau^2 + a^2(\tau)\, d\Omega_3^2) \qquad (5)$$

where $d\Omega_3^2$ is the metric on a unit three-sphere, 1/H is the radius of the euclidean four-sphere, and:

$$\sigma^2 = 2G/3\pi \qquad (6)$$

The action is:

$$I = 1/2 \int d\tau \left[N/a \left(-\left(\frac{a}{N}\frac{da}{d\tau}\right)^2 - a^2 + H^2 a^4 \right) \right] \tag{7}$$

where N is the lapse function(Misner,Thorne,Wheeler 1973).Let us define:

$$h = \det h_{ij} \tag{8}$$

$$\tilde{h}_{ij} = h^{-1/3} h_{ij} \tag{9}$$

$$K = K^i{}_i = h^{ij} K_{ij} \tag{10}$$

where K_{ij} is the extrinsic curvature of S.If we define the functional (Hartle and Hawking 1983):

$$\Phi\left[\tilde{h}_{ij}, K, \phi\right] = \int_0^\infty d\sqrt{h}\, \exp\left[-1/12\pi G \int K\sqrt{h} d^3x\right] \psi(h_{ij}, \phi)$$

$$= \int_C d[g_{\mu\nu}]\, d[\phi]\, \exp(-I^k [g_{\mu\nu}, \phi]) \tag{11}$$

where I^k is the action appropriate for the case in which $\tilde{h}_{ij}$ and K are kept fixed on S(York 1986),we find that the functional ψ given by:

$$\psi[h_{ij}, \phi] = \int_\Gamma d\left[\frac{K}{24\pi i G}\right] \exp\left[1/12\pi G \int K\sqrt{h} d^3x\right] \Phi\left[\tilde{h}_{ij}, K, \phi\right] \tag{12}$$

becomes in our case :

$$\psi(a) = \frac{N_k}{2\pi i} \int_\Gamma dk\, \exp(ka^3 - I^k) \tag{12 bis}$$

where Γ runs from $-i\infty$ to $i\infty$ to the right of every singularity of Φ in the complex k plane,and where $k = \sigma K/9$ .The contour Γ ,parallel to the Im k axis,can be distorted into steepest descent contours(Hartle and Hawking 1983).If $a < 1/H$,the exponent in (12 bis) is stationary when:

$$k = H/3 \sqrt{-1 + \frac{1}{(aH)^2}} = k^* \tag{13}$$

if the part of the four-sphere bounded by a three-sphere is less than a hemisphere.Therefore:

$$\psi \approx \text{const}\, \exp(k^* a^3 - I^{k^*}) = \text{const}\, \exp(a^2/2) \tag{14}$$

if $aH \ll 1$. If $a > 1/H$, we find two complex-conjugate extrema of the exponent in (12 bis):

$$k_+^* = i/3H \sqrt{1 - (aH)^{-2}} \tag{15}$$

$$k_-^* = -i/3H \sqrt{1 - (aH)^{-2}} \tag{16}$$

from which we get:

$$\Psi \approx \text{const} \exp(1/3H^2) \cos((a^2H^2-1)^{3/2} / 3H^2 - \frac{\pi}{4}) \tag{17}$$

Therefore, if $a<1/H$ Ψ is exponentially damped, and this suggests the idea of a classically forbidden region, and if $a>1/H$ Ψ oscillates with a constant amplitude, as it occurs in an allowed region (the so-called Lorentzian region).

3. A MINISUPERSPACE MODEL

In the ADM canonical formalism for general relativity, the metric of the four-manifold M may be locally cast in the form:

$$ds^2 = -(N^2 - N_i N^i) dt^2 + 2N_i dx^i dt + h_{ij} dx^i dx^j \tag{18}$$

where N is the lapse function and N_i is the shift vector. The first-order action for gravitation (York 1986) is, by calling K_{ij} the extrinsic curvature of the three-surface S which divides M into two parts:

$$I_g = 1/16\pi G \int N\sqrt{h}(^3R + K_{ij}K^{ij} - K^2) d^4x \tag{19}$$

If we consider a massive real scalar field which is minimally coupled to the gravitational one, and if we restrict the infinite number of degrees of freedom of matter and gravitation, by requiring that:

$$ds^2 = \sigma^2(-N^2dt^2 + a^2(t)d\Omega_3^2) \tag{20}$$

$$\phi = \phi(t) \tag{21}$$

we find that:

$$I = I_g + I_m = \int (L_g + L_m)dt = -1/2 \int dtNa^3 \Big[\dot{a}^2 N^{-2} a^{-2} - -a^{-2} - \dot{\phi}^2 N^{-2} + m^2 \phi^2 \Big] \tag{22}$$

Therefore:

$$p_a = \partial L / \partial \dot{a} = -a\dot{a}N^{-1} \quad ; \quad p_\phi = \partial L / \partial \dot{\phi} = a^3 \dot{\phi} N^{-1} \tag{23}$$

$$H = p_\phi \dot{\phi} + p_a \dot{a} - L_g - L_m = N/2a^3(p_\phi^2 - a^2 p_a^2 - a^4 + m^2\phi^2 a^6) \quad (24)$$

We must now recall that in a closed universe the time t is just a parameter which may assume arbitrary values according to the choice of N and of N_i, and therefore the wave function ψ does not depend explicitly on time. In particular, by varying the lapse function N at S we push it forward and backward in time (Hawking 1983), and because ψ does not depend on t we have that the functional:

$$\psi = \int_C d[g_{\mu\nu}]\ d[\phi]\ \exp(iI[g_{\mu\nu}, \phi]) \quad (25)$$

must be left unchanged by an infinitesimal translation of N. If the measure in (25) is invariant under translation (Hartle and Hawking 1983), this leads to:

$$0 = \int_C d[g_{\mu\nu}]\ d[\phi]\ \frac{\delta I}{\delta N} \exp(iI[g_{\mu\nu}, \phi]) \quad (26)$$

Furthermore $H_o = \delta I/\delta N$, where:

$$H_o = (16\pi G)\ G_{ijkl}p^{ij}p^{kl} - \sqrt{h}\ {}^3R/16\pi G + \sqrt{h}/2(p_\phi^2\ h^{-1} + m^2\phi^2) \quad (27)$$

$$G_{ijkl} = 1/2h^{-1/2}(h_{ik}h_{jl} + h_{il}h_{jk} - h_{ij}h_{kl}) \quad (28)$$

$$p^{ij} = -\sqrt{h}/16\pi G(K^{ij} - h^{ij}K) \quad (29)$$

$$p_\phi = \sqrt{h}\ N^{-1}\ \dot{\phi} \quad (30)$$

Therefore (26) becomes:

$$0 = \int_C d[g_{\mu\nu}]\ d[\phi]\, H_o\ \exp(iI[g_{\mu\nu}, \phi]) \quad (31)$$

If we now introduce the quantum relations:

$$p^{ij} = -i\ \delta/\delta h_{ij} \qquad p_\phi = -i\ \delta/\delta\phi \quad (32)$$

and if we call $T_{\mu\nu}$ the stress-energy tensor of the field ϕ, (31) becomes:

$$\{-G_{ijkl}\ \delta^2/\delta h_{ij}\,\delta h_{kl} + \sqrt{h}\left[-{}^3R + 16\pi G T_{oo}(\delta/\delta\phi, \phi)\right]\}\ \psi = 0 \quad (33)$$

This is the Wheeler-De Witt equation, and is satisfied also by the euclidean functional integral (4) as one may easily verify. In addition, there are the momentum constraints:

$$\delta\psi/\delta h_{ij}\Big|_i = 8\pi T^{oj}\psi \tag{34}$$

which imply that ψ is the same on three-metrics and matter field configurations that are related by coordinate transformations. The Wheeler-De Witt equation is a zero energy Schrödinger equation; in our case, the Hamiltonian (24) must annihilate the wave function. In order to be able to perform calculations, let us postulate (Hartle and Hawking 1983) that:

$$p_a^2 = -a^{-p}\partial/\partial a(a^p\,\partial/\partial a) \tag{35}$$

$$p = 1 \qquad\qquad \alpha = \ln(a) \tag{36}$$

so that (33) becomes:

$$N/2\,\exp(-3\alpha)\,(\partial^2/\partial\alpha^2 - \partial^2/\partial\phi^2 + V(\alpha,\phi))\psi(\alpha,\phi) = 0 \tag{37}$$

where: $$V(\alpha,\phi) = m^2\phi^2\exp(6\alpha) - \exp(4\alpha) \tag{38}$$

Because of our interest in the Lorentzian region, in which $V>0$ and $|\phi|>1$ (Hawking and Wu 1984), we make the WKB ansatz (we are using $\hbar = c = 1$ units):

$$\psi(\alpha,\phi) = C(\alpha,\phi)\,\exp iS(\alpha,\phi) \tag{39}$$

where $C(\alpha,\phi)$ is a slowly varying function of α and ϕ, from which we get, by neglecting the second derivatives of C with respect to α and ϕ:

$$(\partial^2 S/\partial\alpha^2 - \partial^2 S/\partial\phi^2) + (\partial C/\partial\alpha\,\partial S/\partial\alpha - \partial C/\partial\phi\,\partial S/\partial\phi) = 0 \tag{40}$$

$$(\partial S/\partial\phi)^2 - (\partial S/\partial\alpha)^2 = \exp(4\alpha) - m^2\phi^2\exp(6\alpha) \tag{41}$$

Equation (41) is an Hamilton-Jacobi equation. Let us look for solutions of (41) in the form:

$$S = f_1(\alpha)g_1(\phi) + f_2(\alpha)g_2(\phi) = f_1(\alpha)g_1(\phi)(1 + f_2(\alpha)g_2(\phi)/f_1(\alpha)g_1(\phi))$$

$$= k_1 m\phi\exp(3\alpha)(1 + k_2 m^{-2}\phi^{-2}\exp(-2\alpha)) \tag{42}$$

where k_1 and k_2 are two constants that we can calculate by substituting (42) into (41). From (42) we get:

$$\partial S/\partial\phi = k_1 m\exp(3\alpha) - k_1 k_2 m^{-1}\phi^{-2}\exp(\alpha) \tag{43}$$

$$\partial S/\partial \alpha = 3k_1 m \phi \exp(3\alpha) + k_1 k_2 m^{-1} \phi^{-1} \exp(\alpha) \tag{44}$$

Therefore:

$$(\partial S/\partial \phi)^2 - (\partial S/\partial \alpha)^2 = k_1^2 m^2 \exp(6\alpha) + k_1^2 k_2^2 m^{-2} \phi^{-4} \exp(2\alpha) -$$

$$-2k_1^2 k_2 \phi^{-2} \exp(4\alpha) - 9k_1^2 m^2 \phi^2 \exp(6\alpha) - k_1^2 k_2^2 m^{-2} \phi^{-2} \exp(2\alpha) -$$

$$-6k_1^2 k_2 \exp(4\alpha) \cong -6k_1^2 k_2 \exp(4\alpha) - m^2 k_1^2 (9\phi^2 - 1) \exp(6\alpha)$$

$$\cong -6k_1^2 k_2 \exp(4\alpha) - 9m^2 k_1^2 \phi^2 \exp(6\alpha) \tag{45}$$

which holds when $m\phi \gg 1$, $\phi \gg 1$. The equation (41) is therefore satisfied if:

$$k_1^2 = 1/9 \qquad -6k_1^2 k_2 = 1 \Rightarrow k_2 = -3/2 \tag{46}$$

Therefore, by taking into account also the negative values of ϕ, we have that:

$$S = -1/3\ m |\phi| \exp(3\alpha)(1 - 3/2\ m^{-2} \phi^{-2} \exp(-2\alpha))$$

$$\cong -1/3\ m |\phi| \exp(3\alpha) \tag{47}$$

The relations:

$$p_\alpha = \partial S/\partial \alpha = \partial L/\partial \dot{\alpha} \tag{48}$$

$$p_\phi = \partial S/\partial \phi = \partial L/\partial \dot{\phi} \tag{49}$$

lead then to:

$$d\alpha/dt = m|\phi| \qquad d|\phi|/dt = -m/3 \tag{50}$$

from which we get:

$$|\phi| = |\phi(t=0)| - mt/3 = |\phi_o| - mt/3 \tag{51}$$

$$a(t) = \exp(\alpha(t)) = \exp \int_0^t m|\phi| dt' = a_o \exp(m|\phi_o| t - m^2 t^2/6) \tag{52}$$

where a_o is the initial value of $a(t)$. At the end of the era during which (52) holds, one has:

$$|\phi_o| - mt^*/3 = |\phi(t^*)| = |\phi^*| \tag{53}$$

$$a(t^*) = a_o \exp\left[3\phi_o^{\ 2}/2\ (1 - \phi^{*2} \phi_o^{\ -2})\right] \cong a_o \exp(3\phi_o^{\ 2}/2) \tag{54}$$

if $\phi^* \ll \phi_o$ as we expect.

We have therefore obtained two important results:

(1) We have realized under which conditions there exists an exponential expansion of the early universe;

(2) We find that :

$$\lim_{a\to 0} \psi(a,\phi) = \lim_{\alpha\to -\infty} \psi(\alpha,\phi) = \text{constant} \tag{55}$$

In fact, the semiclassical approximation to the euclidean functional integral (4) is :

$$\psi \approx A \exp(-B) \tag{56}$$

where B is the action of a compact positive-definite solution of the classical field equations which is bounded by a three-sphere with the given values of a and ϕ. The action B vanishes in the limit of small three-geometries, and if we substitute (56) into the Wheeler-De Witt equation obtained after having taken p = 1 in (35), we find that it is consistent to take A = constant when the radius of the three-sphere tends to zero. The relation (55) is due to the fact that ψ has been taken only over compact positive-definite four-metrics. The constant in (55) may be assumed equal to one.

4. THE CASE OF $\lambda\phi^n$ THEORIES

In section 3 I have examined a case (Hawking 1983) in which the potential of the scalar field is $V(\phi) = (m\phi)^2/2$; recently, Linde (1984) has studied inflationary processes driven by a scalar field with a quartic self-interaction. It becomes therefore interesting to study the general case:

$V(\phi) = \frac{\lambda}{2}\phi^n$. Then the Hamilton-Jacobi equation (41) becomes:

$$(\partial S/\partial \phi)^2 - (\partial S/\partial \alpha)^2 = \exp(4\alpha) - \lambda\phi^n \exp(6\alpha) \tag{57}$$

whose approximate solution is (I consider for simplicity only positive values of ϕ):

$$S \cong -1/3\ \lambda^{1/2}\ \phi^{n/2} \exp(3\alpha) \tag{58}$$

By substituting (58) into (48) and (49) we get:

$$d\phi/dt = -n\lambda^{1/2}\phi^{(n/2 - 1)}/6 \tag{59}$$

$$d\alpha/dt = \lambda^{1/2}\phi^{n/2} \tag{60}$$

Let : $\phi_o = \phi(t=0)$ (61)

Therefore:

$$-n\lambda^{1/2} t/6 = \int_{\phi_o}^{\phi} \phi'^{(1 - n/2)} d\phi'$$

$$=\phi^{(2-n/2)}/(2-n/2) - \phi_o^{(2-n/2)}/(2-n/2) \quad (62)$$

for every $n \neq 4$. From (62) we get:

$$\phi = (\phi_o^{(4-n)/2} - (2-n/2)\, n\, \sqrt{\lambda}t/6)^{2/(4-n)} \quad (63)$$

which implies:

$$d\alpha/dt = \sqrt{\lambda}\, (\phi_o^{(4-n)/2} - (2-n/2)n\, \sqrt{\lambda}\, t/6)^{n/(4-n)} \quad (64)$$

$$a(t) = a_o \exp(\sqrt{\lambda} \int_0^t (\phi_o^{(4-n)/2} - (2-n/2)n\, \sqrt{\lambda}\, t'/6)^{n/(4-n)} dt') \quad (65)$$

for every $n \neq 4$.

The case n=2 has been already studied in section 3. If n=3, we get:

$$a(t) = a_o \exp(\phi_o^2 - (\phi_o^{1/2} - \sqrt{\lambda}\, t/4)^4) \quad (66)$$

At the end of the inflationary era one has from (63) :

$$\phi = \phi^* = (\phi_o^{1/2} - \sqrt{\lambda}\, t^*/4)^2 \Rightarrow t = t^* = 4(\phi_o/\lambda)^{1/2}(1-(\phi^*/\phi_o)^{1/2}) \quad (67)$$

Therefore:

$$a(t^*) \cong a_{max} = a_o \exp\phi_o^2 \quad (68)$$

if $\phi^* \ll \phi_o$. Let us now study the case $n > 4$. Let:

$$k = \phi_o^{(4-n)/2} \quad (69)$$

$$b = (n/2 - 2)n\sqrt{\lambda}/6 > 0 \quad (70)$$

$$z = n/(n-4) > 1 \quad (71)$$

Therefore (65) becomes:

$$a(t) = a_o \exp(I(t)) \quad (72)$$

where:

$$I(t) = \sqrt{\lambda} \int_0^t (k+bt')^{-z} dt' = \sqrt{\lambda}/b\, \left((k+bt)^{(-z+1)}/(-z+1) - k^{(-z+1)}/(-z+1) \right) \quad (73)$$

The second term on the right-hand side of (73) is:

$$\lambda^{1/2}\, k^{(-z+1)}/b(z-1) = 3\phi_o^{\,2}/n \qquad (74)$$

Moreover, from(63) we have that, at the end of the inflationary era:

$$\phi = \phi^{*} = (\phi_o^{\,(4-n)/2} + (n/2 - 2)n\lambda^{1/2}t^{*}/6)^{2/(4-n)} \qquad (75)$$

which implies:

$$t^{*} = (12/n(n-4)\sqrt{\lambda}\,)\,\phi_o^{\,(4-n)/2}\,(\,(\phi^{*}/\phi_o)^{(4-n)/2} - 1\,) \qquad (76)$$

If $\phi^{*} \ll \phi_o$, owing to the fact that n is$>$4, (76) becomes:

$$t^{*} \cong (12/n(n-4)\sqrt{\lambda}\,)\,\phi^{*(4-n)/2} \qquad (77)$$

The first term on the right-hand side of(73) becomes then:

$$\lambda^{1/2}\,(k+bt^{*})^{(-z+1)}/b(-z+1)$$

$$= \lambda^{1/2}\,(\phi_o^{\,(4-n)/2} + \phi^{*(4-n)/2})^{(-z+1)}/b(-z+1)$$

$$\cong -3\,\phi^{*2}/n \qquad (78)$$

if $\phi^{*} \ll \phi_o$. From (73), (74) and (78) we find that, at the end of the inflationary era:

$$a(t^{*}) = a_o \exp(3/n\ \phi_o^{\,2}(1-(\phi^{*}/\phi_o)^2)) \cong a_o \exp(3/n\ \phi_o^{\,2}) \qquad (79)$$

If n=4, we can't apply the formula(65); our starting point will be the equations(59) and(60), which become:

$$\phi^{-1}d\phi = -2/3\ \lambda^{1/2}dt \Rightarrow \phi = \phi_o \exp(-2/3\ \lambda^{1/2}t) \qquad (80)$$

$$d\,\alpha/dt = \lambda^{1/2}\phi^2 = \lambda^{1/2}\phi_o^{\,2}\exp(-4/3\ \lambda^{1/2}t) \qquad (81)$$

$$\Rightarrow \alpha(t) - \alpha(o) = 3/4\ \phi_o^{\,2}(1 - \exp(-4/3\ \lambda^{1/2}t)\,) \qquad (82)$$

which implies:

$$a(t) = \exp\alpha(t) = a_o \exp(3/4\ \phi_o^{\,2}(1 - \exp(-4/3\ \lambda^{1/2}t))) \qquad (83)$$

Therefore the cosmic scale factor $a(t^{*})$ at the end of the inflationary era is:

$$a(t^*) = a_o \exp(3/4\phi_o^2(1-(\phi^*/\phi_o)^2)) \cong a_o \exp(3/4\,\phi_o^2) \qquad (84)$$

where $\phi^* \ll \phi_o$ as we expect, and:

$$t^* = 3/2\ \lambda^{1/2} \ln(\phi_o/\phi^*) \qquad (85)$$

We have found that in the case of $\lambda\phi^n$ theories there is an exponential expansion of the cosmic scale factor of the early universe, which attains a value of the order of $a_o \exp(3\,\phi_o^2/n)$ for every n at the end of the inflationary era.

In order to solve the horizon and flatness problems (Guth 1981), the cosmic scale factor must satisfy the inequality:

$$a_{final} \geqslant a_o\ 10^{28} \simeq a_o \exp(65) \qquad (86)$$

which implies in our case:

$$n \leqslant 3\phi_o^2/65 \qquad (87)$$

Therefore a large class of $\lambda\phi^n$ theories seems to be able to drive an inflationary era of the early universe, provided that the initial value ϕ_o is very great. The duration t^* of the inflationary era may still be of order 10^{-33} sec as in the case of the massive scalar field model, provided that the parameter λ is conveniently chosen in (67), (77) and (85).

5. CONCLUSIONS

5.1. Advantages and limits

This approach to cosmological problems has some remarkable advantages. In fact, we have a well-defined proposal for the boundary conditions, which requires that the euclidean four-metrics belonging to the class C of equation (4) are compact. Moreover, we can calculate the semiclassical approximation, as it usually happens when one uses the functional integral, and the functional ψ gives us all the informations that we need. We may therefore calculate the time-evolution of the cosmic scale factor and of the matter field, and we can estimate the density perturbations $\Delta\varrho/\varrho$ (Halliwell and Hawking 1985).

However, there are also various problems. In fact, the underlying idea is that of a closed universe, but unfortunately there is not a strong observational evidence in favour of our universe being closed. The action is not positive-definite, and therefore the definition (4) is in general just a formal one (for models with a positive-definite action, see Horowitz 1985).

We are still unable to solve the operator ordering problem, which arises from the quadratic dependence on the momenta of the Hamiltonian. The formula (35) does not include all possible ambiguities in the operator ordering (for such questions, see Page and Hawking 1986).

We don't know if a fundamental massive scalar field such as that considered in section 3 did really exist.But if we try to mimic its effects by means of quantum corrections to the effective action involving curvature squared terms,we meet serious conceptual problems(Hawking and Luttrell 1984,Hawking 1985).

The minisuperspace models are not the correct ones;we ought to consider the space of all three-geometries and matter field configurations on S,that is the superspace(Fisher 1970,Halliwell and Hawking 1985).Moreover, if ψ corresponds to a family of different classical solutions rather than being peaked around one unique solution,we can ask:which is our universe? Owing to the linearity of quantum mechanics,we may think(Hawking 1983) that the measurements made by intelligent minds in a given universe correspond to the properties of a given classical solution,and the measurements made in another universe correspond to another classical solution,without interference(for a recent study of these problems, one can see Tipler 1986).At last,but not at least,we must remember that in the path-integral approach to quantum gravity one usually accepts the existence of a space-time foam(Wheeler 1963,Hawking 1978).However,we are still unable to prove that the topology of space fluctuates,and serious objections may be raised against this idea(De Witt and Anderson 1986). Perhaps we have at our disposal equations which lead to quite correct results,but which still await a proper interpretation.

5.2. Possible developments

In my opinion it would be very interesting to study the Wheeler-De Witt equation for cosmological models with torsion,by trying to apply the Hartle-Hawking's proposal for the boundary conditions.In addition,we can consider models in which the space-time is not simply connected;therefore the hypersurface which divides the space-time into two parts will have in general more than one connected component.Most physical observables depend only on a single connected component and can be calculated from a density matrix obtained by integrating over the unobserved surfaces(Hawking 1986). It becomes therefore interesting the calculation of the density matrix of the universe.

In simplicial two-dimensional quantum gravity a class of simplicial complexes can be found to which the gravitational action can be extended, for which sums over the class are straightforwardly defined,and for which a manifold dominates the sum in the classical limit(Hartle 1985a,b;1986). Finally,it would be useful to apply the technique of Halliwell and Hawking(1985) for the calculations all over the superspace in the case of more complicated models,and to re-examine the problem of space-time's dimensions(Wu Zong Chao 1985a,b).

ACKNOWLEDGEMENTS

I am very much indebted to professors G.Platania,G.Immirzi,R.De Ritis and M.Abud for having encouraged my work.

REFERENCES

De Witt,B.(1967) Phys.Rev. 160,1113

De Witt,B. and Anderson,A.(1986) Foundations of Physics 16,91

Fisher,A.E.(1970) in Relativity eds.M.Carmeli,S.I.Fickler and L.Witten (Plenum Press)

Gibbons,G.W. and Pope,C.N.(1979) Commun.Math.Phys. 66,267

Guth,A.(1981) Phys.Rev.D 23,347

Halliwell,J.J. and Hawking,S.W.(1985) Phys.Rev.D 31,1777

Hartle,J.B. and Hawking,S.W.(1983) Phys.Rev.D 28,2960

Hartle,J.B.(1984) Phys.Rev.D 29,2730

Hartle,J.B.(1985a) J.Math.Phys. 26,804

Hartle,J.B.(1985b) Class.Quantum Grav. 2,707

Hartle,J.B.(1986) J.Math.Phys. 27,287

Hawking,S.W.(1978) Nucl.Phys.B 144,349

Hawking,S.W.(1983) in Relativity,Groups and Topology II Les Houches 1983, Session XL,eds.B.S.De Witt and R.Stora(Amsterdam:North-Holland)

Hawking,S.W.(1984) Nucl.Phys.B 239,257

Hawking,S.W. and Luttrell,J.C.(1984) Nucl.Phys.B 247,250

Hawking,S.W. and Wu,Z.C.(1984) Phys.Lett. 151B,15

Hawking,S.W.(1985) 'Paper written in honour of the 60th birthday of E.S. Fradkin',DAMTP preprint,September 1985

Hawking,S.W.(1986) 'The density matrix of the universe',DAMTP preprint, April 1986

Horowitz,G.T.(1985) Phys.Rev.D 31,1169

Linde,A.D.(1984) Rep.Progr.Phys. 47,925

Magnon,A. and Ashtekar,A.(1984) Class.Quantum Grav. 1,L39

Misner,C.W.,Thorne,K.S. and Wheeler,J.A.(1973) Gravitation (S.Francisco: Freeman)

Milnor,J.(1962) Morse Theory (Princeton:Princeton University Press)

Page,D. and Hawking,S.W.(1986) Nucl.Phys.B 264,185

Tipler,F.J.(1986) Phys.Reports 137,231

Wheeler,J.A.(1963) in Relativity,Groups and Topology,eds.B.S. and C.M. De Witt(New York:Gordon and Breach)

Wu,Z.C.(1985a) Phys.Rev.D 31,3079

Wu,Z.C.(1985b) Journ.Gen.Rel.Grav. 17,1217

York,J.W.Jr.(1986) Foundations of Physics 16,249

A No-hair Theorem for Inhomogeneous Cosmologies

Lars Gerhard Jensen and Jaime A. Stein-Schabes

Theoretical Astrophysics Group
Fermi National Accelerator Laboratory
Batavia, Illinois 60510

ABSTRACT

We show that under very general conditions any inhomogeneous cosmological model with a positive cosmological constant, that can be described in a synchronous reference system will tend asymptotically in time towards the de Sitter solution, so making the problem of initial conditions less severe. The implications for inflationary scenarios are examined and it is found that after inflation the universe stays isotropic and homogeneous for a very long time.

1986

P. Galeotti and D. N. Schramm (eds.), Gauge Theory and the Early Universe, 343–352.

I. Introduction

The observable universe today seems to be remarkably homogeneous and isotropic on the very large scale and a good cosmological model has been constructed capable of describing its large scale properties very nicely. This is the so called Friedmann-Robertson-Walker cosmology (FRW). It represents a perfectly homogeneous and isotropic space. Regardless of the nice features of this cosmological model, we have to ask why is the universe described by such a model. There are two answers to this question. Either the universe has always been like this, i.e. the initial conditions where such that the universe was and has remained isotropic and homogeneous. Or the universe started in a less symmetrical phase and evolved through some dynamical process to become a FRW today. The former is certainly a very unsatisfactory solution, and from a statistical point of view very improbable. In this paper we shall explore the latter possibility.

The early attempts to solve the problem where based on retaining the homogeneity of space as an essential feature of our universe. One of the reasons for doing so is the fact that all possible geometries describing homogeneous but anisotropic models fall into one of nine classes. These where classified a long time ago and have been extensively studied[1]. These are the so called Bianchi and Kantowski-Sachs models. Early studies of this models showed that some of these eventually (asymptotically) became FRW. However, not all. Nevertheless, to obtain such models, the constraints imposed on the degrees of freedom of the gravitational field are so severe that only a set of measure zero of all initial conditions can give rise to this models.

To relax homogeneity would make the problem mathematically intractable. It would require the knowledge of the general solution of Einstein's equations together with its boundary and/or initial conditions. Clearly a daunting task. However, all is not lost. A way out is provided by the inflationary scenarios[2]. In these models one usually assumes that the universe becomes dominated by a positive vacuum energy, i.e. a cosmological constant $\Lambda > 0$, and for a period of time expands exponentially at the Hubble-rate H= $\sqrt{3\Lambda}$ followed by a reheating period that eventually terminates in a FRW flat, radiation dominated universe, after which it can proceed its evolution in the standard way (see ref. 3. for a review on inflation). If the universe undergoes a period of exponential expansion of more than about sixty Hubble times it is possible to explain the homogeneity and isotropy of our observable universe as well as solve the oldness and monopole problems in a natural way[2].

It has been shown that inflation will take place in almost all Bianchi models with a positive cosmological constant (except perhaps in Bianchi IX)[4] that do not contain vorticity and that once the inflationary phase begins the process of isotropization due to the cosmological constant is very efficient. The end result of this process is to smooth out anisotropies and eliminate curvature. Furthermore it has been shown[5] that for a class of Bianchi cosmologies (the orthogonal models) the effect of anisotropies on scales larger than the horizon will take a very long time to act back on the observable universe. Is worth pointing out that this effect is independent of the Bianchi model and of the initial anisotropy. Despite this result, Bianchi cosmologies are still highly symmetrical, very restrictive models and probably form a set of measure zero among the set of all possible cosmologies. Attempts have been made in the past to generalize this results to nonhomogeneous models. However, the lack of sufficient exact inhomogeneous solutions to Einstein's equations has made this difficult. Some explicit inhomogeneous examples that present the same properties have appeared in the literature[6].

Some of this evidence lead to the belief that some fundamental principle was underlying the process of isotropization and inflation was indeed a very general process. This would have important consequences for the problem of initial conditions, as it would make our presently observed universe much more probable. Some time ago the existence of such a principle was conjectured by Gibbons and Hawking and by Hawking and Moss[7]. It states that cosmologies with a positive cosmological constant would approach the de Sitter solution asymptotically in time, (several alternative formulations of this conjecture have appeared in the literature), this is the so called "no hair" conjecture. Several attempts at proving the conjecture have been made[8], and a general proof has been obtained for homogeneous cosmologies (Bianchi models)[4]. It has also been shown[9] that "general" solutions to Einsteins equations exist which asymptotically approaches the de Sitter solution (at least locally).

Early on it was shown that the conjecture, as it stands was false, as trivial counterexamples can be provided, the most obvious one being the closed FRW model which collapses before it enters an inflationary phase. Nevertheless, the number and diversity of models that did obey this principle lead to the general belief that if not this then maybe a weaker version of the conjecture should be true. In this work we would like to present a proof of the "no hair" conjecture for a very large class of inhomogeneous and anisotropic cosmological models. We will show that given an arbitrary synchronous space-time, a positive cosmological constant, an energy momentum tensor satisfying the weak and strong energy conditions and a constraint on the three curvature of space, then this space-time will evolve towards a de Sitter space-time (at least locally). This result holds in an arbitrary number of dimensions.

We will also present a generalization of earlier work on orthogonal Bianchi models that shows that once inflation takes over and the phase transition is successfully terminated, then the universe remains homogeneous and isotropic for a very long time. This implies that inflation is a a generic feature of almost any universe that contains a positive cosmological constant. Initial conditions of these cosmologies then become almost irrelevant since all the models end up in the same asymptotical state, the de Sitter cosmology.

The paper will be organise as follows. In Section II the metric, field equations and the energy conditions will be given, the conjecture will be formulated and proven, in Section III we will generalize our early result on orthogonal Bianchi models to inhomogeneous models. Section IV will contain some comments and conclusions.

II.The No-Hair Conjecture

We consider Einstein's equations

$$R_{\mu\nu} = T_{\mu\nu} - \frac{1}{2}g_{\mu\nu}T - g_{\mu\nu}\Lambda \tag{2.1}$$

Here $g_{\mu\nu}$ is the space-time metric, $T_{\mu\nu}$ the energy-momentum tensor, and $T = T^{\mu}_{\mu}$. We use the sign conventions $(+,-,-,-)$ and the notation of ref.(10). Greek indices run from 0 to 3 and Latin from 1 to 3. The only assumptions we make about the energy-momentum tensor is that it satisfies (i) the dominant energy condition, this means that $T_{\mu\nu}t^{\mu}t^{\nu} \geq 0$ and $T_{\mu\nu}t^{\nu}$ is non spacelike for all timelike t^{ν}, and (ii) the strong energy condition, that $(T_{\mu\nu} - \frac{1}{2}g_{\mu\nu}T)t^{\mu}t^{\nu} \geq 0$ for all timelike t^{μ}. The dominant energy condition is equivalent to demanding that the energy density is non-negative and the energy flow is causal. All known forms of matter satisfies this condition. (For a perfect fluid it reduces to $\rho \geq |p|$.) The strong energy condition for a perfect fluid reduces to the usual requirement that $\rho + 3p \geq 0$, i.e. a large negative energy density or large negative pressures must be present to violate this condition. We choose to work in a synchronous reference system where $g_{00} = 1$ and $g_{0\nu} = 0$. We shall also introduce a (positive definite) spacial metric tensor $h_{ab} \equiv -g_{ab}$ and define $s_{ab} = \dot{h}_{ab}$. Using this (1) becomes[10]

$$R^0_0 = -\frac{1}{2}\dot{s}^a_a - \frac{1}{4}s^b_a s^a_b = T^0_0 - \frac{1}{2}T - \Lambda$$

$$R^0_a = \frac{1}{2}(s^b_{a;b} - s^b_{b;a}) = T^0_a \tag{2.2}$$

$$R^b_a = -P^b_a - \frac{1}{2\sqrt{h}}\frac{\partial}{\partial t}(\sqrt{h}s^b_a) = T^b_a - \frac{1}{2}\delta^b_a T - \delta^b_a\Lambda$$

Here P_{ab} is the three dimensional Ricci tensor calculated using h_{ab} and $h = det\ h_{ab}$. ($\sqrt{h}$ can be interpreted as the volume element in three-space.) Let us define the volume expansion by $K \equiv \frac{1}{2}\dot{h}/h = \frac{1}{2}s^a_a$. In what follows we make the assumption that the space is open or flat, i.e. that the scalar spatial curvature $P = P^a_a \leq 0$ for all times. Eq. (2.2) then implies

$$-R^0_0 = \dot{K} + \frac{1}{4}s^b_a s^a_b = -T^0_0 + \frac{1}{2}T + \Lambda \tag{2.3a}$$

$$-R^a_a = \dot{K} + K^2 + P = -T^a_a + \frac{3}{2}T + 3\Lambda \tag{2.3b}$$

To proceed and solve (2.3) we must first calculate $s^b_a s^a_b$. If we introduce the trace free part of s_{ab}, $2\sigma_{ab} \equiv (s_{ab} - \frac{1}{3}s^c_c h_{ab})$, we find that

$$s^a_b s^b_a = s_{ab}s^{ab} = \frac{1}{3}(s^a_a)^2 + 4\sigma_{ab}\sigma^{ab} = \frac{4}{3}K^2 + 4\sigma_{ab}\sigma^{ab}$$

Substituting this into (2.3a) we find

$$\dot{K} = \Lambda - \frac{1}{3}K^2 - \sigma_{ab}\sigma^{ab} - (T^0_0 - \frac{1}{2}T) \tag{2.4}$$

Eliminating $\dot{K}$ using (2.3b) this gives

$$\Lambda - \frac{1}{3}K^2 = -\frac{1}{2}\sigma_{ab}\sigma^{ab} - T^0_0 + \frac{P}{2} \tag{2.5}$$

Clearly $\sigma_{ab}\sigma^{ab}$ is non negative and zero only when $\sigma_{ab} = 0$. The strong and dominant energy conditions imply that $T^0_0 - \frac{1}{2}T$ and T_{00} are positive, furthermore $P \leq 0$ so from (2.4) and (2.5) we find

$$\dot{K} \leq \Lambda - \frac{1}{3}K^2 = -\sigma_{ab}\sigma^{ab} - T^0_0 + \frac{P}{2} \leq 0 \tag{2.6}$$

so $K^2 \geq 3\Lambda$. Also, after integration of the first inequality of (2.6) we find that $K \leq \sqrt{3\Lambda}/\tanh(\sqrt{\frac{\Lambda}{3}}(t + t_0(x_c))$, where t_0 only depends on space. (Here we have chosen the positive square root so $K_0 = K(t_0)$ is positive corresponding to an expanding universe.) This implies that asymptotically

$$0 \leq K - \sqrt{3\Lambda} \leq 4\sqrt{3\Lambda}e^{-2\sqrt{\frac{\Lambda}{3}}(t+t_0)}$$

i.e. the expansion rate K of the volume tends to the de Sitter rate of $\sqrt{3\Lambda}$. From (2.6) it follows that $\sigma_{ab}\sigma^{ab}$, T_{00} and $-P/2$ all are suppressed by $8\Lambda e^{-2\sqrt{\frac{\Lambda}{3}}(t+t_0)}$. This means that locally the universe undergoes rapid isotropization, indeed $\sigma_{ab} = 0$ asymptotically implies that

$$\dot{h}_{ab} - 2\sqrt{\frac{\Lambda}{3}}h_{ab} = 0$$

This shows that asymptotically h_{ab} becomes de Sitter, $h_{ab}(t, x_c) = e^{2\sqrt{\frac{\Lambda}{3}}t}\tilde{h}_{ab}(x_c)$, where $\tilde{h}$ only depends on space. Also for consistency we show that the energy momentum tensor decays exponentially: from the dominant energy condition it follows that for all timelike t^ν

$$(T_{0\nu}t^\nu)^2 \geq T_{a\nu}t^\nu h^{ab}T_{b\mu}t^\mu \geq 0 \tag{2.7}$$

Choosing $t^\nu = \delta^{0\nu}$ this shows that

$$T_{00}^2 \geq T_{a0}h^{ab}T_{b0} \geq 0$$

Since T_{00}^2 vanishes faster than h^{ab} this forces T_{0a} to vanish asymptotically. Substituting this back into (2.7) we similarly find that T_{ab} vanishes so all components of the energy-momentum tensor vanish exponentially fast. Using the above it follows that inside any given physical volume V_0 space-time rapidly becomes equal to the vacuum de Sitter space-time: due to the exponential expansion a fluctuation of scale l is rapidly redshifted and "smoothed" over V_0, all anisotropy as well as the energy-momentum tensor decays exponentially. It is important to keep in mind that although space-time becomes de Sitter locally there is no reason for this to happen globally.

We have shown that the "no hair conjecture" is true for a wide class of spatially open and flat cosmological models. The only requirements is the existence of a synchronous reference frame, a positive cosmological constant and an energy momentum tensor that satisfies both the strong and dominant energy conditions.

In particular this applies to the case of homogeneous cosmologies, the Bianchi models. Except for Bianchi IX all these are flat or open, so in the presence of a positive cosmological constant these will all approach de Sitter space, in agreement with previous results[4].

We also wish to point out that our argument holds in higher dimensions: under the conditions given above an $(n+1)$-dimensional cosmology under the influence of a positive cosmological constant will eventually expand at a rate of $\sqrt{\Lambda/n}$ in each of the n spatial directions. This rules out inflation in Kaluza-Klein type theories in which P (the sum of the curvature of the internal space and three space) is negative, since eventually the internal dimensions will expand and become observable. (Although it may be possible to have $P \leq 0$ if one is willing to violate the energy conditions.)

III. Inflationary Cosmology

As we have mentioned in the beginning , the fact that such a large class of cosmologies under the influence of a positive cosmological constant tend to the de Sitter space-time has great importance for the inflationary cosmology scenarios[11]. In these gravity is coupled to a massless scalar field ϕ with a "flat" potential $V(\phi)$. The cosmological constant is fine tuned so that the energy density vanishes at the minimum of V corresponding to the absence of a cosmological constant today. The equation of motion for ϕ is

$$\ddot{\phi} + K\dot{\phi} = -V'(\phi) \tag{3.1}$$

where ϕ is taken to be smooth so that we can neglegt gradient terms in (3.1), and K is given by (2.5). We notice that if $P \leq 0$, then K is greater than the Hubble constant $\sqrt{3\Lambda}$ of de Sitter phase. Therefore the "friction force" felt by the field, $K\dot{\phi}$

is greater than in de Sitter phase making the field roll slower over the potential; in the same token a greater K means faster expansion. We shall assume that initially ϕ is stabilized (by for example initial conditions or thermal corrections) on the "flat" part of V. Then we find that at least as much inflation is produced in the general case with the universe being inhomogeneous prior to inflation than in the usual case with Friedmann-Robertson-Walker (FRW) cosmology prior to inflation (similar arguments were used in ref.(12) for the anisotropic case). The resulting universe is highly homogeneous and isotropic on scales much larger than the horizon and after ϕ has returned to its minimum the universe evolves like the usual FRW model, but without the extra assumption of initial homogeneity or isotropy. We should comment that there in no need to assume that $P \leq 0$ in all of space, in order to have succesful inflation. We just need $P \leq 0$ in some region large enough that surface effects can be neglected, then this region will eventually evolve into de Sitter space and may then become our observable universe. This strongly supports the belief that inflation is a very universal feature largely independent of the initial conditions of the universe.

We have argued that after inflation any inhomogeneities (anisotropies) have been "pushed" far beyond the horizon and the universe evolves almost exactly like the FRW model. However since the horizon volume increases faster than a typical comoving volume in FRW models, then we expect that eventually the inhomogeneities will reappear through the horizon again. Given the number of Hubble times N during which the universe inflates, we shall now estimate how long it takes before the inhomogeneities reappear. To do this we must study the evolution of the inhomogeneities as described by (2.2).

Let us with t_0 define the time of the onset of inflation, and let t_N be N Hubble times later i.e. $t_N = t_0 + \sqrt{\frac{3}{\Lambda}}N$, then found above that the metric satisfies

$$h_{ab}(t_N, x_c) = e^{2N} h_{ab}(t_0, x_c)$$

Using this in the usual expression for the Christoffel symbols, we find that after N Hubble times of inflation the spatial Ricci tensor scales like $P_a^b(t_N, x_c) = e^{-2N} P_a^b(t_0, x_c)$. This gives us the initial conditions for evolution of the Einstein equations (2.2) after inflation. Here we assume as usual an effective reheating after inflation so that all the vacuum energy density is converted into radiation energy density. The energy-momentum tensor after inflation then becomes that of a perfect fluid with $p = \gamma\rho$ ($\gamma = \frac{1}{3}$ for radiation; we are also assuming that there are no cosmic rotation or fluid acceleration). From conservation of the energy-momentum tensor we find that $\rho \propto h^{-\frac{(1+\gamma)}{2}}$, and assuming effective reheating we can find the constant of proportionality,

$$\rho = \Lambda \left(\frac{h_N}{h}\right)^{\frac{(1+\gamma)}{2}}$$

where $h_N \equiv h(t_N, x_c)$. Now let us define new variables $\tilde{h}_{ab}(t, x_c) \equiv e^{-2N} h_{ab}(t, x_c)$, and quantities carrying a tilda are defined as usual but using $\tilde{h}_{ab}$. Then the last equation in (2.2) becomes

$$\frac{1}{2\sqrt{\tilde{h}}} \frac{\partial}{\partial t}(\sqrt{\tilde{h}} \tilde{s}^b_a) = -e^{-2N} \tilde{P}^b_a + \frac{1}{2}(1-\gamma)\Lambda \left(\frac{\tilde{h}_N}{\tilde{h}}\right)^{\frac{(1+\gamma)}{2}} \delta^b_a \tag{3.2}$$

Again, to make this equation more transparent it is advantageous to carry out another change of variables: set $\mu^{-1} = (1-\gamma)\Lambda h_0^{\frac{(1+\gamma)}{2}}$ and define

$$\bar{h}_{ab} = e^{-2N} \mu \tilde{h}_{ab}$$

$$\bar{t} = e^{-2N} \sqrt{\mu} t$$

Using that $\tilde{h}_N = h_0$, then (3.2) becomes

$$\frac{1}{2\sqrt{\bar{h}}} \frac{\partial}{\partial \bar{t}}(\sqrt{\bar{h}} \bar{s}^b_a) = -\bar{P}^b_a + \left(\frac{1}{\bar{h}}\right)^{\frac{(1+\gamma)}{2}} \delta^b_a \tag{3.3a}$$

with initial conditions

$$\bar{h}_{ab} = e^{-2N} \mu h_{ab}(t_0, x_c) \ \text{ at } \ \bar{t} = e^{-2N} \sqrt{\mu} t_N \tag{3.3b}$$

i.e.

$$\bar{h}_{ab}(\bar{t} = 0, x_c) = 0$$

We notice that all scales are eliminated in (3.3) therefore it follows that it will take a time $\bar{t} \simeq 1$ before the curvature term in (3.3) begins to dominate and anisotropy and inhomogeneities becomes important. This corresponds to a cosmic time

$$t_* = e^{2N} \sqrt{(1-\gamma)\Lambda}\; h_0^{\frac{(1+\gamma)}{4}} \tag{3.4}$$

We have here assumed that the universe is radiation dominated after inflation, however a period of matter domination will only increase the time t_* at which the anisotropies and inhomogeneities reappear through the horizon. The reason for this is that for matter the volume increases like t^2 as compared to $t^{\frac{3}{2}}$ for radiation, while the horizon volume increases like t^3 in both cases. This shows that if the universe inflates it will remain homogeneous and isotropic for a very long time. At time t_* the universe again "remembers" initial conditions and may evolve to become inhomogeneous and anisotropic again on observable scales. This result applies in particular to all the open or flat homogeneous cosmologies, the Bianchi I-VIII cosmologies. In the presence of a positive cosmological constant these models will inflate and after inflation remain isotropic until time t_*. This is in agreement with previous and more laborious results obtained for some (the orthogonal) Bianchi models[5].

IV. Conclusions

We showed that inflation is a very universal feature, a process very likely to have occured in the early history of the universe. A modified and slightly different version of the "no hair conjecture" has been proven. This states that given any inhomogeneous and anisotropic space-time that can be written in a synchronous gauge, and with a non-positive three-Ricci scalar, then in the presence of a positive cosmological constant and a fluid whose energy-momentum tensor satisfies the strong and dominant energy conditions, this will evolve towards the de Sitter space-time. In particular we have demonstrated that within our observable universe inflation is capable of smoothing out all inhomogeneities and anisotropies as well as making the space flat, so explaining the nature of our observable universe. We have discussed the relevance of this result to the problem of initial conditions, where we have argued that if the conditions of the conjecture are satisfied the initial condition are not important any more as the final evolution of our model will be independent of them. Clearly this elevates inflation and makes it a much more appealling solution to the question we posed in the introduction. Inflation provides us, for the first time, with a dynamical mechanism by which the universe undergoes a transition from a very inhomogeneous and non-regular past to a nice smooth future. Of course, if the Universe has always been a FRW or something very close to it, then there is nothing to worry about.

Knowing that inflation will take place in a large class of inhomogeneous space-times, we calculated how effective the process of smoothing was. This generalizes our early results on orthogonal Bianchi Models that uses the fact that a similar conjecture had been proven that ensures the existence of an inflationary phase to assess the efficiency of the isotropization. We found that the timescale for anisotropies to act back on the FRW model left by inflation, goes exponentially with the number of e-folds the universe inflates. This is much larger than the age of the universe. Precisely the same happens in the inhomogeneous models. A successfull inflationary phase will produce a very smooth flat radiation dominated FRW model that will stay FRW for a time of order $t_* \simeq e^{2N}\sqrt{\Lambda}$.

We would like to thank Mike Turner and Bob Wald for encouraging conversations and comments. This work was supported by the Department of Energy and NASA.

References

1. M.A.H. Mac Callum in *Cargèse Lectures in Physics* vol.6 edited by E.Shatzman, Gordon and Breach, New York 1973.
2. A.H. Guth, Phys. Rev. **D23**, 347 (1981)
3. M.S. Turner, *The Inflationary Paradigm*, Proceedings of the Cargèse School on Fundamental Physics and Cosmology, edited by J. Audouze and J. Tran Thanh Van, Editions Frontieres, Gif-Sur-Yvette (1985).
4. R.W. Wald, Phys. Rev. **D28**, 2118 (1983) .
5. L.G. Jensen and J.A. Stein-Schabes, Phys. Rev. **D34** (1986)
   M.S. Turner and L. Widrow, *The Bianchi Models and New Inflation* Fermi-Pub 86/49-A25 (1986)
6. J.D. Barrow and J.A. Stein-Schabes, Phys. Lett. **103A**, 315 (1984)
7. G.W. Gibbons and S.W. Hawking, Phys. Rev. **D15**, 2738 (1977).
   S.W. Hawking and I.G. Moss, Phys. Lett. **110B**, 35 (1982).
8. W. Boucher, G.W. Gibbons and G.T. Horowitz, Phys. Rev. **D30**, 2447 (1984).
   W. Boucher, in *Classical General Relativity*, edited by W.B. Bonnor *et al.* (Cambridge University Press, Cambridge, England, 1984).
9. A.A. Starobinskii, J.E.T.P. Letters **37**, 66 (1983).
10. L.D. Landau and E.M. Lifshitz, *The Classical Theory of Fields*, Pergamon Press (1971).
11. A.D. Linde, Phys. Lett. **108B**, 389 (1982).
    A.Albrecht and P.J. Steinhardt Phys. Rev. Lett. **48**, 1220 (1982).
    A.D. Linde, Phys. Lett. **129B**, 177 (1983).
    J. Ellis, D.V. Nanopoulos, K. Olive and K. Tamvakis, Nucl. Phys. **B221** 224 (1983).
12. G. Steigman and M.S. Turner, Phys. Lett. **128B**, 295 (1983).

Constraints on Algebraically Extended Theories of Gravity

Patrick F. Kelly
University of Toronto,
Dept. of Physics,
Toronto ON,
M5S 1A7 CANADA

0. Abstract

The particle spectra for all of the algebraically extended theories of gravity are presented and only those which are free of ghosts and tachyons are suggested for physical viability. Of the much smaller class of theories that pass this hurdle, one in particular is singled out. This model is distinct from previous alternative theories (for instance the NGT of Moffat). The static, spherically-symmetric solutions of this new model are investigated as an aid in determining whether this model has phenomenological interest.

I. Introduction

It is rather ironic that the ideas which later developed into gauge theory emerged from the study of gravity[1], but today the standard model very successfully[2] unites the strong, weak and electromagnetic forces into a cogent framework, while gravity remains aloof. An understanding of gravity is imperative since it is the interaction that governs the large-scale structure and evolution of the universe. General relativity (GR) provides an elegant and appealing theory of gravity and is in excellent agreement with the standard solar system tests[3]. However, most of these tests probe only the linear structure of GR. The ability to account for the perihelion precession of Mercury (a non-linear effect) was one of GR's early triumphs, but the calculation must include the effects of solar oblateness, a parameter not yet known with sufficient precision to unreservedly support GR[4,5]. More seriously, observations of the close binary systems DI Hercules[6] and AS Cam[7] reveal anomalously small periastron precession rates. Classical effects have been unable to account for the large discrepancies between the data and the GR predictions ($> 20\sigma$)[6].

It is then of interest to investigate alternative theories of gravitation. One hopes to construct theories which match the weak field behavior of GR while differing observably in the strong field regime. One such theory has been proposed by Moffat and his collaborators[8] and appears in the literature as NGT–nonsymmetric gravitational theory. The static, spherically-symmetric (SSS) solution in NGT possesses a body-dependent, conserved parameter 'l' (associated with fermion number) as well as the Schwarzschild mass 'm'. The parameter l contributes a retrograde term in the periastron precession calculation and can be fitted to agree with several anomalous binaries[9]. The values of l

P. Galeotti and D.N. Schramm (eds.), Gauge Theory and the Early Universe, 353–360.

obtained by this fit are not in conflict with weak constraints arising from the solar system experiments[10].

In the course of studying the geometrical structure of NGT it was realized that a number of extensions to GR could be produced using a method which was termed "algebraic extension" (AE)[11]. In this article, the AE theories are subjected to tests in order to select those which may be of phenomenological interest. Section two displays the AE theories (there are four plus GR) and establishes notation. In section three, the linearised versions of the theories are computed and required to be free of ghosts and tachyons. A new model, distinct from NGT, is found to circumvent ghost problems by means of an additional gauge invariance [12] in flat space. Section four takes this promising model from section three and with it attempts to determine a physically sensible SSS solution. A rather strange consequence ensues [13] and seems to eliminate this model from contention as a viable theory of gravity.

II. The Programme of Algebraic Extension

AE incorporates an algebraic structure into the geometry of spacetime by allowing tensors in the theory to acquire values in an arbitrary $m+1$ dimensional algebra $\mathcal{A}$ [11]. This corresponds to enlarging the tangent bundle while leaving the base space (4-dim spacetime) unaffected. Denote the generators of the algebra by $\{1, a_i | i = 1 \rightarrow m\}$, and under the operation of 'conjugation' $*: \{1^*, a_i^*\} = \{1, -a_i\}$. The generators compose as

$$a_I \cdot a_J = C^K_{IJ} a_K \quad , \quad I, J, K = 0 \rightarrow m, \quad C^K_{IJ} \in \mathcal{R}. \tag{2.1}$$

The metric tensor is assumed to be 'hermitian' (^ signifies algebra valued quantities),

$$\hat{g}_{\alpha\beta} = 1 g_{(\alpha\beta)} + a_i g^{i}_{[\alpha\beta]}. \tag{2.2}$$

The inverse of the metric tensor is determined by the rule

$$\hat{g}_{\mu\alpha}\hat{g}^{\nu\alpha} = \delta^\nu_\mu = \hat{g}^{\alpha\nu}\hat{g}_{\alpha\mu}. \tag{2.3}$$

Index ordering is always important, as is term ordering for noncommuting algebras. The physical connection is also taken to be hermitian

$$\hat{\Gamma}^\alpha_{\beta\gamma} = 1\Gamma^\alpha_{(\beta\gamma)} + a_i \Gamma^{\alpha\ i}_{[\beta\gamma]}. \tag{2.4}$$

The curvature tensor is defined (as usual) by commuting covariant derivatives[11]

$$\hat{R}^\alpha{}_{\beta\gamma\delta} = \hat{\Gamma}^\alpha_{\beta\delta,\gamma} - \hat{\Gamma}^\alpha_{\gamma\delta,\beta} + \hat{\Gamma}^\lambda_{\beta\delta}\hat{\Gamma}^\alpha_{\gamma\lambda} - \hat{\Gamma}^\lambda_{\gamma\delta}\hat{\Gamma}^\alpha_{\beta\lambda}. \tag{2.5}$$

The most general Lagrangian linear in the curvature is

$$\mathcal{L} = \sqrt{-g}\hat{g}^{\mu\nu}[A_1 \hat{R}^\sigma{}_{\nu\sigma\mu} + A_2(m-3)\hat{R}^\sigma{}_{\nu\mu\sigma}], \tag{2.6}$$

where g is the determinant of $\hat{g}_{\alpha\beta}$ [11]. Insisting that the AE theories satisfy reasonable criteria (algebra/geometry compatibility, reality of physical quantities, and GR in the

Algebra	m	generators	C_{IJ}^{K}	commuting
$\mathcal{R}$	0	1	...	yes
$\mathcal{C}$	1	$1, i$	$i^2 = -1$	yes
$\mathcal{E}$	1	$1, e$	$e^2 = +1$	yes
$\mathcal{Q}$	3	$1, a_1, a_2, a_3$	$a_i \circ a_j = -\delta_{ij} + \epsilon_{ijk} a_k$	no
$\mathcal{H}$	3	$1, a_1, a_2, a_3$	${a_1}^2 = {a_2}^2 = 1, a_1 \circ a_2 = -a_3$	no

Table 1. Algebras

limit of no algebraic structure) [11] limits the possible choice of algebras to five: Real $\mathcal{R}$, Complex $\mathcal{C}$, Hyper-complex $\mathcal{E}$, Quaternion $\mathcal{Q}$, and Hyper-quaternion $\mathcal{H}$.

The $\mathcal{R}$ extension is GR itself. NGT is a particular case of an $\mathcal{E}$ theory. In the approach that follows, the metric and its first derivatives are taken to be the fundamental degrees of freedom and the requirement of geometric compatibility

$$0 = \hat{g}_{\alpha\beta,\mu} - \hat{\Gamma}^{\gamma}_{\mu\alpha}\hat{g}_{\gamma\beta} - \hat{g}_{\alpha\gamma}\hat{\Gamma}^{\gamma}_{\beta\mu}, \tag{2.7}$$

is imposed and implicitly defines the (hermitian) connection in terms of $\hat{g}_{\alpha\beta}$ and $\hat{g}_{\alpha\beta,\mu}$. Invoking compatibility (2.7), it is easily shown that the Lagrangian (2.6) is real under the algebraic operation of conjugation. Field equations follow from a Hilbert (Euler–Lagrange) variation of (2.6).

Several subtle distinctions exist between this formulation of AE theories and NGT. In NGT, the constants A_1 and A_2 are chosen ($m = 1$) such that the term contracted with $\sqrt{-g}\hat{g}^{\mu\nu}$ in (2.6) is hermitian. Thus the Lagrangian is real under conjugation, and field equations are derived using the Palatini variational procedure (where g and Γ are regarded as the fundamental degrees of freedom). As a consequence of the Palatini method, the torsion vector (trace of the torsion tensor) must vanish $0 = \Gamma_\mu = \Gamma^{\alpha}_{[\mu\alpha]}$. NGT regains the lost degrees of freedom through the introduction of a vector torsion W_μ as an independent field in the Lagrangian.

III. Ghost Properties

Despite the fact that these are 'classical' theories, they do predict particle spectra. One criterion used to judge the potential viability of gravitational theories is that the predicted particle(s) not include ghosts and tachyons[14,15]. A ghost particle is one which possesses negative energy and is characterized by having a negative residue at the pole of its propagator. On the other hand, a tachyon propagates faster than light and is distinguished by having a complex mass (the pole of its propagator occurs at a negative value). GR is ghost free since the gauge symmetry associated with general coordinate invariance eliminates any modes other than the spin-2 graviton[16]. NGT was shown to be ghost-free due to the existence of the aforementioned (non-propagating) W_μ field which acts to constrain away ghost modes[17]. A theory examined by Moffat *et al* based on the complex algebra was abandoned when it was found to possess ghosts[18].

In performing the analysis, we first set $A_2(m-3)$ equal to zero † in the general AE Lagrangian (2.6), leaving

$$\mathcal{L} = \sqrt{-g}\hat{g}^{\mu\nu}\hat{R}^{\sigma}{}_{\nu\sigma\mu} = \sqrt{-g}\hat{g}^{\mu\nu}\hat{R}_{\mu\nu}. \tag{3.1}$$

† Recall that in NGT this term is not set to zero.

In order to proceed, we require the linearised (weak field) form of the Lagrangian (3.1). Towards this end, the weak field metric is

$$\hat{g}_{\mu\nu} = \eta_{\mu\nu} + \hat{h}_{\mu\nu} = \eta_{\mu\nu} + 1h_{(\mu\nu)} + a_i h^i_{[\mu\nu]}, \tag{3.2}$$

where $\eta_{\mu\nu} = diag(-,-,-,+)$, and $\hat{h}_{\mu\nu}$ is small. At any given order, indices are raised and lowered with η. To first order, equation (2.3) entails

$$\hat{g}^{\mu\nu} = \eta^{\mu\nu} - \hat{h}^{\mu\nu} \quad , \quad \text{where} \quad \hat{h}^{\mu\nu} = \eta^{\mu\beta}\eta^{\alpha\nu}\hat{h}_{\alpha\beta}. \tag{3.3,4}$$

We introduce the quantities

$$h = \eta^{\mu\nu}\hat{h}_{\mu\nu} = \eta^{\mu\nu}h_{(\mu\nu)}, \tag{3.5}$$

$$h_\mu = h_{[\mu\alpha]}{}^{,\alpha} = \eta^{\alpha\beta}\hat{h}_{[\mu\alpha],\beta}, \tag{3.6}$$

and observe that h is purely real while h_μ has no real part. The connection is determinable to any order in $\hat{h}_{\alpha\beta}$ through the compatibility condition (2.7). To first order then,

$$\Gamma^\alpha_{\beta\gamma} = \frac{1}{2}\eta^{\alpha\rho}(\hat{h}_{\gamma\rho,\beta} + \hat{h}_{\rho\beta,\gamma} - \hat{h}_{\beta\gamma,\rho}) + O(h^2). \tag{3.7}$$

These results, when applied in (3.1), yield the Lagrangian to second order in $\hat{h}_{\alpha\beta}$,

$$\mathcal{L}^{(2)} = \frac{1}{4}\hat{h}^{\mu\nu}\Box\hat{h}_{\nu\mu} - \frac{1}{2}h_{,\nu}(h^{(\mu\nu)}{}_{,\mu} - \frac{1}{2}h^{,\nu}) + \frac{1}{2}h^{(\mu\nu)}{}_{,\mu}h_{(\alpha\nu)}{}^{,\alpha} + \frac{1}{2}\eta^{\sigma\pi}h_\sigma h_\pi, \tag{3.8}$$

where $\Box = \eta^{\alpha\beta}\partial_\alpha\partial_\beta$. Equation (3.8) may be rewritten as

$$\mathcal{L}^{(2)} = \frac{1}{4}h^{(\mu\nu)}\Box h_{(\nu\mu)} - \frac{1}{2}h_{,\nu}(h^{(\mu\nu)}{}_{,\mu} - \frac{1}{2}h^{,\nu}) + \frac{1}{2}h^{(\mu\nu)}{}_{,\mu}h_{(\alpha\nu)}{}^{,\alpha}$$
$$+ \frac{1}{4}\hat{h}^{[\mu\nu]}\Box\hat{h}_{[\nu\mu]} + \frac{1}{2}\eta^{\sigma\pi}h_\sigma h_\pi, \tag{3.9}$$

which illustrates clearly the separation of $\mathcal{L}^{(2)}$ into two pieces

$$\mathcal{L}^{(2)} = \mathcal{L}^{(2)}_{\rm GR} + \mathcal{L}^{(2)}_{\rm skew}. \tag{3.10}$$

The first part, $\mathcal{L}^{(2)}_{\rm GR}$, depends only on the symmetric part of $\hat{h}_{\alpha\beta}$ (*i.e.*, $h_{(\alpha\beta)}$) and is precisely the term that one obtains in GR, as it must be in the AE scheme. Explicitly writing the generators of the algebra in the second piece (see (2.1,3.2)),

$$\mathcal{L}^{(2)}_{\rm skew} = (1C^0_{ij} + a_k C^k_{ij})\{\frac{1}{4}h^{i[\mu\nu]}\Box h^j_{\nu\mu} + \frac{1}{2}\eta^{\sigma\pi}h^i_\sigma h^j_\pi\}. \tag{3.11}$$

Clearly the quantity $\eta^{\sigma\pi}h^i_\sigma h^j_\pi$ is symmetric in ij, less obviously so is the term $h^{i[\mu\nu]}\Box h^j_{\nu\mu}$. The structure constants C^k_{ij} are antisymmetric for all of the algebras under consideration, while the $C^0_{ij} \propto \delta_{ij}$. Thus

$$\mathcal{L}^{(2)}_{\rm skew} = C^0_{ii}\{\frac{1}{4}h^{i[\mu\nu]}\Box h^i_{\nu\mu} + \frac{1}{2}\eta^{\sigma\pi}h^i_\sigma h^i_\pi\}, \tag{3.12}$$

is purely real as again it must be.

There is a complete absence of tachyons in all of the AE theories since neither $\mathcal{L}^{(2)}_{\mathrm{GR}}$ nor $\mathcal{L}^{(2)}_{\mathrm{skew}}$ contain mass terms. In [15] van Nieuwenhuizen examined the general set of antisymmetric real tensor field Lagrangians and has shown that

$$\mathcal{L}^{(2)} = \frac{1}{4}h^{[\mu\nu]}\Box h_{[\nu\mu]} + \frac{1}{2}\eta^{\sigma\pi}h_\sigma h_\pi \tag{3.13}$$

is free of ghosts. In the AE theories then, the sign of the residue at the pole of the propagator depends on the overall sign of $\mathcal{L}^{(2)}$ relative to $\mathcal{L}^{(2)}_{\mathrm{GR}}$. This relative sign is determined for each field by the C^0_{ii} structure constants. In order to have a ghost free theory, all of the C^0_{ii} must be positive. Of the five AE theories, only two, the $\mathcal{R}$ and $\mathcal{E}$ theories satisfy this positivity criterion and hence can be ghost free. The $\mathcal{E}$ algebraic extension corresponds to a real nonsymmetric theory. (Carets are no longer used to denote $\mathcal{E}$–valued quantities in the subsequent analysis.)

The skew Lagrangian (3.13) can be manipulated into the canonical form of the Kalb–Ramond Lagrangian[19]

$$\mathcal{L}_{\mathrm{KR}} = \frac{1}{12}h^{[\nu\mu,\lambda]}h_{[\nu\mu,\lambda]}. \tag{3.14}$$

This Lagrangian is invariant under the gauge transformation

$$h_{[\alpha\beta]} \rightarrow h_{[\alpha\beta]} + 2\varsigma_{[\alpha,\beta]}, \tag{3.15}$$

and the $h_{[\alpha\beta]}$ field is spin-0. Kalb and Ramond originally introduced this Lagrangian to provide a Maxwell–type coupling between closed strings in a dual resonance (string) model. Its interpretation here, as well as the origin of the gauge invariance (3.15) are open questions in this model.

One final consideration needs to be addressed. The Lagrangian (3.1) describes the $\mathcal{Q}$ and $\mathcal{H}$ AE theories in their complete generality, but in the $\mathcal{E}$ case, the term in the Lagrangian (2.6) which was set to zero in (3.1) must, in principle, be included. The contribution of this term to the linearised Lagrangian,

$$\mathcal{L}^{(2)}_{A_2} = 2h^{[\mu\nu]}h_{[\nu,\mu]}, \tag{3.16}$$

is entirely to $\mathcal{L}^{(2)}_{\mathrm{skew}}$ and is not invariant under (3.15). The inclusion of this term would break the gauge invariance of the skew field and would introduce ghosts into the theory, unless another mechanism is available (as in NGT) to eliminate the unphysical modes[17].

The Lagrangian (3.1) in the hyper-complex (real nonsymmetric) case gives rise to a ghost free model of gravity distinct from NGT. In an effort to determine whether this model has phenomenological interest, SSS exact solutions are calculated in the next section.

IV. SSS Solutions For the New Model

A necessary condition for the success of a model of gravity is that the set of solutions to the field equations contain some which can provide a description of the astronomical bodies and systems that are observed. In particular, the model must contain objects that possess the (idealized) properties of stars and our solar system. The simplest situation envisioned is that of a static 'star' (the sun is $> 10^9$ years old), which has perfect spherical symmetry (non-rotating) and is alone in a very large universe. Historically, after tests of self-consistency have been performed on new theories† the next task has been to search for these SSS solutions.

Papapetrou[20] has determined that the SSS metric, expressed in spherical coordinates (r, θ, ϕ, t) has the general form

$$g_{\mu\nu} = \begin{pmatrix} -\alpha & 0 & 0 & w \\ 0 & -\beta & u\sin\theta & 0 \\ 0 & -u\sin\theta & -\beta\sin^2\theta & 0 \\ -w & 0 & 0 & \gamma \end{pmatrix}, \tag{4.1}$$

where α, β, γ, u, and w are functions of r only. The inverse of the metric may be found according to (2.3). The SSS connection is determinable in terms of metric functions from the requirement of geometric compatibility (2.7) [21]. The Γ's are stated in Appendix A of [13], written in the present notation.

Hilbert (Euler–Lagrange) variation of (3.1) with respect to $\sqrt{-g}g^{\mu\nu}$ yields the full vacuum field equations[13,22]

$$\varphi_{\mu\nu} = R_{\mu\nu} - \frac{1}{2}g_{\mu\nu}R + X_{\mu\nu}, \tag{4.2}$$

where $R = g^{\alpha\beta}R_{\alpha\beta}$, and $X_{\alpha\beta} = \Gamma^{\lambda}_{\beta\alpha}\Gamma_{\lambda} - \Gamma_{\beta,\alpha} - \Gamma_{\beta}\Gamma_{\alpha}$. The field equations are hypercomplex hermitian, have the same general structure as the metric (4.1), and $\varphi_{33} = \varphi_{22}\sin^2\theta$. A non-trivial set of field equations is provided by $\{\varphi_{11}, \varphi_{[14]}, \varphi_{44}, \varphi_{22}, \varphi_{[23]}\}$. This set is not completely independent, however, since there exists the freedom to choose a coordinate condition. The completely general field equations could not be solved, but solutions were obtained in three special cases[13]. The two natural special cases are i) $u = 0, w \neq 0$, and ii) $u \neq 0, w = 0$. The third case is where $\gamma(1 - \frac{w^2}{\alpha\gamma})$ is set equal to a constant, c_y say. From the structure of the field equations, and the demand that the solutions become flat space at large r, it is believed that the only solutions which may be of phenomenological interest fall in this category. In the event that $\gamma(1 - \frac{w^2}{\alpha\gamma})$ is not constant, it must go as r^6 for large r.

The general SSS solutions for the metric functions in this, what is believed to be the physical case, are expressible as [13]:

$$\alpha = \left(\frac{ds}{dr}\right)^2(u^2 + \beta^2); \tag{4.3}$$

$$u + i\beta = -\frac{c_0}{4}\sinh^{-2}\left[\frac{\sqrt{ic_0}}{2}(s + s_0)\right], \quad c_0 \neq 0, \tag{4.4}$$

$$= i(s + s_0)^{-2}, \quad c_0 = 0; \tag{4.4a}$$

$$\gamma = c_y\left(1 + \frac{e^{2D}}{u^2 + \beta^2}\right); \tag{4.5}$$

$$w^2 = c_y\left(\frac{ds}{dr}\right)^2 e^{2D}; \tag{4.6}$$

† The new model satisfies geometric (and field theoretic) Bianchi Identities [22].

where s is an arbitrary function of r, c_O is a real constant of integration, s_O is a complex constant of integration, and e^{2D} satisfies

$$\partial_s e^{2D} = c(u^2 + \beta^2) \quad , \quad c = constant. \tag{4.7}$$

Constraints are placed on the values of the constants and the functional form of s by applying boundary conditions.

If the symmetric components are required to approach their flat space values for large r $(\alpha \to 1, \beta \to r^2, \gamma \to 1)$, then $c_y = 1$ and $s + s_O \sim \frac{-1}{r}$. Equation (4.7) could only be integrated in the case $c_O = 0$, and at large r,

$$e^{2D} \sim \frac{c}{r} r^3 + c_D \quad , \quad c_D = constant, \quad c_O = 0. \tag{4.8}$$

The only a priori constraint placed on the skew components is that they not grow asymptotically at a rate greater than r^2. In the $c = 0$, and $c \neq 0, c_O = 0$ cases, both u and w tend to zero as $r \to \infty$.

The asymptotic argument is further refined by insisting that the symmetric components go over into the Schwarzschild solution for large r. When this limit is taken, the $c = 0$ solutions do not contain an identifiable 'mass'. (The 'mass' being related to the coefficient of the $\frac{1}{r}$ term in the expansion of γ). For $c \neq 0$, the equations (4.5,6) become

$$\gamma \sim 1 + \frac{c/3}{r}, \tag{4.9}$$

$$w^2 \sim \frac{c/3}{r(1 + \frac{\alpha_O}{r})} \sim \frac{c/3}{r}. \tag{4.10}$$

We note that in this theory the 'mass' arises in a fundamentally different manner than in the Schwarzschild form in GR. In order to recover the Schwarzschild form in (4.9) we are forced to choose $\frac{c}{3} = -2m < 0$. However, this is disasterous since with this choice, (4.10) implies that w is complex and the theory has ghosts. This unseemly result will also occur if $c \neq 0$, and $c_O \neq 0$ since the e^{2D} terms in (4.5,6) enter with the same sign.

This is perplexing behavior: positive mass SSS solutions require that the model possess ghosts; while conversely, models which are ghost free have negative mass SSS solutions!

V. Remarks

Of the four non-trivial algebraically extended theories of gravity, the three which are based on the $\mathcal{C}$, $\mathcal{Q}$, and $\mathcal{H}$ algebras predict ghost particles. Without an additional mechanism to eliminate the ghosts these theories must be discarded.

The $\mathcal{E}$ algebra alone is able to satisfy the no-ghost criterion. It is able to accomodate (at least?) two distinct models, NGT and the model studied here, which are ghost free albeit for rather different reasons. Of the two models, it appears that only NGT is able to produce a physically sensible SSS solution. The failure of the new model to do so suffices to eliminate it from consideration as a viable alternative theory of gravity. There remains the possibility that the $w = 0$ case may be amenable to physical interpretation but this is unlikely. (Of course the penultimate test of the phenomenological viability of this theory would be a PPN expansion.) Hence, of all of the theories of this type that have been investigated in any detail, NGT is the only one which has survived the various tests that have been performed. The range of alternative theories has been reduced considerably but whether NGT is the unique theory of this kind with phenomenological applicability is still unknown.

VI. Acknowledgments

This work was supported in part by the Natural Sciences and Engineering Research Council of Canada. Further, I would like to thank R. B. Mann, J. W. Moffat, C. Morningstar and C. Reader for many interesting discussions.

VII. References

1. H. Weyl, Sitzungsber. d. Preuss. Akad. d. Wiss., 465, (1918) ; H. Weyl, Ann. d. Physik **59**,101, (1919) .
2. See for example *Proc. 1985 Int. Symp. on Lepton and Proton Interactions at High Energies,* ed. M. Konuma, K. Takahashi, (Kyoto University: 1986) .
3. C. M. Will, *Theory and Experiment in Gravitational Physics,* (Cambridge: CUP), 1981.
4. R. Reasonberg, *Proc. 9th Int. Conf. on Cosmology and Astrophysics,* ed. P. Bargmann and V. de Sabbata, (New York: Plenum), 1986.
5. L. Campbell, J. C. McDow, J. W. Moffat, D. Vincent, Nature **305**, 508, (1982) .
6. F. P. Guinan, F. P. Maloney, Astron. J. **90**, 1519, (1985) .
7. Kh. F. Khaliullin, V.S. Kozyreva, Astrophs. Space Sci. **94**, 115, (1983) .
8. J. W. Moffat, Phys. Rev. D **19**, 3554, (1979) ; J. W. Moffat, *Proc. 7th Int. School of Gravitation and Cosmology,* ed. V. de Sabbata, (New York: Plenum), 1981 ; J. W. Moffat, Found. Phys. **14**, 1217, (1984) ; J. W. Moffat, J. Math. Phys. **25**, 347, (1984) ; G. Kunstatter, R. Yates, J. Phys. A **14**, 847, (1981) ; G. Kunstatter, J. W. Moffat, J. Malzan, J. Math. Phys. **24**, 886, (1983) .
9. J. W. Moffat, Astrophys. J. Lett. **287**, L77, (1984) ; J. W. Moffat, Can. J. Phys. **64**, 178, (1986) .
10. J. W. Moffat, Phys. Rev. Lett. **50**, 709, (1983) .
11. R. B. Mann, Class. Quantum Grav. **1**, 561, (1984) .
12. P. F. Kelly, R. B. Mann, Class. Quantum Grav. **3**, 705, (1986) .
13. P. F. Kelly, R. B. Mann, *U. of Toronto Preprint,* in preparation .
14. E. Sezgin, P. van Nieuwenhuizen, Phys. Rev. D **21**, 3269, (1980) .
15. P. van Nieuwenhuizen, Nucl. Phys. B **60**, 478, (1973) .
16. S. Weinberg, *Gravitation and Cosmology,* (Wiley), 1972 .
17. R. B. Mann, J. W. Moffat, Phys. Rev. D **26**, 1858, (1982) ; H. P. Leivo, G. Kunstatter, P. Savaria, Class. Quantum Grav. **1**, 7, (1984) .
18. See [17] and R. B. Mann, J. W. Moffat, J. G. Taylor, Phys. Lett. **97B**, 73, (1980) .
19. M. Kalb, P. Ramond, Phys. Rev. D **9**, 2274, (1974) ; E. Cremmer, J. Scherk, Nucl. Phys. B **72**, 117, (1974) .
20. A. Papapetrou, Proc. Roy. Ir. Acad. Sci. **A 52**, 69, (1948) .
21. M. A. Tonnelat, *Einstein's Unified Field Theory,* tr. R. Akerib, (Gordon and Breach), 1966 ; D. N. Pant, Nuovo Cimento B **25**, 175, (1975) .
22. P. F. Kelly, M. Sc. Thesis, University of Toronto, unpublished 1985.

ON KALUZA-KLEIN THEORIES

Vinicio Pelino
Istituto Astronomico
Università di Roma I
Via Lancisi 29
00161 Roma Italy

ABSTRACT. In this paper we briefly review the use of extradimensions in the context of modern physics. Thereafter we investigate the problem of the world's dimensionality, proposing a statistical theory of the spatial dimensions in the early history of our universe.

> The main question is if
> the Lord had a choice when
> He created the world.
> A. Einstein

The search for a unified theory of all fundamental interactions has led physicists to consider seriously the idea that our world can have more than four dimensions.* In the standard Kaluza-Klein approach(1) isometries of the compact manifold M_k associated with the our observed universe M_4, manifest themselves as gauge interactions in M_4; in this way the true world has the structure $M_4 \cdot M_k$. Therefore, the study of topological properties of M_k is very important in order to have a realistic theory in the expanded world. For example it is very easy to see that the eigenvalues of the Klein-Gordon operator in M_k, give the spectrum of besonic particles in M_4. In fact, for a massless scalar lar field Φ in $M_D = M_4 \cdot M_k$ the K-G equation can be written:

$$\Box\Phi = 0$$

Using the ansatz $\Phi_D = \Phi_4 \cdot \Phi_k$ this equation becomes:

$$(\Box - M^2)\Phi_4 = 0$$

P. Galeotti and D. N. Schramm (eds.), Gauge Theory and the Early Universe, 361–365.

where $M^2\Phi = \square\Phi_k$. In this context the use of index theory is very helpful for the study of physical operators[2]. The main problem in this kind of "particle physics" approach is the search for constraints on the number of dimensions of the compact manifold whose isometry group must contain the standard SU(3) × SU(2) × U(1) gauge group. In the next paragraph we will briefly see how this problem is discussed in supergravity and superstrings theories.

The "particle physics" approach.
In supergravity theories tnere is a relation between the number of gravitinos N and the number of dimensions D of the whole spacetime;[3] it is:

$$\text{High N low D} = \text{High D low N} \qquad (1)$$

The non-existence of particles having spin higher than 2 puts a constraint on max D, in fact it can be shown by reason of supersymmetry[4] that D = 11. Therefore, by (1) we have the equivalence between the two theories (N = 8, D = 4) and (N = 1, D = 11). We note that the (N = 1, D = 11) theory is a pure gravity theory, this means that it is not necessary to postulate the separate existence of Yang-Mills fields, they are a consequence of gravity. The big trouble of this very fascinating theory is the problem of chirality. In fact we cannot define Weyl spinors on an odd dimensional manifold, and this fact makes it impossible to have a complex representation of the particles spectrum. In superstrings theory the quantization of the supersymmetric action leads to the result: D = 10. An external gauge field must be introduced in the structure of the theory, and in order to make it a nomaly-free we must have SO(32) or the more promising heterotic group $E_8 \times E_8$. We note that both of these groups have dimension 496, and it could be thinkable to embed them in the general coordinate group of a 496 dimensional compact manifold[5], in order to have a pure Kaluza-Klein theory, in this way our world will have 506 dimensions!

The "cosmological approach".
In the "particle physics" approach the structure of the world as a product manifold $M_4 \cdot M_k$ is taken as an assumption. Therefore we are not able to explain the reason for which we live in a three expanded spatial dimensions. (Incidentally, it is remarkable that also three families of particles seem to well describe all the matter in the universe!). We will not describe here the amazing properties of 3-spatial geometry in relation to the existence of living structures and solar systems (see for example the nice paper "Dimensionality" by Barrow[6]). In our opinion the subject of dimensionality must be considered as an initial conditions problem for our universe. In fact we believe that our world is the final result of a dynamical process involving an arbitrary number of initial dimensions[7]. Supported by this idea we have studied the vacuum Einstein equations on a D = d + 1 arbitrary dimensional Torus, using for the scale

factors the following ansatz: $a_k = e^{\alpha nt}$ with $k = 1, \ldots, d$, $\alpha_n \in \mathbb{C}$. In this way the Einstein equations $R_{1n} = 0$ are reduced to the following algebraic system

$$\begin{cases} \sum_n \alpha_n^2 = 0 \\ \sum_{i \neq n} \alpha_1 \alpha_n = 0 \end{cases} \tag{2}$$

In such a system we are free to choose d-2 free variables $\alpha_j = 0, \pm 1$ $(j = 1, \ldots, d - 2)$, obtaining $d(d - 1)/2$ solutions, one of which is the trivial "static" one $(0, \ldots, 0)$ with this choice the behaviour of the scale factors for the Torus is given by $\mathrm{Re}\alpha$; we can have $d\uparrow$ expanding $(\mathrm{Re}\alpha > 0)$, $d\downarrow$ contracting $(\mathrm{Re}\alpha < 0)$, d^0 static $\mathrm{Re}\alpha = 0$ dimensions. Of course as $t \to \infty$ the contracting dimensions disappear, we call such a process Dimensional-decay. Moreover, we consider the set of solutions for the system (2) as universes having a certain probability to happen in an instantanic transition[8] from a spacetime foam of compact manifolds of arbitrary initial dimensions. Introducing a quantity, defined instability factor

$$I = (d_0 - d_f)^{-1} \tag{3}$$

where d_0 and d_f are respectively the initial and final spatial dimensions of the Torus, we consider the set of D-decay solutions as formed by elements of different statistical weight, and write the probabilities as:

$$\begin{cases} \mathbb{P}[V(d_0 \to d_f)] = 1/1 + A \sum_{k \neq 0} e^{-dk} \\ \mathbb{P}[V(d_0 \to d_f)] = Ae^{-df}/1 + A \sum_{k \neq 0} e^{-dk} \end{cases} \tag{4}$$

where the former is related to d_0 conservation (the trivial solution), while the latter is affected by D-decay process, the quantity A is given by:

$$A = e^{d_0}(e^{d_0} - 1) \tag{5}$$

At this point we put three constraints on our model:

i) Einstein equations must be formulated also in d_f;

ii) Gravitations must exist on d_f;

iii) D-decay must be isotropic (this means $|\mathrm{Re}\alpha_n| = 1$).

As consequence of it we have

$$d_f = d\uparrow \tag{6}$$

and $d_f \neq 1, 2$ because in 2D gravity the Einstein tensor vanishes identically, and in 3D there are no gravitons[9]. Finally in the limit $d_0 \to \infty$ we have:

$$\begin{cases} \mathbb{P}[V(d_0 \to d_0)] = 0 \\ \mathbb{P}[V(d_0 \to d_f)] = (e - 1)e^{-d_f} \end{cases} \tag{7}$$

Therefore, last considerations seem to indicate $d_f = d\uparrow = 3$ as the best probability that a universe has to happen in the whole range of d_0. Unfortunately it seems that the ansatz can be applied only for complex hermitian metrics:

$$ds^2 = g_{i\bar{k}}\, dz^i d\bar{z}^k \tag{8}$$

By the way, we could suppose a universe having a complex metric structure, (see for example the powerful applications of Calabi-Yau manifolds to superstrings[10]).

Acknowledgement. The author would like to thank Prof. F. Occhionero for his kindly interest on this work. Also, he would like to thank Prof. E. Kolb for the very useful conversations on extradimensions.

*See Kolb lecture in these proceedings.

1) Duff, Nilsson, Pope, Phys. Rep. Feb. 1986.
2) E. Witten, in Shelter Island II Ed. Jackiw et al. 1984.
3) P. van Nieuvanhuizen in "Relativity, Group, Topology", Les Houches 1984.
4) W. Nahm Nucl. Phys B 135 (1978) 149.
5) M. J. Duff, B. E. W. Nilsson, C. N. Pope, Phys Lett. 163 (B) 1985, 343.
6) J. D. Barrow, Phil. Trans. R. Soc. Lond. A310 (1983) 337.
7) V. Pelino to appear in "Proceedings of the 26° Liège International Astrophysical Colloquium" 1986.
8) A. Vilenkin Nucl. Phys B 252 (1985) 141.
9) R. Jackiw, Nucl. Phys B 252 (1985) 343.
10) P. Candelas, G. Horowitz, Astrominger, E. Witten in "Geometry, Anomalies, Topology" World Scientific Publishing 1985.

THE INTERGALACTIC MEDIUM

Xavier Barcons
Departamento de Física Teórica, Universidad de Santander, 39005 Santander, Spain
and
Institute of Astronomy, Madingley Road
Cambridge CB3 OHA, England

1. INTRODUCTION

The fact that the process of galaxy formation must not necessarily be 100% efficient, raises the possibility that some portion of matter in the Universe lies in the intergalactic space. The aim of this seminar is to show that a hot electron - proton Intergalactic Medium (IGM) can exist, though some points are still unclear.

The starting point for the determination of the IGM features is the diffuse X-Ray Background (XRB) in the energy range from 3 to 300 keV. The spectrum of this radiation looks like thermal bremsstrahlung of a hot ($T > 10^8$ K) plasma (Marshall et al. 1980).

As no other known resolved X-Ray sources show the spectral shape of the XRB (except for an ad hoc distribution of Active Galactic Nuclei), a hot IGM provides a natural explanation for this background. Taking into account that the XRB is very isotropic (less than 2% of anisotropies in angular scales of 5 degrees x 5 degrees; Shafer & Fabian 1983) we infer that the hot IGM must be very uniformly distributed.

In Section 2, the evaluation of the IGM parameters (density, temperature,...) is discussed. In Section 3 we show that the existence of intergalactic absorption clouds (Lyman α clouds) does not exclude this model for the IGM if these clouds are non spherical. Section 4 summarizes our conclusions.

2. THE X-RAY BACKGROUND AND THE INTERGALACTIC MEDIUM

The best data of the XRB in the range 3 - 300 keV are reported in Marshall et al. (1980) and Gruber et al. (1984). By fitting the XRB with the bremsstrahlung of the IGM we could in principle obtain its temperature and density, but a number of points have to be taken into account.

P. Galeotti and D. N. Schramm (eds.), Gauge Theory and the Early Universe, 367–371.

The first one is the expansion of the Universe. Due to this fact, the IGM was denser and hotter in the past and therefore the radiated power was higher.

The second point is that for a temperature, at present, of about 40 keV (Marshall et al. 1980), the plasma is just slightly relativistic, but in the past relativistic corrections were more important. This has, as a consequence, that relativistic thermodynamics should be taken into account in order to evaluate how the IGM adiabatically cools with the expansion of the Universe (Barcons & Lapiedra 1985). On the other hand, electron - electron bremsstrahlung must be considered at relativistic temperatures (Stepney & Guilbert 1983) in addition to electron - proton bremsstrahlung (Górecki & Kluźniak 1981).

A third point to be considered is that of the distribution functions. The electron - electron relaxation timescale τ_{ee} and the electron - proton coupling time τ_{ep} fulfil $\tau_{ee} < H_0^{-1} << \tau_{ep}$. Therefore electrons and protons do not probably have the same temperature, but electrons are thermalized by themselves. Consequently, protons are assumed to be at rest and a relativistic equilibrium distribution is adopted for the electrons.

The last point to be emphasized is concerned with the heating of the IGM. As heating mechanisms are far from clear, we assume an instantaneous heating at a redshift z_m. This value can not be less than 2 because no significant portion of neutral hydrogen is seen between QSOs and us. On the other hand, the values of z_m much greater than 4 are not allowed because the interaction of the IGM with the microwave background through inverse Compton scattering will steepen too much the bremsstrahlung spectrum in relation with the observed XRB spectrum (Guilbert & Fabian 1986).

According to these points we evaluate the received energy at a given frequency per unit time unit solid angle and unit frequency band by integrating the emitted spectra along the line-of-sight.

Comparison of these theoretical spectra with the XRB shows that the temperature, at present, is about 25 keV (Barcons 1986; Guilbert & Fabian 1986). The density depends on z_m and adopting $q_0=½$ and $H_0=55$ km sec^{-1} Mpc^{-1} we find a density parameter for the IGM $\Omega_{IGM} \equiv nm_{pr}/\rho_{cr}$ running from 0.39 ($z_m=2$) to 0.21 ($z_m=4$).

This last result seems to be discouraging, because standard theories of primordial nucleosynthesis (Yang et al. 1984) demand $\Omega_{baryon} < 0.19$. However, clumping of the IGM makes bremsstrahlung more efficient and the above values for Ω_{IGM} are probably higher than the real ones (Guilbert & Fabian 1986).

3. LYMAN α CLOUDS

Lyman α clouds are intergalactic structures containing some neutral hydrogen which absorbs the UV Lyman α lines from QSOs (see Sargent et al. 1980 for a review). The typical column density contributed by a single cloud is about 10^{14} cm^{-2} and line widths suggest that the temperature is about 10^4 K. The enormous pressure exerted by the IGM on Ly α clouds demands that these must have densities ~ 1 cm^{-3} at a typical redshift $z \sim 2$.

On the other hand, from the observations of double QSOs, where common absorption lines are found (Young et al. 1981; Foltz et al. 1984), it seems that either clouds have a transverse size of 10 kpc or they are clumped on these scales. As there is some evidence for clumping, Ly α clouds probably have a transverse size smaller than 10 kpc.

The problem arises if we consider that the integrated UV radiation from QSOs provides the only source of ionization, because the flux is about 10^{-20} erg sec^{-1} cm^{-2} Hz^{-1}. If clouds are in equilibrium this will provide a very low ionization fraction at a density of $1cm^{-3}$, and from the observation of typical column densities we would conclude that their longitudinal size is about 10^{-4} pc. This probably means that clouds are non spherical but a far from equilibrium state is also suggested (Barcons & Fabian 1986).

On the non sphericity of Ly α clouds there is some independent evidence from the distribution of column densities at high redshift (Carswell et al. 1986). These distributions are not well fitted with spherical clouds unless an ad hoc distribution for their radii is assumed. However, if we assume that clouds form a sample of spheroids with eccentricity $e > 0.999$, the distribution of column densities is explained in a natural way (Barcons & Fabian 1986). This does not necessarily mean that Ly α clouds are spheroids, but there is a range in the observed column density distribution which seems to be distinctive of thin clouds.

As a conclusion, it seems that Ly α clouds are non spherical, and that their existence does not necessarily exclude the proposed model for the IGM.

4. CONCLUSIONS

Although some progress has been made supporting the existence of a baryonic IGM, more work is needed concerning other related problems. The main question is related with the nature of the sources which have reheated the IGM at a so high temperature, because about 1% of the rest mass of galaxies is needed to reach a temperature of 25 keV for the IGM. Together with this point there is the problem of the origin and shape of Lyα clouds. All of these points need more study.

Finally, we mention an independent test which can be used to see whether or not the hot IGM exists. If the IGM is ionized, it should produce dispersion effects on electromagnetic waves. Concretely, radiowaves with different frequencies will be delayed in a different way and therefore, by looking at an (extragalactic) periodic source or explosion, the IGM could be detected through this method (Haddok & Sciama 1965; Barcons & Lapiedra 1985).

Our conclusions can be summarized as follows:

-A hot uniform IGM responsible for the XRB should have $T \sim 25$ keV and $\Omega_{IGM} > 0.2$, but clumping of the gas can reduce this last factor.

-This hot IGM has probably been heated at $3 < z_m < 4$ but more work is needed on heating processes.

-The existence of Lyman α clouds does not rule out the above model if clouds are non spherical, which is, on the other hand, suggested by the observed distribution of column densities.

ACKNOWLEDGEMENT

I am grateful to A.C. Fabian for his collaboration in working out some of the results presented here.

REFERENCES

Barcons, X. and Lapiedra, R. 1985. Ap. J. **289**, 33

Barcons, X. 1986. 'X-Ray Radiation from the Intergalactic Plasma'. Ap. J. (In the press)

Barcons, X. and Fabian, A.C. 1986. 'Are Lyman α clouds non spherical?' Mon. Not. R. Astr. Soc.(In the press)

Carswell, R.F., Webb, J.K., Baldwin, J.A. and Atwood, B. 1986. 'Lyman Line Absorbing Clouds at Very High Redshift'. Ap. J. (Submitted)

Foltz, C.B., Weymann, R.J., Roser, H.J. and Chaffee, F.H. 1984. Ap. J. **281**, L1

Górecki, A. and Kluźniak, W. 1981. Acta Astronomica **31**,457

Gruber, D.E., Rotschild, R.E., Matteson, J.L. and Kinzer, R.L. 1984. MPE Report **184**, 129

Guilbert, P.W. and Fabian, A.C. 1986. 'A hot Intergalactic Medium?' Mon. Not. R. Astr. Soc. (In the press)

Haddok, F.T. and Sciama, D.W. 1965. Phys. Rev. Lett. **14**, 1007

Marshall, F.E., Boldt, E.A., Holt, S.S., Miller, R.B., Mushotzky, R.F., Rose, L.A., Rotschild, R.E. and Serlemitsos, P.J. 1980. Ap. J. **235**, 4

Sargent, W.L.W., Young, P.J., Boksenberg, A. and Tytler, D. 1980. Ap. J. Suppl. **42**, 41

Shafer, R.A. and Fabian, A.C. 1983. 'The (An)Isotropy of the X-Ray Sky' in Early Evolution of the Universe and its present Structure. Ed. by G. Abell and G. Chincarini. IAU **104**, 33

Stepney, S. and Guilbert, P.W. 1983. Mon. Not. R. Astr. Soc. **204**, 1269

Yang, J., Turner, M.S., Steigman, G., Schramm, D.N. and Olive, K.A. 1984. Ap. J. **281**, 493

Young, P.J., Sargent, W.L.W., Boksenberg, A. and Oke, J.B. 1981. Ap. J. **249**, 415

RELATIONS BETWEEN ASTRONOMICAL PARAMETERS FOR THE UNIVERSE WITH COSMOLOGICAL CONSTANT AND RADIATION PRESSURE

M. Dąbrowski
Astronomical Observatory of the Wrocław University
University of Wrocław
ul.Kopernika 11
51-622 Wrocław
Poland

J. Stelmach
Institute of Theoretical Physics
University of Wrocław
ul.Cybulskiego 36
50-205 Wrocław
Poland

ABSTRACT. A purpose of the talk is a derivation of the so called redshift - magnitude formula for the relation between luminosity distance and redshift of a galaxy in the open and closed Friedman universe with cosmological constant, radiation pressure and matter at zero temperature. Similar problem had been already treated by several authors (Mattig 1958, Kaufman and Schucking 1971, Kaufman 1971). However, our generalization consists in taking the radiation pressure into account. Due to using of usual mathematical apparatus - elliptic Weierstrass functions the final result takes closed form.

1. INTRODUCTION

The standard cosmological Friedman models are usually based on two simplifying assumptions concerning the cosmological constant and the radiation pressure. For the most part one neglects both. In such cases one treats the substance filling the universe as a pressureless dust (Friedman 1922, 1924, Lemaitre 1931, de Sitter 1932, Heckman 1942, Stabell and Refsdal 1966). The reason for this approach is the observation that at the present stage of the evolution the matter plays the dominant role (Mc Vittie 1965, Weinberg 1972, Narlikar 1983) and the pressure of the chaotic motion of galaxies and the cosmological constant are negligible. It turned out that taking the pressure effects into account did not lead to any qualitative changes in Friedman's solutions (Robertson 1933, Tolman 1934, Stabell 1968).The result has not essentially changed even if one considered the radiation pressure.

As far as solution of the Friedman equation with pressure does not

P. Galeotti and D. N. Schramm (eds.), Gauge Theory and the Early Universe, 373–382.

offer particular difficulties, this is not the case when the cosmological constant is present (Einstein 1917, Friedman 1932, Sandage 1961, Stabell and Refsdal 1966). However, analytic solution of the most general form of the Friedman equation with cosmological constant and radiation pressure is possible using elliptic functions and was discussed by several authors (Edwards 1972, 1973, Coquereaux and Grossmann 1982, Dąbrowski and Stelmach 1986a). In spite of the conviction that the cosmological constant is at the present epoch almost equal to zero the appropriate models are still examined mainly due to the predictions of quantum theory or the inflationary ideas (Guth 1981, Linde 1982).

A direct motivation for the present work is a relation between the apparent luminosity of galaxies and a redshift in the Friedman universe without the cosmological constant and the pressure, given by Mattig (Mattig 1958). Mattig's formula had been next generalized to the case with cosmological constant and applied to some numerical calculations (Refsdal 1966, Stabell and Refsdal 1966, Solheim 1966, Refsdal, Stabell and de Langer 1966). A particularly elegant form of these formulae has been obtained with the aid of the elliptic functions (Kaufman and Schuking 1971, Kaufman 1971).

A purpose of the present talk is a generalization of Mattig's formula for the relation between luminosity distance and redshift of a galaxy in the open and closed Friedman universe with cosmological constant, radiation pressure and matter at zero temperature. Since, throughout the paper, the matter is treated as a pressureless dust the new formula, formally, cannot be extrapolated to enough early stages of the universe. Therefore acceptable values of redshifts are bounded from above. Similarly to Kaufman and Schucking we use elliptic Weierstrass functions. The problem can be solved by application of the fractional linear transformation of the general Friedman equation into the canonical Weierstrass form. The method was used by Coquereaux and Grassmann and allowed to find explicit solutions of the Friedman equation with positive curvature, cosmological constant and radiation pressure. Similar calculations for the open universes were also performed (Dąbrowski and Stelmach 1986a).

2. BASIC NOTIONS

Our principal assumption is that the universe is homogeneous and isotropic. In consequence the metric is of the Robertson-Walker type

$$ds^2 = dt^2 - R^2(t)\left[\frac{dr^2}{1-kr^2} + r^2(d\Theta^2 + \sin^2\Theta)d\phi^2\right] \tag{1}$$

where $R(t)$ is a scale factor and $k=0,\pm 1$ is a curvature index. Moreover, the Einstein equation simplify to the form obtained by Friedman

$$\dot{R}^2 + k = \frac{8\pi 6}{3}\rho R^2 \tag{2}$$

where ρ is energy density. Generally ρ is time dependent. Hence in order to facilitate handling with the above equation one usually splits the energy density into three parts

$$\rho = \rho_m + \rho_r + \rho_v \tag{3}$$

where

ρ_m - energy density of matter
ρ_r - energy density of radiation
ρ_v - energy density of vacuum playing role of the cosmological constant

Now we assume that there is no interaction between matter and radiation at the considered stage of the evolution. In other words no part of the matter changes into radiation and vice-versa. It means that as well as matter as the radiation live independetly in the universe, which is relevant to the galactic era. Thus the energy conservation

$$\frac{d}{dt}(\rho R^3) = - p \frac{d}{dt} R^3 \tag{4}$$

has to be satisfied for each ingredient separately giving

$$\rho_m = \frac{C_m}{R^3} \tag{5a}$$

$$\rho_r = \frac{C_r}{R^4} \tag{5b}$$

where C_m and C_r are some constants up to now not specified. Putting in addition

$$\rho_v = \frac{\Lambda}{8\pi G}$$

we get the Friedman equation in a form

$$\frac{\dot{R}^2}{R^2} = \frac{C_r}{R^4} + \frac{C_m}{R^3} - \frac{k}{R^2} + \frac{\Lambda}{3} \tag{6}$$

suitable for our further considerations

Now we introduce the following astronomical parameters

$H(t) \equiv \frac{\dot{R}}{R}$ - Hubble constant

$q(t) \equiv - \frac{R\ddot{R}}{\dot{R}^2}$ - deceleration parameter

$\sigma_m(t) \quad \frac{4\pi G\rho_m}{3H^2}$ - matter density parameter

$$\rho_r(t) \equiv \frac{4\pi G\rho_r}{3H^2} \quad \text{- radiation density parameter} \tag{7d}$$

$$D(t_o) \equiv \frac{R^2(t_o)}{R(t)} r \tag{7e}$$

- luminosity distance between a galaxy which emitted a light ray at cosmic time t and an observer who detected the light at the moment t_o.

$$z \equiv \frac{R(t_o)}{R(t)} - 1 \quad \text{- redshift for this galaxy} \tag{7f}$$

Now we shall find the relations between these observational parameters. Since they all correspond to the moment of the observation $t=t_o$ for simplicity we denote them with the index zero.

$$H_o \equiv H(t_o), \quad q_o \equiv q(t_o), \quad \sigma_{mo} \equiv \sigma_m(t_o), \quad \sigma_{ro} \equiv \sigma_r(t_o) \tag{8}$$

Some of the mentioned relations may be easily obtained just by writing down the Friedman equation for $t=t_o$ and using the definitions of the astronomical parameters

$$H_o^2 = H_o^2(2\sigma_{ro} + 2\sigma_{mo}) + \frac{\Lambda}{3} - \frac{k}{R_o^2} \tag{9}$$

or by differentiating both sides of the Friedman equation with respect to t and then by putting $t=t_o$

$$\frac{\Lambda}{3} = H_o^2(2\sigma_{ro} + \sigma_{mo} - q_o) \tag{10}$$

Now I pass to the main topic of the talk - calculation of the generalized Hubble's law.

3. REDSHIFT MAGNITUDE FORMULA

The starting point of the calculation is the formula for the luminosity distance - magnitude measured by astronomers. I remind that it is expressed in terms of the redshift z, today's value of the scale factor R_o and the parameter r which is the radial coordinate of the observer in the coordinate system connected with the galaxy. Since neither R_o nor r are observable they should be calculated and expressed by σ_{ro}, σ_{mo}, H_o, q_o and z. In order to do that I define usual in this context new coordinate χ instead of r

$$r \equiv S(\chi) = \begin{cases} \sin\chi , & k=1 \\ \chi , & k=0 \\ \mathrm{sh}\chi , & k=-1 \end{cases} \tag{11}$$

Now applying this variable χ and the identity (9) we find the formula for the luminosity distance in a form

$$D_o = \frac{1+z}{H_o} \left[\frac{k}{4\sigma_{ro}+3\sigma_{mo}-q_o-1}\right]^{\frac{1}{2}} S(\chi) \tag{12}$$

The nontrivial point of the calculations comes now and consists in expressing $S(\chi)$ in terms of observational parameters. We perform suitable calculations passing to new dimensionless variables and parameters. We define

$$d\tau \equiv \frac{dt}{R} \tag{13a}$$

$$\Lambda_c \equiv \frac{4}{9C_m^2} \tag{13b}$$

$$T \equiv \Lambda_c^{-\frac{1}{2}} R^{-1} \tag{13c}$$

$$\lambda \equiv \Lambda / \Lambda_c \tag{13d}$$

$$\alpha \equiv C_r \Lambda_c \tag{13e}$$

Then the Friedman equation takes a polynomial form

$$\left(\frac{dT}{d\tau}\right)^2 = \alpha T^4 + \frac{2}{3}T^3 - kT^2 + \frac{\lambda}{3}, \tag{14}$$

with α and λ given by the formulae

$$\alpha = \frac{2\sigma_{ro}}{9\sigma_{mo}^2 k} (4\sigma_{ro} + 3\sigma_{mo} - q_o - 1) \tag{15a}$$

$$\lambda = 27\sigma_{mo}^2 k^3 \frac{2\sigma_{ro}+\sigma_{mo}-q_o}{(4\sigma_{ro}+3\sigma_{mo}-q_o-1)^3} \tag{15b}$$

In order to find χ we come back to the formula defining the line element in the curved space-time. We choose such a coordinate system that the light ray comes to us at constant values of Θ and ϕ. Then the distance parameter χ may be calculated by the integration of the expression

$$d\chi = -\frac{dt}{R} = -d\tau \tag{16}$$

using the Friedman equation in a new form

$$\chi = \int_{T_o}^{T} \frac{dT}{(\alpha T^4 + \frac{2}{3}T^3 - kT^2 + \frac{\lambda}{3})^{\frac{1}{2}}} = \tau - \tau_o \tag{17}$$

The reduced temperatures T and T_o corresponding to the time of the emission and to the time of the detection of the light ray may be easily expressed in terms of the observational parameters (Dąbrowski and Stelmach 1986b)

$$T_o = \frac{3\sigma_{mo}k}{4\sigma_{ro} + 3\sigma_{mo} - q_o - 1} \tag{18a}$$

$$T = T_o(z + 1) \tag{18b}$$

We notice now that from the qualitative point of view the problem is already solved since the right hand side of the formula (17) is expressed only in terms of astronomical parameters. The fundamental problem lies now in the explicit calculation of the integral. In some special cases the integration is easy. It happens if the expression

$$Q \equiv \alpha T^4 + \frac{2}{3} T^3 - kT^2 + \frac{\lambda}{3} \tag{19}$$

possesses at least one double root, for example in the case of a vanishing cosmological constant. Such cases are called degenerate and shall be discussed later on. However, in the general case the integration cannot be performed in an elementary way and a theory of elliptic functions should be used (Tricomi 1937, Abramovitz and Stegun 1964). Following already mentioned paper (Coquereaux and Grassmann 1982) we use elliptic Weierstrass functions which appear when the so called fractional linear transformation is applied

$$T(\tau) = \frac{\frac{1}{4}Q'(T_j)}{\mathcal{P}(\tau) - \frac{1}{24} Q''(T_j)} + T_j \ . \tag{20}$$

T_j is a root of the algebraic equation

$$Q(T) = 0 \tag{21}$$

and $Q'(T)$ and $Q''(T)$ are the first and the second derivatives of $Q(T)$ with respect to T.

The Friedman equation takes now the canonical Weierstrass form

$$\left(\frac{d\mathcal{P}}{d\tau}\right)^2 = 4\mathcal{P}^3 - g_2\mathcal{P} - g_3 \tag{22}$$

where g_2 and g_3 are some invariants

$$g_2 = \frac{k^2}{12} + \frac{\alpha\lambda}{3} \tag{23a}$$

$$g_3 = 6^{-3}(k^3-2\lambda) - \frac{k\alpha\lambda}{18} \tag{23b}$$

The solution of the above equation is called Weierstrass elliptic function and is given by the series

$$P(\tau) = \frac{1}{\tau^2} + \frac{g_2}{20}\tau^2 + \frac{g_3}{28}\tau^4 + \ldots \tag{24}$$

In order to find χ it is convenient to calculate $P(\chi)$ at first and then to take the inverse function P^{-1}.
For this purpose we use helpful formula for P-function

$$P(\chi) = P(\tau-\tau_o) = -P(\tau) - P(\tau_o) + \frac{1}{4}\left[\frac{P'(\tau) + P'(\tau_o)}{P(\tau) - P(\tau_o)}\right]^2 \tag{25}$$

$P(\tau)$ and $P(\tau_o)$ may be now reexpressed in terms of reduced temperatures T and T_o with the aid of the fractional linear transformation

$$P(\chi) = -\frac{1}{4}Q'(T_j)\left(\frac{1}{T-T_j} - \frac{1}{T_o-T_j}\right) - \frac{1}{12}Q''(T_j)$$

$$+ \frac{1}{4}\left[\frac{(T_o-T_j)^2 Q^{\frac{1}{2}}(T) + (T-T_j)^2 Q^{\frac{1}{2}}(T_o)}{(T_o-T)(T-T_j)(T_o-T_j)}\right]^2 \tag{26}$$

We notice that the right-hand side depends now only on the obserbational parameters by means of α,λ,T and T_o. However, for practical applications the formula is useless because it depends also on T_j which in a general case is not easy to be found.Since $P(\chi)$ as a function of observational parameters can not depend on the choice of the root T_j, elimination of T_j in the above formula should be performed.

I will not discuss the technical details of this elimination (Dąbrowski and Stelmach 1986b). Instead I focus myself on the result which reads

$$P(\chi) = \frac{k}{4\sigma_{ro}+3\sigma_{mo}-q_o-1}\left\{\frac{4\sigma_{ro}+3\sigma_{mo}-q_o-1}{6} - \frac{z+2}{2}[\sigma_{mo}+\sigma_{ro}(z+2)]\right.$$

$$\left. + \frac{1}{4z^2}\left[1+[2\sigma_{ro}z^2(z+2)^2+\sigma_{mo}z^2(2z+3)+q_o z(z+2)+(z+1)^2]^{\frac{1}{2}}\right]^2\right\} \tag{27}$$

χ parameter calculated from the above expression as an inverse function of $P(\chi)$ is just the desired magnitude which has to be inserted into the formula for the luminosity distance-so called generalized Hubble's law.

$$D_o = \frac{z+1}{H_o}\left(\frac{k}{4\sigma_{ro}+3\sigma_{mo}-q_o-1}\right)^{\frac{1}{2}} S\left\{P^{-1}\ \frac{k}{4\sigma_{ro}+3\sigma_{mo}-q_o-1}\left\{\frac{4\sigma_{ro}+3\sigma_{mo}-q_o-1}{6}\right.\right.$$

$$- \frac{z+2}{2}\left[\sigma_{mo}+\sigma_{ro}(z+2)\right] + \frac{1}{4z^2}\left[1+\left[2\sigma_{ro}z^2(z+2)^2\right.\right.$$

$$\left.\left.\left.\left. + \sigma_{mo}z^2(2z+3) + q_o z(z+2) + (z+1)^2\right]^{\frac{1}{2}}\right]^2\right\}\right\} \qquad (28)$$

4. DEGENERATE CASES

At the end of my talk I would like to show how obtained by us redshift magnitude formula goes over into the usual Hubble's law.

Earlier I mentioned about degenerate cases, i.e. the cases when the elliptic functions degenerate to elementary ones. It happens for example when the cosmological constant vanishes

$$\lambda=0 \Leftrightarrow q_o=2\sigma_{ro} + \sigma_{mo} . \qquad (29)$$

Then our formula simplifies (Dąbrowski and Stelmach 1986b)

$$D_o = \frac{\sigma_{mo}z+(\sigma_{mo}+2\sigma_{ro}-1)\{[2\sigma_{ro}z^2+2z(2\sigma_{ro}+\sigma_{mo})+1]^{\frac{1}{2}} - 1\}}{H_o[\sigma_{mo}^2 + 2\sigma_{ro}(2\sigma_{ro}+2\sigma_{ro} - 1)]} \qquad (30)$$

Moreover putting $\sigma_{ro}=0$ (radiation is negligible) we recover known formula of Mattig

$$D_o = \frac{\sigma_{mo}z + (\sigma_{mo} - 1)\{(2z\sigma_{mo} + 1) - 1\}}{H_o\sigma_{mo}^2} \qquad (31)$$

Assuming now that the space is flat ($k=0 \Leftrightarrow \sigma_{mo}=\frac{1}{2}$) we get

$$D_o = \frac{2}{H_o}(z + 1 - \sqrt{z+1}) \qquad (32)$$

Finally for small redshifts we can expand the square root into series and we find the usual Hubble law

$$z = H_o D_o . \qquad (33)$$

5. CONCLUSIONS

The main result of my present talk was finding of the general, explicit formula for the luminosity distance of a galaxy as a function of the astronomical parameters in open and closed Friedman models with cosmological constant and radiation pressure. I remind that the original formula of Mattig include neither pressure nor cosmological constant. Generalization of Kaufman and Schucking consisted in taking the cosmological constant into account. We have made one step further by considering the radiation pressure.

Although the assumption that the matter is a zero temperature dust noninteracting with radiation renders impossible to examine early stages of the universe (where the expansion is not adiabatic) it seems that our formula allows to test cosmological models, at least at sufficiently late stage of the evolution, with more precision. We notice that coefficient σ_{ro} (responsible for the radiation) appears always with the highest power of z. It means that for sufficiently large values of redshifts the radiation cannot be neglected. Testing the old universe requires considering of the full formula.

Other general relations between astronomical parameters are still calculated and shall be published elswhere.

ACKNOWLEDGMENTS

One of the authors J.S. with to thank the Organizers of the "International School of Particle Astrophysics" for covering the staying expenses in Erice, for perfect organization of the School and for giving the possibility to deliver the present seminar.

REFERENCES

Abramovitz, M., and Stegun, L.A. (1964),*Handbook of Mathematical Functions*, Dover, New York.
Coquereaux, R., and Grossmann. A. (1982).*Ann.Phys*.N.Y. 143,296.
Dąbrowski, M.P., and Stelmach, J. (1986b). *Astron.J.*(in press).
Edwards, D. (1972), *Mon.Not.R.Astr.Soc.*159,51.
Edwards, D. (1973). *Astrophys.Space Sci.*24,563.
Einstein, A. (1917), *Sitzber.Preuss.Adad.Wiss.*142.
Friedman, A. (1922), *Z.Phys.*10,377.
Friedman, A. (1924), *Z.Phys.*21,326.
Guth, A. (1981), *Phys.Rev.*D23,347.
Heckman, O. (1942), *Theorien der Kosmologie*, Springer-Verlag , Berlin.
Kaufman, S.E., Schucking, E.L. (1971),*Astron.J.*76,583.
Kaufman, S.E. (1971), *Astron.J.*76,751.
Lemaitre, G. (1931), *Mon.Not.R.Astr.Soc.*91,490.
Linde, A.D. (1982), *Phys.Lett.*108B,389.
Mattig, von W. (1958),*Astr.Nachr.*284,109.
Mc Vittie, G.C. (1965), *General Relativity and Cosmology*,University of Illinois Press, Urbana.
Narlikar, J.V. (1983). *Introduction to Cosmology*, Jones and Barlett Publishers, Inc., Boston.

Refsdal, S. (1966), *Mon.Not.R.Astr.Soc.* **132**,101.
Refsdal, S., Stabell, R., and de Lange, F.G. (1966),*Mem.R.Astr.Soc.* **71**, 143.
Robertson, H.P. (1933), *Rev.Mod.Phys.* **5**, 62.
Sandage, A. (1961),*Astrophys.J.* **133**, 355.
Sitter de W. (1932),*Proc.Akad.Wetenech*,Amsterdam, **35**,596.
Solheim, J.E. (1966),*Mon.Not.R.Astr.Soc.* **133**, 321.
Stabell, R., and Refsdal, S. (1966),*Mon.Not.R.Astr.Soc.* **132**, 379.
Stabell, R. (1968), *Mon.Not.R.Astr.Soc.* **138**, 313.
Tolman, R.C. (1934),*Relativity, Thermodynamics and Cosmology*, Oxford at the Clarendon Press.
Tricomi, F. (1937), *Funzioni Elittische*, Nicola Zanichelli Editore, Bologna.
Weinberg, S. (1972), *Gravitation and Cosmology*, John Wiley and Sons,Inc. New York - London - Sydnay - Toronto.
Dąbrowski, M.P., and Stelmach, J. (1986a) *Ann.Phys.N.Y.* **166**, 422.